ANGLICANISM

STEPHEN NEILL

ANGLICANISM

Fourth edition

OXFORD UNIVERSITY PRESS
NEW YORK

Contents

Preface

I am glad that this book, which has been out of print for a number of years, is once again to appear on the market, in revised and somewhat shortened form, and to become available to readers. The immense rise in the cost of book production has made it impossible to carry out quite such a complete revision as both author and publishers would have desired. In a small number of cases, paragraphs have been marked with an asterisk as an indication that the information supplied in them is now out of date. For more recent information on these matters the reader may consult the *Epilogue*.

In the twenty years since *Anglicanism* was first written, many things have happened in the Anglican world; a steady stream of books and articles has issued from the press. I cannot claim to have kept fully abreast of the flood, though I have done my best. But I do know the Anglican Communion a great deal better than I did in 1956. In the intervening years it has been my privilege to travel extensively, to encounter Anglican churches everywhere from the Arctic Ocean to New Zealand, and from New York to Tokyo in both directions, to minister in almost two hundred Anglican dioceses, and to become personally acquainted with almost all the outstanding leaders of the Anglican fellowship during this period.

The result of these experiences has been to strengthen my attachment to the Anglican way of worshipping God and the Anglican manner of living the Christian life. The Anglican Communion has obvious weaknesses. I could wish that in some ways it were other than it is, and that it would learn to make better use of the immense spiritual resources that have been committed to it by the Lord of the Church. Other Communions have their obvious excellences, theological and liturgical. Yet this is a fellowship in which it is possible for me to proclaim all that I believe to be true, and in which I am not required to teach anything which I believe to be untrue. Why then should I wish to exchange it for any other? In this part of the fellowship of the people of Christ I am content to live and to die.

S.N.

Oxford, April, 1976

CHAPTER I

Beginnings

No man knows when the Gospel of Jesus Christ was first preached in the British Islands; but there is reason to think that no long period elapsed between the Resurrection and the origins of a Church in England. Tertullian at the beginning of the third century claimed that parts of Britain unreached by the Romans had become subject to the law of Christ. Tertullian is a rhetorical writer, and there may be some element of exaggeration in the claim. But there is no reason to doubt that it is substantially correct. Christianity travelled fast along the trade routes and with the army; there would be nothing surprising in the presence of Christian soldiers on the Roman Wall in the second century A.D.

Archaeological evidence is as yet scanty; but what there is – a villa chapel in Kent, a later church at Silchester, the presence here and there of the 'Chi-Rho' symbol – suggests a fairly wide diffusion of the Gospel in Roman Britain. 'We may picture it as already deeply rooted in the populations of the shrunken and impoverished towns, and making its way by degrees among the peasantry. And . . . we may conjecture that it had already taken root over much of the highland zone.'[1] Three British bishops were present at the Council of Arles in A.D. 314.

But with that early Romano-British Christianity the Church of England has only the most tenuous of connexions. The invasion of the heathen Angles, Saxons, and Jutes broke up the settled life of towns and villas, and seems to have destroyed with it the earlier organization of the Church. For more than a century Britain ceased to be part of the civilized world.

Yet the connexion does exist. For before its decline British Christianity had driven out in three directions to the

1. R. G. Collingwood and J. N. L. Myres: *Roman Britain* (1936), p. 272.

conversion of the Celtic world. In the fifth century Illtud pressed forward into the mountains of Wales, and founded a great monastery, the first home of Welsh Christianity. Before the end of the fourth century Ninian had returned from Rome where he had been educated, had built his great monastery Candida Casa at Whithorn in Galloway, and had begun the long and difficult task of evangelizing the Picts. In 432 Patrick, who like most other famous Irishmen was not Irish, crossed to Ireland, and gave just a generation of human life to its conversion. With the breakdown of the Roman Empire, culture found its last refuge in the most distant of the Western Islands, flourished gloriously in Ireland, and in time flowed back in missionary enterprise and educational effort to the continent of Europe. It is the reflection of this mellow culture that meets us in the pages of the Venerable Bede.

This Celtic Christianity was very different from that of Roman times, when the centre of Christian life had been the city, with the bishop who was the chief pastor of the city church. In these Celtic lands there were no cities. The centre of everything was the monastery. Of course there were bishops, as in all other Churches, but there was no regular diocesan system. The great leaders were the heads of the famous monasteries.

From Ireland the Gospel crossed again to Scotland with Columba's foundation of Iona (563). From Iona, 'the chief of nearly all the monasteries of the northern Irish and of all the monasteries of the Picts', Aidan came in 634 to convert the northern English. This too was a monastic enterprise, marked by extreme austerity of life and by a most attractive simplicity of character. Few things in ecclesiastical history are more charming than Bede's descriptions of the holy men of that primitive time; as this of Cuthbert, that 'he was wont chiefly to resort to those places, and preach in such villages, as being seated high up amid craggy uncouth mountains, were frightful to others to behold, and whose poverty and barbarity rendered them inaccessible to other teachers . . . he would often stay a whole month, before he returned home, continuing among the mountains to allure that rustic people by his preaching and example to heavenly employments'.

In the meantime the most famous of all the missionary embassies arrived in England. Pope Gregory, rightly called the Great, bethought him of the lost and distant province, and sent Augustine and his company of monks to Canterbury (597). This was not so completely a pioneer enterprise as is sometimes imagined. Bertha the wife of King Ethelbert was already a Christian, and there may have been, even in Kent, surviving fragments from the ancient British Churches. Nevertheless, this was a memorable happening. From now on, from north and south, the Christianization of Britain went steadily forward, though with many setbacks and hindrances through the incursions of the Danes, until it was completed, in so far as the conversion of any country can be said to be completed.

Historians have rightly stressed the wisdom of Pope Gregory's famous letter to Augustine, in which he advises him not necessarily to impose upon the newly-formed Church of the English all the usages and customs with which he had been familiar at Rome: 'For things are not to be loved for the sake of places, but places for the sake of good things. Choose, therefore, from every Church those things that are pious, religious and upright, and when you have as it were made them up into one body, let the minds of the English be accustomed thereunto.'

There were thus three streams in early English Christianity – the Romano-British, the Celtic, and the Roman. Because of this, and because the directly Roman strain was the latest to arrive, some Anglicans not very favourably disposed towards the see of Rome have argued as though the Church in England had as it were a primitive and inherent right to independence of Rome. But such argument rests upon a complete misunderstanding of the way men thought in those early centuries. At that time every Christian in the West believed that there was only one Church and that he formed part of it. Within that one Church there might be controversy over details. The Celtic Churches were conservative and clung to practices that had gone out of fashion at

Rome. Most of these problems seem to us, at this distance of time, very trivial; but Churches become attached to trivial things, and do not like to have them altered. These Churches claimed the right to hold fast to their own traditions. But this did not constitute any claim to independence *from* the Church; it was only a difference of opinion as to the measure of freedom that might be claimed by a local Church *within* the fellowship of the one Church.

Uniformity and variety, centralization and independence – there is a kind of ebb and flow in the life of the Church between these poles. Much of the history of the English Church before the Reformation can be summed up in terms of the tensions between these two ideals. Frequently it was the ecclesiastics who were in favour of uniformity, which in those times meant closer dependence on the see of Rome; whereas the monarchs on the whole were sturdy defenders of a measure of national indepeadence. In any case the idea that a local Church could be completely independent of the one Church, that it could exist for itself without maintaining fellowship with all the other Churches in the world, had not in those days dawned in the mind of anyone in the West.

The first notable champion of a closer dependence of the English Church upon Rome was that stormy petrel of the seventh century, Wilfrid of Ripon. Wilfrid (*c*. 634–709) was certainly one of the greatest men of his age, but it is hard to find him a lovable character. He had the zeal and passion of the true evangelist; but he was liable to be blinded by the identification of his own interests or prejudices with the good of the Church, and to mar his advocacy of good causes by stirring up the hostility of those who might have been his friends. His place in our story is related to his advocacy of the supremacy of Rome as against the claim to independence maintained by the Celtic Churches.

The question at issue was one of practical, though not of theological, significance – the method of calculating the date at which Easter is to be observed. This still causes difficulty in areas where the Eastern and Western Churches co-exist, since they still differ in their calendar (though not in the same manner as the Celtic and Roman Churches of old)' and it may happen that one group of Christians is celebrating the joys of Easter, while another, perhaps within the same family, is still observing the austerities of Lent. What was highly characteristic was the manner in which the problem was handled in Britain. The King, Oswy of Northumbria, convened a Synod, presided over it, and in the light of the discussion gave the decision by which the practice of the Church was to be determined. The Synod was held, in 664, at a place called Streanaeshalch, which has usually and probably rightly been identified with Whitby. After the Celtic case had been put forward, Wilfrid replied, laying the greatest stress on the authority of Peter, and asking: 'though your fathers were holy, do you think that their small number, in a corner of the remotest island, is to be preferred before the universal Church of Christ throughout the world?' The King, apparently impressed by this argument, gave his decision in favour of the usages of Rome, 'lest, when I come to the gates of the kingdom of heaven, there should be none to open them, he being my adversary who is proved to have the keys'.

The only date that any Englishman can be quite sure of remembering is 1066 – and not without reason, since the Norman conquest marked an epoch in the history of Britain and of the West. England became, and remained for more than four centuries, unquestionably a part of Europe. The change could not be without its influence on the life of the English Church; one of its effects was the far closer integration of that Church in the general life of the Church in the West. The Anglo-Saxon Church has suffered the denigration that is the usual fate of the vanquished; in recent times greater justice has been done to its merits, and to the outstanding men who served it. But it was an isolated Church. There can be little doubt that the closer relationship with Rome, which followed on the Norman Conquest, was of advantage both to the English Church and to the nation with the spiritual care of which it was charged.

This was the period of the first great reform of the Western Church after its emergence from the confusion of the Dark

Ages. The central figure in this reform was Hildebrand, Pope Gregory VII, who although he became Pope only in 1073 had for ten years before that date been the dominant power in Rome and the inspirer of papal policies. With an intense passion for righteousness, and a conviction that righteousness could be done only if the power of kings and rulers was made subject to that of the Church, Gregory desired at every point to assert the pre-eminence of the spiritual power, and to organize the clergy as a disciplined and obedient army, through whom the papal decisions could be carried into effect. Such was the Pope with whom William the Conqueror had to deal.

The Conqueror appointed as his Archbishop of Canterbury a man who was pledged to the reforming ideals of the Pope, but at the same time was capable of working loyally with his king. Lanfranc (c. 1005–1089) was an Italian, who had become prior of Bec in Normandy, and then abbot of William's great monastery at Caen. It seems that he was very unwilling to accept the heavy charge of Canterbury; but history has judged him to have been one of the greatest in its succession of Archbishops. Much in his administration still remains obscure, but the main lines of it are clear. Most of the Anglo-Saxon bishops were quickly got rid of and were replaced by foreigners. A remarkable series of Church Councils was held; through these the primacy of Canterbury over the other bishoprics was firmly established, and Lanfranc's ideas of discipline were spread throughout the Church. The first serious attempt was made to impose on the clergy that celibacy which was one of the favourite ideas of Hildebrand, but had never been regarded as a necessary part of the clerical life in England. Above all, Lanfranc brought with him from the Continent the Canon Law, that vast confused body of Church law, discipline, and practice, which had grown up over the centuries and was still growing. Both the law itself, and its introduction in England, were very far from complete; yet from now on the Englishman was subject to two jurisdictions; that of the King and that of the bishop, with endless complications where the boundaries between the one and the other were not clearly defined, but with a general obligation resting on the secular power to support the judgement of the Church.

Lanfranc was the Pope's man, as well as the King's, and this may seem to suggest a complete subordination of the English Church to the papacy. But, as in all things English, we run at once into the element of paradox. From a very early date England had been known as the specially devoted child of the papacy; but this was not held incompatible with vigorous resistance to individual Popes. No English church-man has ever defended the freedom of his Church with greater vigour than William the Conqueror.

William had special reasons for being grateful to the Pope, who had supported the dubiously legal enterprise which had made him King of England. But when the Pope tried to make capital out of this, in order to claim that England owed the Pope feudal allegiance and to demand fealty of William, the Conqueror flatly refused; he had made no promise of fealty, none of his predecessors had ever stood in such a relationship to the Pope, neither would he. He forbade any papal letters to be received in England without his permission. He forbade the proposal of any legislation in any Council of the English Church without his prior consent. He forbade his bishops to go to Rome without his permission, even though they had been expressly and personally sum-moned by the Pope.

None of these things was ever forgotten. These practices, imposed by the Conqueror through the sheer force of his personal authority, came to be collected and to be known as the 'ancestral customs' of the kingdom. It was believed that they had permanently regulative power on the relationships between the Church of England and Rome.

The most famous of all the conflicts between King and Church was that between Henry II and Thomas à Becket, made familiar, by Mr T. S. Eliot's admirable *Murder in the Cathedral*, to many people who have otherwise little acquaintance with the history of the English Church. Henry and Thomas were both difficult people to deal with, and it is clear that each had the gift of exasperating the other beyond bearing. The main cause of their quarrel was one in which it may seem to us that reconciliation ought to have been possible – the extent and nature of the King's authority over clergymen who had committed crimes. Henry wished to strengthen the authority of his own courts, and it must be admitted that he had a not unreasonable case; the greatest of all authorities on English affairs in the Middle Ages, F. W. Maitland (1850–1906), wrote that 'one of the worst evils of the later Middle Ages was the benefit of clergy'.[1] But Henry's insistence on what he believed to be the ancient customs of the realm seemed to Becket an infringement of the liberty of the Church, and he determined to resist, if need be to the death.

When, on 29 December 1170, four knights, taking literally some rash words of Henry, murdered Becket in his own cathedral, the whole of Christendom shuddered at the sacrilege. Henry, though he disclaimed responsibility, did exemplary penance, and bared his back to the scourges of the monks of Canterbury. Within little more than two years, on 21 February 1173, Becket was canonized by the Pope. His shrine in Canterbury became one of the richest and most splendid in Christendom, and uncounted multitudes of pilgrims made their leisurely way through the Weald of Kent 'the holy blisful martyr for to seke'. It might seem that the Church had won hands down.

Certainly Henry made some concessions, and was unable to carry through all the reforms that he had had in mind.[2] But he does not seem, in reality, to have lost much of the control that an English king felt that he must exercise upon the Church. In the Middle Ages, bishops were such important people in the life of the nation – great nobles, leading

1. 'Originally the privilege of exemption from trial by a secular court allowed to or claimed by clergymen arraigned for felony; in later times the privilege of exemption from the sentence, which, in case of certain offences, might be pleaded on his first conviction by everyone who could read.' *Oxford English Dictionary*, s.v. Clergy.

2. On this see Z. N. Brooke: *The English Church and the Papacy* (1931), pp. 190–214. Especially p. 212: 'As a result of the concessions of Henry II, canon law becomes at last completely valid in this country; in the full practice of the law, the English Church is able to take its place with the rest of the Church.'

administrators as well as spiritual princes – that no king felt that he could allow the choice of them to fall into other hands than his own. He regularly had to make a bargain with the Pope about the appointment, and there were the canonical rights of the cathedral chapters to be borne in mind. But in 1173, the very year of Becket's canonization, Henry dealt with the vacancy in the bishopric of Winchester, one of the most important of all the English sees, by sending the following missive to the monks who had the right of election: 'Henry, King of the English etc. to his faithful monks of the church of Winchester, greeting. I order you to hold a free election, but, nevertheless, I forbid you to elect anyone except Richard my clerk, the archdeacon of Poitiers.' Apparently in nearly eight centuries little has changed in the control of the Crown over the election of the bishops of the English Church.

It is exceedingly difficult for modern man to think himself back into the mind of medieval man, and to understand his attitude to his Church. Two special points of difficulty may be mentioned, because of the long shadows which they cast on all the subsequent history.

First, then, the medieval view of property and of the obligations attaching to it. The Church possesses property, mostly land, which it holds under a great variety of tenures, and from which it supports those who do it service. There is so much property, and so much work to be done. It is almost self-evident to modern man that one who holds an office should do the work which is attached to that office, and for which he is being paid. Nothing could be further from the mind of medieval man. One who held an office was certainly responsible to see that the work was done; but if he could persuade a deputy to do it for a tenth of the income, retaining nine-tenths himself, there was nothing in that to which anyone could reasonably take objection. As the Church of England was saddled with the remnants of this system until well on in the nineteenth century – indeed is not altogether clear of it today – it is important to understand how the system worked.

To us it seems illogical that, at the end of the Middle

Ages, several bishops of Worcester should be Italians resident in Rome, who never visited England.[1] But this was really a convenient method by which the King made sure of having a friendly agent at Rome, such as on occasion it might be very much to his advantage to have. And the absence of the bishop did not mean that the work was not done; there were always impecunious Irish bishops about the place, and titular bishops without sees, who could be relied on to perform the necessary episcopal acts; and the surviving episcopal registers show with what diligence and regularity the official work of the dioceses was carried on.[2]

Nor must it be supposed that this system always worked to the disadvantage of Englishmen. It was possible for an English archdeacon to be at the same time chancellor of the cathedral of Valladolid in Spain. Cardinal Wolsey was Bishop of Tournai in Belgium before ever he became Archbishop of York, and went there in 1513 to be enthroned. And had not the English cardinal Nicholas Breakspear become Pope in 1154 as Adrian IV, the only Englishman ever to hold that dignity?

Nevertheless there was a feeling in England – and this grew stronger with the increasing centralization of affairs in Rome, and the increasing claims of the Popes to exercise direct authority everywhere – that there were too many foreigners in England, and too many foreigners abroad who were drawing English revenues for which they did no work. And indeed the figures are astonishing. At York in the fourteenth century less than half the cathedral appointments were held by Englishmen; at Salisbury fourteen out of twenty-nine non-resident canons were foreigners. It is not surprising that in 1307 the parliament of Carlisle protested against the 'unbridled multitude' of papal 'provisions', by which unwanted foreigners were thrust into good positions in the English Church.

1. From 1497 to 1535, when Hugh Latimer became bishop. Cardinal Campeggio, appointed by the Pope to try the case of Henry VIII against Catherine of Aragon, was Bishop of Salisbury, 1525-35.

2. Professor Knowles gives a long list of such episcopal helps in *The Religious Orders in England*, Vol. 11 (1955), pp. 372-5.

One result of the growing discontent was the passing in 1351 of the Statute of Provisors, repeated in 1390, which laid down that elections and presentations to benefices should be free; and that, if the Pope 'provided' to any appointment, that should be forfeited to the Crown. It is most unlikely that this statute was ever literally applied – the system of provisions supplied the king with far too many convenient bargaining counters. But it is sometimes useful to have a statute on the books, even if it is not applied; and here was a weapon that could be used, if ever papal aggression became a matter of serious controversy.

Even more difficult for modern man to understand is the medieval churchman's love of litigation. The ordinary Englishman's idea is to have as little to do with the law as possible, and the great majority of people have never been in a court of law in their lives. To the medieval European litigation was a major interest, without which he would have felt his life much the poorer. A distinguished modern scholar has written of the monks as litigious capitalists, who were quite prepared on occasion to forge documents to protect their interests and property. Gregory VII spoke and wrote often of 'justice', and this has often been interpreted in terms of a high ideal of moral righteousness; but in many contexts what Gregory was really concerned about was nothing more than the question of who had the legal right on his side. 'At the time when Gregory became Pope we are entering an age of intense ecclesiastical litigation. Churchmen, forbidden to take the sword, exposed their bodies to dangers greater than those of battle in long and trying journeys in pursuit of justice, in the defence of the rights, dignities and prerogatives of their churches, in the vainglorious desire for prestige or the dour assertion of principle.'[1]

Rome was recognized as the fount of law for the Christian world, and a steadily increasing volume of business poured

1. R. W. Southern: *The Making of the Middle Ages* (1953), p. 146. Mr G. H. Cook, in *The English Medieval Parish Church* (1954), pp. 25–6, gives interesting details of the great series of lawsuits between the parishioners and the monks of Abingdon in the fourteenth century.

into the papal chancery. This did not mean that every suit was heard in Rome (though it might happen to leading ecclesiastics to spend years on end at the papal court awaiting the deliverance of judgement in their suits), for the Pope could appoint deputies in any country to hear cases on his behalf. But the system did mean a great enhancement of papal power, and a great enrichment of papal officials. In the end the practical English reacted violently against the system.

The first Statute of Praemunire, enacted in 1353, did not mention Rome at all; but it did lay down 'that anyone drawing the king's subjects out of the realm on pleas the cognizance whereof belongs to the King's courts, or impeaching the judgements given in those courts, shall have a day appointed to answer his contempt, and if he fail to do so shall be put out of the king's protection, his lands and goods forfeited, and his body imprisoned at the king's pleasure'. A later Act of 1393 declared that the Pope by his actions had caused the laws of the realm 'to be defeated and avoided at his will, in perpetual destruction of the sovereignty of the king'. These acts in no way whatever indicate an inclination of the English people to separate themselves from the great Church of the West,[1] and in fact their influence on legal proceedings was astonishingly small. But once again this was a useful statute to have on the books, and it is not difficult to imagine what kind of use could be made of it by an autocratically-minded king.[2]

1. Dr W. A. Pantin has rightly stressed this point: 'Nothing could be more clear than that the fourteenth-century English Church was very consciously part of the universal Church, in ecclesiastical government and in its intellectual and spiritual life. And this was achieved without any diminution of national characteristics .... This combination of genuine Englishry with being part of an international Church and culture is one of the most interesting features of medieval England.' *The English Church in the Fourteenth Century* (1954), p. 5.

2. The concluding words of Miss K. N. Wood-Legh's careful *Studies in Church life in England under Edward III* (1934) are: 'At any rate the degree of control which the State was able to exercise over the affairs of the Church was an important factor in the life of the times, and if, as there is every reason to believe, it was maintained throughout the Middle Ages, it must have made the changes which followed the breach with Rome less startling to the ordinary man than they otherwise would have been.'

In matters of practice the English might show signs of restiveness; in matters of faith they lived, by medieval standards, lives of unimpeachable orthodoxy. One of the first appearances of Britain in Church history was in the person of the blameless monk Pelagius, who in the early years of the fifth century more than met his match in controversy with the great Augustine of Hippo concerning the dark problems of predestination and free will, and has given his name to the characteristic English heresy, Pelagianism. After that nothing of the kind happened in Britain for a thousand years – and then appeared John Wyclif (d. 1384). Wyclif has often been called 'the Morning Star of the Reformation', but the use of the term can hardly be justified by the facts. His writings reveal him as a bitter, angry, disappointed man, with a harsh and narrow mind.[1] His intellect moved wholly within the limits of medieval scholastic logic; but his ruthless application of this logic led him by degrees to question almost everything that the medieval Church held sacred – the right of the Church to hold property, pilgrimages and the veneration of the saints, the authority of the episcopate, the doctrine of transubstantiation in the Mass.

Similar, though perhaps less radical, attacks had been made by other medieval writers; but one thing in Wyclif is genuinely new – the basis on which his criticisms were made. It was his aim to restore the Scriptures to their position as the unique and sole authority for life and doctrine in the Church. It is difficult to determine exactly what connexion Wyclif himself had with the version of the Scriptures that commonly bears his name. One translation of the whole Bible seems to have been completed in his lifetime, and there can be little doubt that the work was carried out through his inspiration and in part under his direction, though it is not possible to be sure to what extent he himself was responsible for the rough, vigorous, and still for the most part intelligible, English in which the translation is phrased. What is important is that Wyclif bequeathed this principle of the

1. Professor Knowles writes of 'a mind to which personal affection was a stranger', and 'an embittered spirit and a mind, perhaps, pathologically obdurate and inflexible'. *Op. cit.*, p. 101.

sole authority and general accessibility of the Holy Scriptures to that Lollard movement, through which Wycliffite
ideas continued to circulate among the population for the
next century and a half. 'Thei seien thus, that what euer
man or woman wole be meke in spirit and wole preie God
helpe him, schal without faile undirstonde ech partie of
holy scripture.' From the point of view of the medieval
Church this was heresy; but we encounter here an idea that
was taken up again at the time of the Reformation, and it is
not surprising that the Reformers believed themselves to be
reaping where Wyclif had sown.[1]

If we could be carried back six hundred years in time,
what would be the first thing that would strike us about the
Church in England? The answer would, I think, certainly
be its enormous power.

Perhaps one-third of the whole land-surface of England
had come, as parish glebe, through royal grants, or legacies
to the religious houses, into the hands of the Church in one
or other of its aspects. Almost all the greatest and most splendid buildings in the country were ecclesiastical in character.
The great churchmen were among the richest subjects in
the country, and until the end of the Middle Ages almost
monopolized government and administration.

Ecclesiastics formed a numerous and prosperous community. It has been reckoned that the clergy accounted for
rather more than two per cent of the population. Of course
not all clergymen were priests; many of them had taken
only minor orders. But it seems likely that on the average
one in every hundred grown men in the country was a
priest.

The Church was ubiquitous. In the fourteenth century,
the average population of a parish was between four and five

1. The weighty opinion of Professor Norman Sykes is that 'Wyclif's
successors, generally called the "known men", or the "just, fast men",
supplied a continuous and unbroken link between medieval Lollardy
and sixteenth-century Lutheranism in England ... and they did their
work so well that when Lutheranism reached England in the reign of
Henry VIII it seemed at first sight but a continuation of Lollardy'.
The English Religious Tradition (1953), p. 14.

hundred; but perhaps the majority of the people lived in villages with less than two hundred inhabitants. The church bell could be heard everywhere. The festivals of the Church were the festivals of the village; and in days when most men were illiterate, these feasts were also the village calendar – do we not still speak of the Michaelmas holidays? In such small communities there is little that remains unknown, and both secular and Church authorities maintained a diligent watch on everything that everybody did, such as makes our temporary wartime restrictions on liberty seem by comparison a happy holiday of freedom. The parson knew just how much he could claim by way of tithe, and was prepared to go to considerable lengths to extract his due. It was one of the great merits of Chaucer's Poor Parson that 'full looth were him to cursen for his tithes'. In the background was always the archdeacon's court, the loathed instrument for the collection of fines and fees. Any man who committed such a flagrant offence as eating meat on a fast-day could hardly hope to escape observation. Excommunication could carry with it very serious consequences. Even prominent people could be subjected to public, painful, and, to our minds, humiliating penances; as for example, to walk round Taunton market-place, clad only in a shirt and breeches, 'and when he comes to the middle of it stand still for a time at the discretion of the chaplain robed in a surplice, and with a whip in his hand who follows him'.

It is difficult to tell just how well each part of the religious system was working at any given period. From every century come the bitter complaints of the reformers; but these must always be treated with a measure of caution. One scandalous priest in a hundred is news; the ninety and nine who faithfully and quietly do their duty remain unrecorded and unnoticed. In our own day it would be possible to draw a highly unflattering picture of the Church, in which no detail would be false but which would represent a considerable distortion of the whole. It seems likely that the general level of the English Church remained fairly steady from the twelfth to the fifteenth century, though with a certain decline in devotion that gathered speed rather rapidly as the Middle

Ages reached their close. But within this picture of general stability, certain changes in emphasis do appear.

The great period of English monasticism had come to an end by the middle of the thirteenth century. Before that date the monasteries had produced many great leaders and some saints. They had covered the face of England with buildings of marvellous beauty, and they had done much to enrich the country by transforming its methods of agriculture and trade. But gradually life and heroism departed. It was not so much that monks and nuns led scandalous lives – it was much more that they had become very ordinary and secular. Perhaps Bishop Stubbs' judgement on the monasteries was a little too severe – that 'their inhabitants were bachelor country gentlemen, more polished and charitable, but little more learned, or more pure in life than their lay neighbours'. But it was a friendly critic who wrote of the nunneries shortly before the Reformation: 'I am bound to say that the description of nunneries as houses of religion, even in the purely technical sense, was and had long been inaccurate. The regular discipline which religion in that sense implies, the strict observance of the rule, had vanished, enclosure was disregarded, worldly desires were freely gratified and the life of the nuns hardly to be distinguished from that of the world around them.'[1] And our most learned authority of all, Professor Knowles of Cambridge, has told us that in the fifteenth century the monasteries, and indeed the Church as a whole, did not produce any towering example of holiness, such as alone can bring an organization in decline back to the way of creative renewal.[2]

1. A. H. Thompson: *Visitations in the Diocese of Lincoln, 1518–31* (1940), Vol 1. p. lxxxvii.

2. *The Religious Orders in England*, Vol. 11 (1955), p. 364: 'The fifteenth century saw no great English churchman and no English saint. When all has been said that can be said of causes and influences, in things of the spirit it is, humanly speaking, the men that matter, and they were not forthcoming. Whatever there may have been of silent and hidden sanctity – and this, *ex hypothesi*, may equally well be assumed or neglected in any age – no Englishman arose in the fifteenth century to show his countrymen the truth and the charity of Christ, which alone would have been able to make the dry bones live, or to see himself, and convey to others, the fullness of meaning of the First Commandment.'

The coming of the Friars in the thirteenth century brought a new force into English religious life. The Friars supplied some of the best scholars and thinkers at the universities, and the most frequent and popular preachers of the next three centuries. But they too had been affected by the desires and the cares of the world, and what specially offended contemporary sentiment was the contrast between their profession of utter poverty and their gradually increasing wealth. Chaucer's Friar is among the most unpleasant of his pilgrims, cunning, sycophantic, and corrupt:

> Curteis he was and lowly of servyse,
> Ther was no man nowher so vertuous.
> He was the beste beggere in his hous.

In the fourteenth century there is evident a turning away of interest, as measured by gifts and benefactions, from the monasteries to the parish churches. This was a right movement – ultimately the life of the Christian fellowship is in the parish, and in the parish church which is its centre. Here English Christianity had its greatest treasure of all. The movement for church-building has never ceased, and continues in our own day; it attained its greatest glory in the thirteenth and fourteenth centuries. Almost every village had its church, generally following one of the main traditions of architecture, yet almost every one marked by some felicitous adventure in originality. Even after Puritan destructiveness and the horrors of nineteenth-century restoration have done their worst, what is left to us is of wonderful splendour, the greatest of all the legacies of the Ages of Faith.

It is hard to form any general picture of the pastors who ministered in these sanctuaries. Many of them were men of humble origin and very inadequate erudition. The great tithes of many parishes were held by a monastery or some lay impropriator; in consequence, then as now, many of the parochial clergy were wretchedly underpaid. There are many complaints of irregularity and scandalous living. The pious fifteenth-century writer John Myrc would hardly have found it necessary to give to the clergy the advice:

> Taverns also thou must forsake
> And merchandise thou shalt not make,
> Wrestling and shooting and such manner game
> Thou might not use without blame,

unless there was at least some probability that the advice would be needed. But on the whole the medieval parish priest emerges better from the sober conclusions of historical research than from the gloomy pictures drawn by the satirists of his own day. The best known of all is, of course, Chaucer's Poor Parson:

> This noble ensample to his sheepe he yaf
> That first he wroghte and afterward he taughte . . .
> A better preest I trowe that nowher noon ys;
> He waited after no pompe and reverence,
> Ne maked him a spiced conscience,
> But Cristes loore, and his Apostles twelve,
> He taughte, but first he folwed it hym selve.

It is probable that this portrait, like so many of the others in the Prologue to the *Canterbury Tales*, was drawn from life. Chaucer would not so have drawn it, unless he had been sure that it would be recognized as reasonably typical and not merely exceptional.

In the fifteenth century another change becomes notice-able. Less money goes to the churches, and more to the foundation of chantries. A chantry was a chapel endowed to provide for a priest to sing masses in perpetuity, usually for the soul of the founder or for those of his kin. In some cases whole groups of chantry priests lived together in a collegiate body, as at what is now Manchester Cathedral. Often the chantry was a chapel within an existing church, such as the great chantry of William of Wykeham in Winchester Cathedral. This new tendency marked a certain turning away from the corporate to the individual element in religion, from the fellowship of worship in the congregation to preoccupation with the destiny of the individual soul.

This was natural in a time when life was becoming increasingly difficult and uncertain, and when a certain failure of nerve and confidence becomes perceptible.[1] The Churches laid increasing stress on the terrors of death, of the judgement to come, and of that hell where 'they that be dampnet to hell, thay styntyn never to cry and yelle: "Woo ys hym that theyre shall go"'. In this mood William Dunbar (d. 1520) writes his *Lament for the Makers*, where every verse ends with the refrain *Timor mortis conturbat me*, 'the fear of death troubles me':

> The state of man does change and vary,
> Now sound, now sick, now blyth, now sary,
> Now dansand mirry, now like to die:
> *Timor mortis conturbat me.*

> Unto the Death gois all Estatis,
> Princis, Prelatis, and Potestatis,
> Baith rich and poor of all degree:
> *Timor mortis conturbat me.*

No Christian ought to write in such terms as these; we have come a long way indeed from Paul's desire to depart and to be with Christ, which is very much better. But, if no man could be sure that he was saved, and if the best that he could hope for was an almost endless and agonizing period in purgatory, and if that period could be shortened through the offering of the sacrifice of the Mass on his behalf, it might well seem wise to make such a disposition of his earthly goods as would give him the greatest ease, or at least reduce his dis-ease, in the life to come.

Measured by that infallible thermometer of spiritual life, the expansion of the Church through missionary work, the fifteenth is the gloomiest of all the centuries. It is the only one in which hardly any growth seems to be traceable. But all was not dark. Even the fifteenth century had its saints,

1. J. Huizinga, in *The Waning of the Middle Ages* (1924), has laid stress on the element of strain and tension which becomes apparent in the declining days of the Middle Ages, and persisted into the sixteenth century.

and produced one of the most famous of all Christian classics, the *De Imitatione Christi* (*c*, 1450). But this book itself is symptomatic of a tendency to an inwardness in religion, which might bear little relation to the routine and the organization of the visible Church. Late medieval religion in England is distinguished by the remarkable succession of the English mystics – Richard Rolle of Hampole, *The Cloud of Unknowing*, Margery Kempe, and many others. All perfectly orthodox in intention, these mystics sought their consolation not in pantheistic imaginings, but in intense devotion to the passion of the crucified Saviour.

> My dere-worthly derlyng, sa dolefully dyght,
> Sa straytly upryght streyned on the rode;
> For thi mykel mekenes, thi mercy, thi myght,
> Thow bete al my bales with bote of thi blode.[1]

Good men of that epoch might well seem to themselves to find valid reasons for turning away from the visible Church, since that Church had indeed come to sorry estate. From the beginning of the Great Schism in 1379, there had been two, and sometimes three, Popes in Christendom, each claiming the full allegiance of all the faithful on pain of everlasting damnation. It was hard for the ordinary believer to discern the glorious Bride of Christ through these outward deformities. A new and hopeful beginning was made in the Council of Constance (1415), and the Conciliar Movement of which it was the expression. Good and earnest men, and some not so good and earnest, strove to reform the Church in head and members, and to curb the autocracy of the papacy by giving the Church the outlines of a constitutional form of government. England, as an integral part of Western Christendom, was naturally represented by some of its bishops at these great gatherings of the Church. The Councils did succeed in restoring the unity of the papacy and so of the Church; but the reforming movement died away in frustration, and no great spiritual impetus was released to stem the decay of inner Christian life.

1. Richard Rolle: *A Song of the Passion.*

In 1453 the fall of Constantinople before the Turks shook the whole of Christendom, and removed the last bulwark of defence between the Christian West and the victorious advance of Islam. For two centuries Constantinople had been only a shadow of itself, but it had been there. From now on, one Christian province after another was lost to Europe, until in 1683 the armies of the Turks were encamped at the gates of Vienna.

Throughout the fifteenth century, and with increasing impetus, the great movement known as the Renaissance was liberating the human spirit, and opening out before it new roads of discovery and experience. But in none of its earlier forms did the Renaissance bring much comfort to a starved and impoverished Christendom. It might express itself in the somewhat cloudy neo-Platonism of such men as Pico della Mirandola (1463–94) in Italy; in that exaggerated regard for pre-Christian classical antiquity which among other things suggested the rewriting of the rather barbarous Latin of the Church's service books in the classical rhythms of Cicero; or in frank scepticism and a return to pagan sensuality; hardly ever in a rediscovery of the hidden powers and principles of Christian living.

And yet at the end of the century the Western Church was still a fabric of impressive unity and solid power. The ordinary man in England as elsewhere complained of his Church and criticized it, but still he loved it and depended on it. The Mass was the centre of the life of the parish; few parishioners would be absent on Sunday, and a surprising number of Christians in England found their way to church even on weekdays. Probably the ordinary man understood little of the teaching of the Church – late medieval preaching was directed more to the emotions than to the mind. Often his behaviour in Church was deplorable. Yet the thought of being excluded from the Church still had power to horrify him, and he expected, when life ended, to be laid to rest with its prayers. The man of 1500 could see no further into the future than we. But certainly he would have been utterly incredulous had he been told that within fifty years the stately unity of Western Christendom would have been brought irreparably to an end, and that the last years of the century would see separated Churches locked in mortal and apparently irreconcilable conflict with one another.

If it was a new world that was born in the sixteenth century, the pangs through which it came to the birth were very long and exceeding bitter.

CHAPTER 2

Reconstruction

THE Reformation of the English Church in the sixteenth
century is something about which everyone living in an
English-speaking country, and particularly in Britain, has to
make up his mind. The evidences of it, and of its remoter
consequences, shout at him from almost every street, in the
presence of Churches calling themselves by many and
various names, agreed only in regarding themselves each as
the true and legitimate heir of the Christianity of the New
Testament.

Five distinct points of view can readily be distinguished.

There are those who hold that the Reformation was a
thoroughly bad thing, which ought never to have happened,
and of which no good can be said. This is naturally the view
of nearly all Roman Catholics, and of some sentimental
idealizers of the Middle Ages.

Some think that the Reformation was on the whole a bad
thing, but that it had certain redeeming features, such as
the gift to the people of public worship in a tongue that they
could understand. This is broadly the position of the Anglo-
Catholic wing of the Anglican Churches.

Others believe that the Reformation was on the whole a
good thing, but that it was marred by some unnecessary
violence, and that some of the traditional good things that it
rejected might better have been retained. This is roughly
the position that is maintained in this book.

Many people have held and hold that the Anglican
Reformation was the greatest thing that ever happened in
the history of the English people, that it found the exact
middle point between not going far enough and going too
far, and that the Church of England is suspended on the
perfect point of balance between 'the meretricious gaudi-
ness of the Church of Rome, and the squalid sluttery of a

fanatical conventicle'. This was the view of almost all Anglicans between about 1633 and 1833, and of almost all Evangelical Anglicans up till the present time.

Lastly, some think that the Reformation was a good thing, but that it would have been even better if it had gone a great deal further than it did. Some Anglicans have held this view, among them the late Queen Victoria. It is almost universally held by continental Protestants, and is naturally shared by the majority of English Nonconformists.

One or two other major points must be made clear before we come to Henry VIII and his six wives.

The first is that, though we speak often of 'the Reformation' in the singular, there were as a matter of fact no less than six distinct types of Reformation of the Church in the sixteenth century.

There was, first, the immense work of renewal that took place in the Roman Catholic Church, and is often called the Counter-reformation, because to a large extent it took place as a reaction against the Protestant movements in various countries. This was conservative, in the sense that it did not ask any fundamental theological questions but codified Roman Catholic doctrine as it had taken shape at the end of the Middle Ages. From it the Church of Rome emerged with its doctrine more sharply defined, its organization improved, its discipline strengthened, and the power and authority of the Pope greatly enhanced. The theological movement was accompanied by a great renewal of devotional life, and a tremendous missionary effort, the Jesuits taking the lead, which in a century had carried the Roman faith to the four corners of the earth.

Next, there was the movement which, beginning in Germany with Luther, spread quickly to Scandinavia. This was based on a rediscovery of the Bible and of the doctrine of justification by faith, which we shall later return to consider in more detail. Luther, although a reformer, was still in many ways a medieval man. At some points his thought was revolutionary, but at others he was surprisingly conservative. English travellers, familiar with the bareness of most nonconformist chapels in England, are often astonished to find in Germany Lutheran churches that have retained almost their medieval aspect, with lights and crucifixes on the altar, many of the ancient elements of the liturgy, and even the chanting of the canticles to Gregorian plainsong.

The Reformation which, under the leadership of Zwingli and Calvin, took hold of Switzerland, France, Holland, and Scotland, was more radical, and more intellectual, than that of Luther. Equally firm in its insistence on the Bible as the sole authority, it took as its starting-point the sovereignty of God; and, as a practical principle of reformation, the view that anything which is not expressly commanded in the Bible ought to be rejected from the Church. As a result it departed very far from the ancient traditions of the doctrine, the worship, and the organization of the Church.

All these three types of Reformation believed themselves to be maintaining the essential continuity of the Church. This is interestingly illustrated by the researches of Professor Jaques Courvoisier into the question of the date at which the Church of Geneva became 'Protestant'. The surprising answer is 'after the revolution of 1848'. Up till that time, the Church had been officially simply 'the Church of Geneva'; that Church had certainly changed its form, but it still claimed to be the Church that had existed in Geneva since the earliest Christian centuries.

The fourth type, now to be considered, was far more radical than any of the three so far mentioned. The people called 'sectaries' or 'Anabaptists' denied that there had been any continuity of the Church. Their movement was not the reformation but the re-birth of the Church. They were as opposed to Lutherans and Calvinists as to Roman Catholics; on all these organized Churches they could only call down the doom of Babylon: 'Down with it, down with it, even to the ground.' These folk rejected infant baptism, and denied every claim on the part of the State to be concerned in the affairs of the Church. The Church was not 'the State at prayer';[1] it was the gathered company of the elect believers in Christ. The holders of these tenets were suspected of being

1. The Church of England has been described as 'the Conservative Party at prayer'.

political revolutionaries, and the extravagances of some of their followers lent colour to this suspicion. They were bitterly persecuted in every country of Europe in which they made their appearance, and almost as cruelly by Protestants as by Roman Catholics.[1]

The fifth reformation was that of the humanists. The father of this movement, though he himself never left that Roman Church which he rather ruthlessly criticized, was Desiderius Erasmus (1466–1536), the great Dutch scholar and humanist. In his Christian writings there is strong emphasis on the ethical element in Christianity and on Jesus Christ as our example, less understanding of Jesus as our Saviour and of the doctrines that are bound up with recognition of man's need for a Saviour. In the hands of such followers of Erasmus as the Italian humanists Jacob Acontius and the Sozzini, this movement became unmistakably unorthodox; it denied the unique divine Sonship of Jesus Christ and the doctrine of the Trinity. By 1645 the word 'Socinian' had passed into the English language as the modern equivalent of the ancient 'Arian'.

The Anglican Reformation is related to all these other movements, but it is not to be identified with any of them. Influences from all of them have played upon it. At one point or another it seems to approach rather close to each of them in turn. The tendency, therefore, has been to attempt to explain Anglicanism in the light of something that is not Anglican. Continental Protestant scholars have always tended to interpret the Anglican tradition as a continuation and extension of the ideas of Erasmus. As a scholar and an educationist, Erasmus, who had taught Greek at Cambridge (and wrote rather scathingly of English accommodation and English beer), exercised a deep influence on English thought and practice; it is almost startling that there are by contrast so few traces of his influence on the development of English religion and Anglican theology. Roman Catholics tend to class Anglicans with other 'Protestants', and thereby to make impossible for themselves any understanding of what

1. In England twenty-four Anabaptists were tried in one single day in 1535, and fourteen were sentenced to be burned alive.

Anglicanism really is. We cannot begin to understand it unless we are prepared to accept it as something unique and unlike anything else. If we do so, we may be able to grasp something of its genius, to understand its situation in the complicated map of Christendom, and perhaps to believe in the special vocation that seems to have been accorded to it amid the many and various Churches of the Christian world.

Although we have so much evidence about the sixteenth century, and although so much has been written on it, we still do not know exactly what happened at the Reformation. What proportion of the clergy and people genuinely accepted it? How many slipped into it through sheer indifference? To what extent was its success ensured by coercive action on the part of the State and nothing else? Careful research is gradually throwing light on these problems, but much remains to be done. Traditionally the Reformation has been interpreted as a purely religious movement. We can now see that the new nationalism, and the new type of monarchy, of which Machiavelli's *The Prince* (*c.* 1513) is the charter, played a much more than negligible part in its development. But political changes do not produce religious revolutions. More recently, stress has been laid on economic factors: the Reformation was a part of the shift of power from the ancient aristocracy to the thriving burgher class in the cities and to the smaller landed gentry. But simple Protestants were not so simple as to allow themselves to be burned alive just in order to make safe the enrichment of covetous laymen by the spoils of the Church. We must recognize and allow for all these other factors. But in the end we must come back to the religious question, which presented itself to the best men and women of that age as supremely important.

What is often overlooked is that the Reformation is part of the laymen's revolution. For centuries the Church had dominated every part of the nation's life, even its military activity. Now the laymen were determined to bring that domination to an end. The appointment in 1529 of Sir Thomas More (1478–1535) as Chancellor of the kingdom was a sign of the change in the times. It is part of the greatness of More that, as England's first great lay statesman other than a king, he was the forerunner of the modern prime minister; and as the medieval man, willing to give his life in opposition to the changes in religion, he became the ancestor of the modern Roman Catholic saints and martyrs. Even later, it was not impossible for an ecclesiastic to be also a statesman. For a time under Charles I the estimable William Juxon, Bishop of London, was Treasurer of the realm ('I judge you, my Lord of London, the fittest since you have no children', said the king)[1]. But this was an exception. The Church in England has always exercised great influence in the life of the nation; since the Reformation it has never been able to exercise unquestioned and unchallenged authority.

In the reign of Henry VIII, three factors tending towards a change in religion were at work – the desire of the king for a male heir, the growth of national and anti-clerical feeling, and the spread of Lutheran ideas. We shall consider each of these in turn.

As a very young man Henry had been married to the widow of his brother Arthur, Catherine of Aragon, a pious Spanish lady a good many years older than himself. Catherine had borne him many children, but they had all died in infancy, except for one daughter, the Princess Mary. It was uncertain whether a queen could rule in England.[2] It is hard for us, in these days of untroubled constitutional succession, to realize how vital it was in the period of personal rule that succession to the throne should be determined in advance without any possibility of doubt. Henry desired desperately to have a son and heir. Then his conscience got to work. Had he ever really been married to Catherine? Was not the death of so many children a judgement of God on what was really an incestuous union? Was he not really free to marry whom he would, and in particular a lady named

1. This was the same Juxon to whom on the scaffold the king uttered the mysterious word 'Remember'.
2. England was not, like some continental countries, under the Salic law, which excluded women from the succession; but the nation's only experience of female rule was the disastrous attempt of the Empress Matilda, in the reign of Stephen, to make herself queen.

Anne Boleyn? Henry was not an unconscientious man. He had a sensitive conscience; the only trouble was that his conscience so often told him that what he wanted to do was right.

We commonly speak of Henry's divorce. But this is inaccurate. What Henry wanted was not a divorce – that is a declaration that a marriage once made had now been broken – but a nullity, a declaration that, owing to a canonical impediment to marriage, no real marriage had ever taken place at all. Henry had a much better case than has often been admitted in modern times. This was, apparently, the first occasion in history on which a dispensation had been given by the Pope for the marriage of a man with his deceased brother's widow. Many eminent scholars sincerely doubted whether even the Pope could give such a dispensation. It seemed to run counter to an express command of Leviticus – does it not stand written in Leviticus xx. 21: 'If a man shall take his brother's wife, it is impurity . . . they shall be childless'? And was this not very nearly the fate that had befallen Henry? In any case, Popes had in the past done far worse things than that for which Henry was asking. A great many Englishmen knew that in 1152 the Pope had separated Eleanor, Queen of France, from her husband Louis VII, after fourteen years of marriage and the birth of children, really on no ground other than that of incompatibility of mind (though it was declared that they were within the prohibited degrees), and that within three months she had married Henry of Anjou, later King Henry II of England.

Nothing could be further from the truth than to represent the Pope as standing for the sanctity of Christian marriage and Henry as trying to debase it for the furtherance of his own lusts. If the Pope had from the first moment told Henry plainly that the dissolution of his marriage could in no circumstances be considered, he might have restrained Henry from the step that he was bent on taking, and at least he would have retained the respect of Christendom. But this was far from being the Pope's attitude. As H. G. Wells so truly remarked, Englishmen objected to the Pope not because he was the religious head of the Church, but because he was not. It was plain to all men that the Pope was longing to accord to Henry the nullity for which he asked, if only he could find some means of doing so without offending the Emperor Charles V, who was Queen Catherine's nephew, and of whom he was mortally afraid. He was prepared to go almost to any length, even apparently to the length of permitting Henry in these special circumstances to commit bigamy. But Henry was determined to have a marriage which would leave no doubt as to the legitimacy of his heirs, and nothing less than unquestioned freedom to marry again would satisfy him. So the case dragged on from court to court, to the scandal of Christendom. Henry was getting older year by year, and in 1529 no decision in his case had yet been reached.

Then someone – perhaps it was Thomas Cromwell, but we do not know – put into Henry's mind the question whether it was necessary for him to go to the Pope at all. In civil or criminal matters, no appeal from the king's courts could lie to any court beyond the seas, as for instance to the courts of the Holy Roman Empire. Was there any reason for an appeal in an ecclesiastical case to lie beyond the ecclesiastical courts of the land, and to be heard by a foreign bishop? This raised a weighty question – who has the right to declare and execute the law in England? One of the results of the Renaissance of the fifteenth century was that men's attention had been directed back to the original texts of Roman law, and in particular to the great lawgiving emperor Justinian in the sixth century. There men saw a prince who was *fons utriusque juris*, the source of the law of the Church as well as of the law of the State. It was the business of the patriarchs and bishops to put the law into effect, but the law that they administered was the law of the Christian emperor, and it was under his authority that they administered it. Charlemagne in the West, at the beginning of the ninth century, had handled the affairs of the Church in exactly the same way. William the Conqueror, as we have seen, had prevented any appeals passing from England to Rome. Was Henry any less potent as a prince than they?

Here we can see the significance of the affirmation, in the

preamble to the Act in Restraint of Appeals (1532), that
'this realm of England is an *empire* . . . governed by one
supreme head and king . . . institute and furnished by the
goodness and sufferance of Almighty God, with plenary
whole and entire power pre-eminence authority prerogative
and jurisdiction to render and yield justice and final determination to all manner of folk . . . in all causes . . . without
restraint or provocation to any foreign princes or poten-
tates'. We have already stressed the legal cast of the
thinking of our forefathers. This was a legal question. What
is an empire? It is a realm, which is wholly independent
legally (and that meant, to Henry and his advisers, in the
law both of Church and State) of every other realm. But, if
this was so, if Henry was the new Justinian, what became of
the Pope's claim that he was the supreme judge of Christendom, and that he and he alone had the final voice in all
ecclesiastical causes? Henry answered roundly that this
was an usurped jurisdiction; former Popes had made no
such claim, and it had not been admitted by earlier English
kings; it was an abuse that had crept in in the times of
ignorance – but now the times of ignorance had passed
away.

It may be questioned whether Henry was right in his interpretation of history. At least he must be given credit for
believing what he affirmed, and for being convinced that he
was in fact restoring those good old customs of his ancestors,
which in course of time had become obscured. There is no
reason to question the verdict of one of the most learned of
English historians, E. A. Freeman: Henry VIII, he wrote,
had no thought of 'setting up a new Church, but simply of
reforming the existing English Church. Nothing was further
from the mind of Henry VIII or of Elizabeth than the
thought that either of them was doing anything new.
Neither of them ever thought for a moment of establishing
anything at all. In their own eyes they were not establishing
but reforming; they were not pulling down or setting up, but
putting to rights.'[1]

1. E. A. Freeman: *Disestablishment and Disendowment* (2nd ed., 1885)
p. 66.

In the light of his new discovery Henry claimed to be the Supreme Head, under Christ, of the English Church. It is essential to understand what is, and what is not, included under this term. No English sovereign has ever claimed the right or the authority to perform spiritual functions in the Church. The term 'Supreme Head' was primarily legal in its application. And the pointed question that could be asked of the king's subjects was this – 'Has the Pope any jurisdiction in England?' 'Jurisdiction' here does not mean some general spiritual influence or control or power of direction. The word is to be taken in its exact sense; it means the authority to declare what the law of the Church in England is, and to put that law in execution through the Pope's own courts. Henry denied that the Pope had any such power. To say that any foreign potentate has authority or rule in England is treachery to the existing government. That is why Henry declared that to refuse him the title of Supreme Head of the Church was *treason*, a diminution of the authority of the king in his own dominions.

Of all the clerics in the land the bishops naturally stood in closest relation to the Pope. In 1534 Henry made a pained discovery: 'We thought that the clergy of our realm had been our subjects wholly. But now we have well perceived that they be but half our subjects, yea and scarce our subjects. For all the prelates at their consecration make an oath to the Pope clean contrary to the oath they take to us, so that they seem his subjects and not ours.' This could not be allowed to continue, and Henry would himself take better order. Basing himself on the privilege, maintained by the founders of abbeys and their heirs, of naming the person to be elected to the headship of an abbey, Henry declared that he would now, as sole heir of the founders of the English bishoprics, take sole responsibility for choosing the men that were to fill them. He would send down to the chapter a single name, and the chapter must with 'all speed and celerity' elect the candidate proposed to them by the King, under the heavy penalties of the Act of Praemunire. In practice this did not differ very much from what had happened in the past; English kings had usually got the bishops that they wanted, though skilful negotiations with the Pope had often been necessary. Nor was the English Post-Reformation practice as different as is often supposed from that prevailing in Roman Catholic countries. In Austria, till the end of the first World War, the bishops were in fact the nominees of the Emperor. In the France of the Third Republic up to 1905, they were in fact the nominees of a probably anti-clerical Minister of the Interior. Nevertheless there were grave disadvantages in the autocratic system introduced by Henry, and that system weighs heavy on the Church of England at the present day.

When an English bishopric is vacant, the Queen takes action under the Act of 1534, and sends down one name only to the chapter of the diocese concerned. The chapter meets and, after solemnly invoking the guidance of the Holy Spirit, proceeds to elect the candidate, in the choice of whom the Church as such has had no voice whatever. No man can become a diocesan bishop in the Church of England without having to submit to what can hardly be regarded as an edifying piece of ceremonial. It is sometimes maintained that the Church has some control over the appointment, and that the chapter can refuse to elect the royal nominee. If, in the course of 442 years, any chapter had once refused to elect, there would be some force in the argument. It is a happy thing that all the other Anglican Provinces have freed themselves from the past, and that it is only on the forty-three dioceses of the English provinces that the hand of Henry VIII still rests so heavily.

Nothing in the long history of the separation of the Church of England from the Church of Rome is more astonishing than that the King encountered so little opposition to his plans. Only two among the greatest men in the kingdom refused to take the Oath of Supremacy, and so perished as martyrs on behalf of the old regime.[1]

John Fisher, Bishop of Rochester (c. 1469–1535), whom even the unsympathetic chronicler Edward Hall described as 'a man of very good life', was known as a strong supporter

1. We must not, however, forget the Carthusians of London, and other less outstanding martyrs.

of Queen Catherine, and a stout upholder of the old ways. At the beginning of the breach with Rome he had been in trouble for remarking in the House of Lords: 'My Lords, you see daily what Bills come here from the Commons house, and all is for the destruction of the Church. For God's sake see what a realm the kingdom of Bohemia was; and when the Church went down, then fell the glory of the kingdom. Now with the Commons is nothing but *down with the Church*, and all this meseemeth is for lack of faith only.' When Fisher was already in prison for refusal to take the oath, the Pope appointed him as Cardinal; and this indiscreet act made inescapable a doom which probably was already certain. He was beheaded on 22 June 1535.

Sir Thomas More was one of the greatest men and greatest Christians of his day. He is fortunate in having been the subject of one of the earliest and best of British biographies,[1] and so it comes about that we know him better than almost any other man of his day. We are able to see him in the calm and gaiety of his family life, in the austere devoutness of his religion, in the firm resolution with which he endured long imprisonment and uncertainty, and the quiet courage – one is tempted again to write 'gaiety' – with which he met his death. More has suffered from the adulation which in some quarters has been poured on him. He was not without his faults. He could be a harsh and unfair controversialist. Like other men of his time, he believed in the persecution of heretics and the suppression of heresy by violence. But he was a good man.

As the friend of the 'Oxford Reformers', Erasmus and Dean Colet of St Paul's (c. 1467–1519), and as the author of *Utopia*, More was well aware that there was a great deal in the medieval Church that stood in need of reformation. He was prepared to go a long way with Henry in his plans, and to expound them to others. But when it came to the Oath, he found that he could not in conscience take it. Perhaps, more far-sighted than the King, he saw that the moderate reconstruction of the Church which Henry had in mind would quickly turn itself into a revolution, and would be

1. That by William Roper, More's son-in-law, first printed in 1626.

carried to lengths which would have horrified Henry himself, if he could have imagined them. He saw the line on which he must stand, and from that line nothing would move him. Everyone wanted to save him. Henry himself expressed the deepest regret at his death, and the court wore mourning for a fortnight after his execution. But this was one of those tragic cases in which what one man saw as religious obligation was interpreted by another in terms of political necessity; and neither could give way. To Henry, More had become simply a treasonable subject, and so he must die. More met his death sweetly and without rancour. Is there on record any nobler utterance of a man under sentence of death than More's last words to his judges at the end of his trial: 'More have I not to say, my lords, but that like as the blessed Apostle St Paul, as we read in the Acts of the Apostles, was present and consented to the death of St Stephen, and kept their clothes that stoned him to death, and yet be they now both twain holy Saints in heaven, and shall continue there friends together for ever, so I verily trust, and shall therefore right heartily pray, that though your lordships have now here in earth been Judges to my condemnation, we may yet hereafter in heaven merrily all meet together, to our everlasting salvation. And thus I desire Almighty God to preserve and defend the king's Majesty, and to send him good counsel.'

We must raise once again the question how it came about that the King met with so little resolute opposition in the carrying out of his plans. Amid the confusions and complexities of those crowded years, it is possible to identify at least six factors as having worked in favour of the King.

There was, first, the plain fact that Henry was a resolute and dangerous person. In those days personal rule was a very real thing. If Henry had made up his mind on anything, those who opposed him knew that they were jeopardizing their lives by doing so. When the Duke of Norfolk remarked to More, 'By God body, master More, *Indignatio principis mors est*', More had quietly replied, 'Is that all, my lord? . . . Then in good faith there is no more difference between your grace and me, but that I shall die today, and you to-morrow.' But not all shared More's quiet equanimity and courage.

Secondly, quite apart from Henry's personal qualities, the sixteenth century was a time when kingship came into its own. England had lived through the dreadful miseries of the Wars of the Roses, and knew what it was to have no firm and settled government. What most men want is peace and quiet, and they will put up with a great deal from a government, provided that it is fairly competent and reasonably just. The commons of England had not been taught to think for themselves; they had been taught to obey. Resistance to the King was resistance to the will of God. If the King's grace will have it this way rather than that, then for most of them that was sufficient reason.

Thirdly, the critical spirit of the Renaissance had undermined men's implicit confidence in the traditional and the familiar. Much of the Pope's claim to universal dominion had been based on the Forged Decretals, that remarkable collection of papal documents from the time of the apostles onwards, of which some were genuine and ancient, but the majority had been produced in France during the ninth century. Men asked how it came about that, though the basis of the claim was now known to be false, the claim still continued to be made.

Fourthly, propaganda had begun to work with a power such as it had never had in earlier days, and the King was in complete control of all the printing presses. The printing press had been Luther's great ally. All his works were printed in astonishingly large editions, and sent all over Europe, not yet nationally divided by the lack of a common tongue. The English Reformation would have taken on a quite different form, if the press had not been there to supply cheap Bibles and prayer books. Sir Thomas More estimated that in his day half the grown men in England could read. This may have been an exaggeration, and certainly the percentage of women was very much smaller. But still there was a large and increasing reading public, and the King could see to it that they read what was good for them.

During these troubled years no book was allowed to appear in England unless it was favourable to the King's cause. One of the favourites was the *Defensor Pacis* of Marsilius of Padua, which was at this time translated into English.[1] Marsilius, an Italian who became Rector of the University of Padua in 1312, held startlingly modern views on the problems of politics and religion, of State and Church. He believed that the whole edifice of papal power was a perversion of the true principles of Christianity and a menace to the peace of Christendom. The Church is the whole body of believers, and the supreme authority in Christendom should be not the Pope, but a general council of the Church. Priests must have no coercive powers or the right to impose excommunication. Nor must they own property; they may have the use of it, but donations made in favour of the Church still remain the property of the donors – a doctrine which Henry was to find particularly useful in his dealings with the monasteries. Altogether Marsilius is an author for whom the King is likely to have had a special affection – and this was the kind of reading material that he thought suitable for his subjects.

Fifthly, the new power of nationalism was spreading throughout Europe, and was stronger in the island kingdom than it had as yet become elsewhere. Bernard Shaw is writing of course anachronistically and yet perceptively, when he makes his Inquisitor in *Saint Joan* give the name 'nationalism' to one of the new heresies that he sees threatening the future of the world. It is clear that the majority of Henry's bishops, the ablest among them, Stephen Gardiner of Winchester (*c.* 1493–1555) included, believed it possible to combine the freedom of a national Church with a purely spiritual overlordship of the Pope in Western Christendom. They did not desire any change in the doctrines of the Church – the authority of the Pope had not yet been defined

1. This translation, by Marshall, published in 1535, omitted, according to Dr Previté-Orton, about one-fifth of the work; it was badly done and severely edited in the interests of monarchy. The first complete modern translation into any European language seems to be the English version of Dr Alan Gewirth, published at Chicago in 1956.

as a doctrine – they did desire limitations on the power that a foreign bishop could exercise in the realm. They were mistaken – in both directions; both because the Pope would not accept any diminution of his claims, and because Protestantism would sweep away so many of the familiar land-marks. And when 'the universal pastor of the faithful' can be referred to as a foreign bishop, the battle of the moderate conservatives has been in point of fact already lost.

Finally, and most important of all, Henry could not have put his plans into execution unless anti-clerical feeling had become very strong in the land. The clergy had made themselves hated – by the vexatious claims and exactions of the Church courts, through the overweening magnificence and arrogance of Wolsey, through their failure in so many cases to attend to their spiritual duties. When the opportunity came to bring the clergy down, very few among the laity were found ready to stand up in their defence. It was the Commons who did the King's work, and in the Commons' house the clergy were not directly represented. 'If Henry VIII had not led the Reformation Parliament, it would not have achieved the Parliamentary Reformation. But equally, if – *per impossibile* – Henry had attempted to use this, or any other of his parliaments, not to fortify, but to weaken, the State against the Church, the layman against the cleric, he would have found Parliament uncooperative to the point of hostility. His "faithful commons" did what he asked them to do, not simply because he asked them to do it, but because it was what they themselves would have done if they, and not he, had been responsible for shaping policy.'[1]

Up to this point, we have not yet encountered anything that can properly be called reformation, or anything that is of fundamental importance for Anglicanism. Henry had carried out immense changes in the organization of the medieval Church, and much of what he did has been of continuing importance in the life of the Church of England,

1. S. T. Bindoff: *Tudor England* (1950), The Pelican History of England, Vol. 5, p. 98.

though not in the other Anglican Churches. But in all that
we have so far considered, there has not been a single trace
of any new religious initiative, or of any new power in the
preaching of the Gospel. It is only when we turn to the
third of our three great factors, the infiltration of new ideas
from the Lutheran Reformation in Germany, that we are at
grips with creative forces that, for good and ill, have permanently moulded that strange and beautiful thing called
Anglicanism. It is certain that Lutheran books were being
read in England not later than 1521.

The intellectual centre of the new movement was Cambridge. There a group of earnest young men used to meet, in
order to read and discuss the Lutheran books, at the White
Horse Tavern, which in consequence acquired for itself in
the University the name of 'Germany'. This contact between Luther and the English mind was the real beginning of
the English Reformation. What did the readers find in
Luther that stirred them to new ideas, to impatience, and
in certain cases to readiness for revolution?

The first answer is that they learned from Luther to read
the Bible in a new way. Luther has become a figure of controversy, and many people think of him only as a polemical
writer and an iconoclast. But this is fundamentally to misunderstand him. Luther was primarily a great preacher,
and a great teacher of the Bible. For thirty years he regularly preached and lectured. Many of his sermons have at
last been translated into English, but they are still little
known in the English-speaking world. This accounts for
many misunderstandings. Discarding the use of allegorical
and mystical interpretation, he went straight to the text, and
in particular to the original text of the New Testament,
which since the publication of Erasmus's Greek Testament
in 1516 was readily available to anyone who could read
Greek, and asked what it meant. Under this direct handling
the simple words of the Gospels took on new life. Using
plain, racy speech, with constant allusions to current situations, Luther directed his appeals steadily, not to the emotions, but to the minds and to the wills of his hearers. Those
who heard and read him felt themselves to be entering into

a new world. By his own translation of the Bible into German Martin Luther laid the foundations of modern German prose literature.

We do not know to what extent the medieval Christian was familiar with the Bible; differing views are held on the subject by the experts. Certainly the Bible had been accessible to the scholars. But they had read it within the limits of certain fixed categories of interpretation, and that meant that they had seen certain things and had been blind to others. What men learned from Luther was to read it with a different set of preconceptions, and, when they did so, they felt that they were reading it for the first time. They entered into a personal relationship with it, and felt that it was a living word of God spoken directly to their own souls. Nicholas Ridley, later Bishop of London and a martyr (c. 1500–55), shall be our witness to what this new experience meant to them 'Farewell, Pembroke Hall [Cambridge], of late mine own College, my cure and my charge. . . . In thy orchard (the walls, butts, and trees, if they could speak, would bear me witness), I learned without book almost all Paul's Epistles, yea and I ween all the canonical Epistles, save only the Apocalypse. Of which study, although in time a great part did depart from me, yet the sweet smell thereof, I trust, I shall carry with me into heaven: for the profit thereof I think I have felt in all my lifetime ever after.'[1]

Among other things that Luther had rediscovered in the Bible was the apostolic doctrine of justification by faith. What is meant by this formidably technical term? The great problem for one who knows himself to be a sinner is to discover how he may commend himself to God, and how he can be sure that he has been accepted by God. The answer of the New Testament is that the question has been wrongly framed; man has no need to commend himself to God; it is God Who commends Himself to man: 'God commendeth his own love toward us in that, while we were yet sinners, Christ died for us' (Rom. v. 8). The doctrine of justification is

1. Quoted in H. C. G. Moule: *Bishop Ridley on the Lord's Supper* (1895), p. 8.

simply the theological expression of the words spoken as a criticism of Jesus Christ in the Gospels: 'The man receiveth sinners and eateth with them.' What had the sinners done that they should be acceptable to Jesus Christ? The answer is that they had done nothing – they had simply allowed themselves to be found by the good Shepherd, who had come seeking them. How do we know that this is true of us also? Because God always comes to meet us in the Cross and resurrection of Jesus Christ.

Justification means the establishment of a new and permanent relationship between God and the sinner. It is that act in which God in Christ declares Himself to be for all eternity gracious to the sinner who trusts in Him, regardless of that sinner's deserts. It is the act in which the sinner renounces for all eternity every dependence on himself, on what he can do, on his own merits, and commits himself wholly, unreservedly and unconditionally to the forgiving mercy of a gracious God, whom he has come to know in the death and resurrection of Jesus Christ.

Of course this doctrine can be perverted in an 'antinomian' sense; if God will forgive us, whenever and however much we sin, why should we bother not to sin? St Paul himself was well aware of this, and complained of those who interpreted his teaching as meaning 'Let us continue in sin that grace may abound', or 'Let us do evil that good may come'. This was to St Paul so evident an example of sin against the Holy Spirit that he does not stay to argue with the detractors; he dismisses them with the indignant remark: 'Whose damnation is just.' Luther would have replied that no one can fall into this error, if he has the most elementary understanding of what faith means. For faith in the New Testament means the total self-commitment of man to God in trust, in adoring gratitude, in love, and in humble obedience. If any of these elements is lacking, so-called faith is not faith in the true sense of the term. What Luther had in fact done, and this was perhaps his greatest service to theology, was to restore the true relationship between faith and works. Works are not to be understood as good deeds which man does in order to commend himself to God or to make satisfaction for his sins. The idea of merit on the human side is wholly and for ever excluded. Good works are those deeds which the Christian cannot refrain from doing, because he is constrained by the love of Christ, now dwelling in his heart by faith.

In point of fact, there has rarely been any connexion between Reformation doctrine and antinomian practice. The Churches which have most faithfully proclaimed justification by faith are also those which have maintained the most exacting, and sometimes puritanical, moral standard. For we shall not understand the Reformers at all, unless we realize that what these men were primarily concerned about was holiness. They had tried the medieval way, and had not found that it gave them the victory that they sought. They believed themselves to see the results of its failure in the low level of Christian life around them, even among the clergy. They were convinced that the new understanding of the Gospel, with its appeal to the whole man, mind and conscience and will, could bring about that inner reformation that to them was more important than any change in ritual or in the organization of the Church.

They read in the New Testament that the obligation to holiness rests on the whole people of God, not on one part of it only. For them there was no division of the world into 'religious' (in the sense of monastic) and secular; there might be differences of vocation (though this in the enthusiasm of their new views they were inclined to deny), but there could be no two standards of achievement. The layman no less than the priest is called to 'perfect holiness in the fear of God'. This is the ideal that is solemnly and soberly set forth in the address to the candidates in the Anglican form for the Ordering of Priests: 'Your bounden duty to bring all such as are or shall be committed to your charge, unto that agreement in the faith and knowledge of God, and to that ripeness and perfectness of age in Christ, that there be no place left among you, either for error in religion, or for viciousness in life.' This is a tremendous ideal, and Anglican history is strewn with the wrecks of our failure to attain to it. But which is better – to set before ourselves a lofty ideal, and to fail to reach it, or to acquiesce in the acceptance of a lower ideal?

This, and nothing else, is the famous and much misunderstood Reformation doctrine of 'vocation'. It means that the layman also is called to be a saint; that the place in which he must work out his saintliness is the home, the bank, the factory, the dock, the field; and that, if he has understood his vocation, he can be sure that God will be as much with him there as He is with the priest saying the services in Church. It is not necessary to labour this point for the Anglican, since John Keble has expressed it for him perfectly in one of the hymns that he knows best:

> We need not bid, for cloistered cell,
> Our neighbour and our work farewell,
> Nor strive to wind ourselves too high
> For sinful man beneath the sky;
> The daily round, the common task
> Will furnish all we need to ask –
> Room to deny ourselves, a road
> To lead us daily nearer God.

With all these rediscoveries the Reformers recovered also something of the joy of the early Christians, and above all of their confidence in the face of death. Late medieval religion had tended on the whole to be a gloomy thing. The best that a Christian could hope for, even if he had managed to escape the endless pains of hell, was to awake in a Purgatory little less terrible than hell. The spontaneous joy of the Gospel had been hidden under layers of anxiety and fear.[1] How different is the spirit of comfort and good hope that breathes through the Anglican service for the Burial of the

1. This may be illustrated by a quotation from the good Bishop John Fisher, to whom we have already referred. It is clear that in his understanding the pains of Purgatory, while they last, *separate us* from God. Even in this life pain 'will not suffer the soul to remember itself, much less therefore it shall have any remembrance abiding in torments, for cause also the pains of purgatory be much more than the pains of this world. Who may remember God as he ought to do, being in that painful place? Therefore the prophet saith *quoniam non est in morte qui memor fit tui* [in death there is none that hath remembrance of thee].' Sermon on Psalm VI.

Dead. There is still the profound sense of sinfulness ('We meekly beseech Thee, O Father, to raise us from the death of sin unto the life of righteousness'), the same acceptance of death as a sign of the judgement of God on sinful humanity; but death is swallowed up in victory; 'Almighty God . . . with whom the souls of the faithful, after they are delivered from the burden of the flesh, are in joy and felicity.' If there are to be 'Requiem Masses' in the Church, the liturgical colour for them can only be white, the colour for Easter.

One of the young men who read Lutheran books at the White Horse Tavern was named Thomas Cranmer. This had a good deal to do with the future of the Church of England. In 1529 Cranmer was useful to the King in the matter of his 'divorce'. On 23 August 1532, Archbishop Warham of Canterbury died, and Henry decided that Cranmer should be his successor. Cranmer, at the time absent in Germany, knew something of the burdens that would be laid on his academic and unwilling shoulders, and did his utmost to refuse the proffered honour. But Henry usually got what he wanted. What he got in this case was the Archbishop that he needed. The best thing that can be said on behalf of Henry is that he never ceased to respect and protect Cranmer, and that Cranmer came to love his royal master so much that, when the King died, he allowed his beard to grow in sign of perpetual mourning for him. What Cranmer got was in the end martyrdom. But in the intervening twenty years he had done more than any other one man to make the Church of England what it is today. We have no English Luther or Calvin ('and a good thing too', many Anglicans would be inclined to add); we have as our chief reformer the man who had a greater genius for liturgical worship than any other of whom we have record in the whole history of the Church.[1]

Few men who have made a mark in history have been more cruelly traduced than Cranmer. He has been called pusillanimous, a time-server, a hypocrite, and so on. In recent years we have come nearer to an equitable judgement upon him, but it is doubtful whether the Church of England

1. The classical liturgies of the fourth century all bear the marks of genius; but they are all anonymous.

has yet come to appreciate to the full the greatness of a very
great man. This is not to say that Cranmer made no mis-
takes, or that he was entirely superior to the follies and prejudices of his age. From the point of view of those who
regard the English Reformation as an act of arrant folly
and a crime, Cranmer is the principal villain of the piece.
In a situation of extraordinary difficulty, when the majority
of the bishops were against him and the King's Council was
capable of framing a plot against his life,[1] he sometimes
concealed his objectives and moved tortuously towards
them. Yet those who call him timid forget that he was the
only man who had the courage to plead with Henry for his
injured daughter, the Princess Mary, the only man who
wrote to Henry on behalf of Thomas Cromwell after the all-
powerful minister's fall. Those who have read Cranmer's
writings are unlikely to doubt the splendid integrity of his
mind and character – and all this combined with such
meekness that it was said of him, 'If you do my Lord of
Canterbury an injury, you will make him your friend for
life'.[2]

Cranmer was a typical Cambridge don, a man, that is,
who liked to look on every side of every question, and was
slow to make up his mind on any. But once it was made up,
it was made up for good. It is very characteristic of him that
he has left it on record that, though as early as 1525 he
began to pray that the Pope's jurisdiction should be brought
to an end in England, it was not till twenty years later that
he began seriously to question the current doctrine of the
nature of the presence of Christ in the Holy Communion.
By the highest standards he was not a very learned man; he
had not, that is to say, the immense erudition of Calvin or
Bishop Andrewes; but he was widely read in the Fathers and
in the ancient liturgies of the Church, and he had a cool,

1. This business is set forth at great length in Shakespeare's *King Henry
VIII*, Act v, Scene 1.
2. Shakespeare, who elsewhere refers to 'the virtuous Cranmer', gives.
this version of the saying:

> Do my Lord of Canterbury
> A shrewd turn, and he is your friend for ever.

clear, patient, questing mind. He enjoyed the society of men who were better scholars than himself, and had the art of persuading them to put the treasures of their scholarship at his disposal.[1]

What then were Cranmer's Reformation ideals, and how did he attempt to put them into practice? We may pass over concerns of merely historic interest, such as his share in the affairs of the King's various marriages, and his tacking and veering in response to the changing moods of royal policy, and concentrate on those things that have left a permanent mark on the life of the English Church.

First, then, it is to be noted that Cranmer, like the other Reformers, had fallen in love with the Bible. But his love took a particular form. He believed that the Bible was the living word of God to every man, and that it comes with the greatest power when unaccompanied by any human gloss, comment, or exposition. He was convinced that, if his fellow countrymen could be induced to read the word of God, or, if illiterate, to hear it read, it would in course of time make its way into their hearts and consciences. It was only in the next reign that Cranmer was able to provide his Church with a lectionary; when he was able to do so, he made the Church of England in a day the greatest Bible-reading Church in the world. In no other Church anywhere is the Bible read in public worship so regularly, with such order, and at such length, as in the Anglican fellowship of Churches. In making such provision Cranmer was laying heavy demands on his Englishmen, and reposing great confidence in them. But in that too he was the typical Anglican – Anglicanism is a form of the Christian faith that demands and expects a great deal from ordinary people.

At another point Cranmer's outlook has laid its impress on Anglican theology. Almost all Churches agree that the ordained minister is a 'minister of the Word and Sacraments'. In all the Churches of the Lutheran and Reformed Com-

---

1. On this, see the excellent account in G. W. Bromiley: *Thomas Cranmer, Theologian* (1956), pp. 1–11. Henry's own estimate is summed up in his remark to Stephen Gardiner: 'My lord of Canterbury is too old a truant for us twain.'

munions, 'minister of the Word' means a man who preaches; in the Church of England, the minister of the Word is primarily the man who is authorized to *read* the word of God in the congregation. This implies no depreciation of preaching, but that is another and a separate vocation. The distinction is preserved in the curiously ungainly phrase, pronounced at the most solemn moment in the Ordering of Deacons: 'Take thou authority to read the Gospel in the Church of God, and to preach the same, if thou be thereto licensed by the Bishop himself.'

One of the great moments in the life of Cranmer arrived in September 1538, when the government ordered that every parish in the country should purchase a Bible of the largest volume in English, to be set up in every church, where the parishioners might most commodiously resort to the same and read it. To make plain how this could come about we must go back a little in our story.

We have seen that, since the end of the fourteenth century, the Wycliffite Bible had been circulating quietly in manuscript among the people. The inaugurator of the new era of the English Bible was William Tyndale (*c.* 1492–1536), a Gloucestershire man, who had studied both at Oxford and at Cambridge, had learned Greek and Hebrew, and made it his life's work to give his fellow-countrymen the Bible in their own tongue.

Tyndale began his work in England, but the atmosphere of suspicion and oppression made life dangerous for a would-be translator, and he was wise to escape to the Continent. His translation was completed in Wittenberg, the home of Luther, in 1524, and appeared from the press at Worms in 1526. Tyndale's later years were spent in Amsterdam, where the immunities of the English Merchant Adventurers protected him, and he was able to continue his translation of the Old Testament up to the Second Book of Chronicles. But in 1535 he was lured from sanctuary by an *agent provocateur*, and arrested by the authorities. After eighteen months of imprisonment he was first strangled, and then burnt, at Louvain in October 1536. From the beginning the blood of the English martyrs is on the English Bible.

Tyndale worked from the Greek and Hebrew, with the best helps then available, such as Luther's German version, and the Latin translation which accompanied the Greek Testament that Erasmus had produced in Basel in 1516. He had better luck than most pioneers in translation; the best tribute to his genius as a translator is the fact that, when the lessons are read in church in England today, for all the many revisions that have intervened between Tyndale and our time, at least seventy per cent of what we hear is read exactly as Tyndale set it down more than four hundred years ago.[1]

This is not to say that Tyndale's version is perfect. At times he misunderstood his original. At times his choice of words appeared to be determined by controversial interest rather than by considerations of scholarship, as when he rendered *presbyteros* by 'senior', and *ecclesia* by 'congregation'. Tyndale would probably have answered that it is impossible to separate words from their associations. We can correctly render the Latin word *poenitentia* in English either by 'penance' or by 'repentance'; if we wish the reader to understand, we must avoid the word which will suggest to him a meaning different from that which we believe the original to have conveyed. In some cases later usage has supported Tyndale, in others it has gone back to the apparently more obvious word. At times Tyndale carried his principle that the Bible should speak to ordinary men and women rather too far. The statement that Paul 'sailed away from Philippi after the Easter holidays' (Acts xx. 6) sounds much more like a twentieth than a sixteenth-century translation; and when he writes of those who 'chop and change with the word of God' (2 Cor. ii. 17), all later translators have found it advisable to return more nearly to the dignity of the original. Still worse, Tyndale unlike Cranmer was not prepared to allow the word of God to do its own work, but felt it necessary to garnish his margins with notes, in which none of the pungency of the controversialist was

1. 'Nine-tenths of the Authorized New Testament is still Tyndale, and the best is still his', says Mr J. Isaacs, in *The Bible in its Ancient and English Versions* (1940), p. 160, but not in the New English Bible.

lacking. The Church authorities could hardly be expected to welcome a version in which the similarities between the Bishop of Rome and the antichrist were made so very plain.[1]

As was natural, the introduction of such books into England and the reading of them by English subjects was forbidden. The government was slow to take to the burning of heretics' bodies; it had no inhibitions about the burning of books. On Quinquagesima Sunday 1526 Wolsey himself presided over one burning at St Paul's; another in October of the same year led the Pope's emissary Cardinal Campeggio to write that 'no holocaust could be more pleasing to God'. Archbishop Warham was reduced to the method of himself purchasing the forbidden books, and expended the then gigantic sum of £62 9s. 4d. on this futile expedient. But the English people would have the Bible, and nothing would stop them. Protestants have almost always been good organizers; many of the London merchants were already convinced Protestants, and with their help forbidden literature entered the country in ever-increasing quantities; though Canon Maynard Smith's remark that 'it is to Big Business that we principally owe our version of the New Testament' is perhaps more epigrammatic than precise.

If people were to be prevented from reading the Bible in an undesirable translation, the only remedy was to provide them with an official and satisfactory version, and to this the government in the end made up its mind. But which version? That was the question.

Tyndale held the field, but not unchallenged by rivals. The first rival was Miles Coverdale, later Bishop of Exeter, who in 1535 completed a version based on the Latin Vulgate tempered by Luther's German Version of the original. Coverdale was much inferior to Tyndale as a scholar, but he had a sensitive ear for rhythm, and this is preserved in the

---

1. e.g. On Numbers xxiii. 8, 'How shall I curse whom God curseth not?'; margin: 'The Pope can tell how'. Deut. xxiii. 18: 'Neither bring the hire of an whore nor the price of a dog into the house of the Lord thy God'; margin: 'The Pope will take tribute of them yet, and bishops and abbots desire no better tenants'.

Prayer Book Version of the Psalms, the original Coverdale revised by Coverdale himself.

In the end Cranmer's choice fell on yet a third version, which in truth was no version. In 1537 one 'Thomas Matthew', who was in reality John Rogers, later a martyr for the faith under Queen Mary, had produced and dedicated to the king an English Bible, which was in fact simply Tyndale, as far as Tyndale had gone, supplemented by Coverdale for those parts which Tyndale had not lived to translate. This, shorn of the offensive comments and introductions, and somewhat further revised by Coverdale, was the volume that appeared in 1538, printed in Paris, since English printing was not yet good enough for such an enterprise, and set up by royal authority as the 'Great Bible' for the reading of the English people. This marked a real revolution. Many older people would perhaps have agreed with the old Duke of Norfolk: 'I never read the Scripture, nor never will read it. It was merry in England afore the new learning came up: yea, I would all things were as hath been in time past.' But now the people as a whole would have the Bible. Six copies were set up in Old St Paul's, and the people crowded eagerly into the cathedral all day long to listen to any who could and would read with an audible voice.

A further step forward was taken when in February 1543 the Convocation of Canterbury ordered that 'every Sunday and holy day throughout the year the curate of every parish church, after *Te Deum* and *Magnificat*, should openly read to the people one chapter of the New Testament in English without exposition, and when the New Testament was read over, then to begin the Old'.

But this was very far short of what Cranmer hoped to see. He had become convinced that every part of public worship should be in a language understanded of the people. Here he had lighted on that principle which more than any other drew a clear line between reformed and unreformed Churches. In all the reformed Churches of the world, whether they called themselves Catholic or Protestant, the aim, was that the people should 'pray with the spirit and . . . pray with the understanding also . . . sing with the spirit and . . . sing with the understanding also' (1 Cor. xiv. 15); and therefore that everything in worship should be in the tongue which they familiarly used at home. The use of an unknown tongue may add to the sense of mystery; it does not tend to the enlightenment of the understanding.

Cranmer's first great opportunity came in 1544,[1] when in view of 'the miserable state of all Christendom' 'plagued with most cruel wars', he was permitted to 'set forth certain godly prayers and suffrages in our native tongue'. So was published 'An exhortation unto prayer . . . to be read to the people afore processions. Also a Litany with the suffrages to be said or sung in the time of the said processions.' The Litany, the first form of public worship to be officially authorized in the English language, had come into being. If proof were needed of its surpassing merit, nothing more need be cited than the fact that after more than four centuries it is still used throughout the world in almost exactly the form in which it came from Cranmer's pen.

This first great work exhibits all Cranmer's excellences as a liturgiologist. He did not strive for originality – the original sources of much of the Litany can be pointed out – but on almost everything that he used Cranmer impressed the marks of his own genius. The petitions in the Litany are not just strung or flung together – there is a strict logical order of thought, as any reader can ascertain for himself by analysing them.

Cranmer had no gift at all for writing verse; his efforts, as he himself admitted to the King, lacked 'grace and facility', and others might feel inclined to put the matter rather more strongly. But he could wield English prose as hardly any other master of the English language has ever done. Perhaps no other English writer, except Tennyson, has ever had so precise an apprehension of the exact value of every single English vowel; the blending of vowel sounds in such a majestic phrase as 'Almighty God, who alone canst order the unruly wills and affections of sinful men' is a

1. The matter had already been discussed in Convocation on 21 February 1542.

perpetual delight to the ear. Cranmer had carefully studied the *Clausula*, the rhetorical rules for the termination of a sentence, and knew the difference between a rising and falling rhythm. Thus when he came to render the ancient *Kyrie Eleison*, he did not translate it literally, 'Lord, have mercy', but wrote 'Lord, have mercy upon us', thus retaining the length of the original and the necessary falling rhythm. The building up of the clauses and sentences has been so carefully done that the alteration of a single word can be ruinous. In the petition 'In all time of our tribulation, in all time of our wealth, in the hour of death and in the day of judgement', the American Prayer Book, on the ground that the word 'wealth' in the sense of 'well-being' is somewhat archaic, reads 'in all time of our prosperity'. The rhythm has been irretrievably destroyed.

The Litany has only one major defect. It was written in time of trouble, and the suffrages with which it concludes have a plaintive, sorrowful note, which renders them a little unsuitable for general use. The explanation of this is that here, as in so many things, we have only a broken arch of the great fabric planned by Cranmer's splendid genius. It was in his mind to produce a complete *Processional*, with services for many occasions, joyful as well as sorrowful. We are left wondering what other riches might have been ours, if the plan had ever been completed.

It is a little difficult to guess what may have been passing through the mind of Henry VIII in the last ten years of his reign, and what his attitude to the development of the Reformation really was. He must have known that about half his bishops held 'Lutheran' views, but he continued to respect and to protect both Cranmer and Gardiner. He did not hold with the marriage of the clergy; but he must have known, when he chose Cranmer for the Archbishopric of Canterbury, that he was already a married – in fact a twice-married – man; perhaps he thought that the regular married life of an Archbishop of Canterbury was no worse than the open concubinage in which Wolsey had lived as Archbishop of York. Sometimes Henry was negotiating with the German princes, and sometimes he was not; Lutheran influence went up and down with the political barometer. It must have been very difficult for his subjects to remember under how many Articles they were living. The Six Articles of 1539 marked perhaps the furthest limits of 'Catholic reaction', but they were never put into force with their full rigour. Henry continued to burn 'Lutherans' and to hang 'Papists', sometimes together, which was thought ill of; but never in large numbers, and never to such an extent as to lose the goodwill of the majority of his subjects. But, most significantly, he had permitted his only son to be brought up by Protestant tutors; he can hardly have doubted what would happen immediately upon his own death.

Much had been done. The Pope's authority had been rejected. The monasteries had been dissolved. The revenues of the chantries had been placed at the King's disposal. The great shrine of Thomas à Becket at Canterbury had been destroyed. But Mass continued to be said, as it always had been; and the official religion of England continued to be 'Catholicism' without the Pope. But the main preoccupation of this strange, perplexing man, as of all Tudor sovereigns, was the unity and prosperity of his people. When Henry for the last time prorogued Parliament in person, on Christmas Eve 1545, he delivered, 'so sententiously, so kingly, so fatherly,' an astonishing oration on Christian charity and unity. Papist, Lutheran, Anabaptist, he averred, were names devised by the devil for severing one man's heart from another. He regretted that that 'most precious jewel the word of God is disputed, rhymed, sung and jangled in every alehouse and tavern'; Christian charity was faint among them, and God little served. All this he urged them to amend. 'Then may I justly rejoice that thus long I have lived to see this day, and you by verity, conscience and charity between yourselves may in this point, as you be in divers others, accounted among the rest of the world as blessed men.'

But the unity was not to be as easily restored as Henry hoped. England was not the only country in the world in which the face of religion was being changed, and the rivalry of the various forms of Reformation was already developing.

In 1536, the young French humanist John Calvin (1509–64), not yet the head and dictator of the Church of Geneva, produced the first edition of his *Institution of the Christian Religion*. This great work is one of the outstanding classics of the Christian world, and still continues to influence the hearts and consciences of men. Calvin was the oracle of a Reformation better thought-out, more logical, in some ways more exacting and ruthless than that of Luther. Scotland was to become the second home of Calvinism. The Church of England was to fall for a time so deeply under its influence as to give some foundation for the statesman's remark that the Church of England had 'a Popish liturgy and Calvinistic articles'. By the end of Henry's reign, Calvin was already the theological guide of many English churchmen.

For years the Protestants had been committing their cause to a General Council of the Church. At last, after endless delays, the Pope's Council was opened at Trent on 13 December 1545. On 8 April 1546, a small group of about thirty bishops, almost all Italians, changed the whole course of the history of the Church. They affirmed that the Church receives with equal veneration (*pari veneratione*) the Holy Scriptures of the Old and New Testament, and the traditions *written and unwritten*, which have been preserved in the Church since the time of the apostles.[1] This was new doctrine, and involved a rejection of the whole Catholic tradition of the Church from the beginning. The trouble was in the little word *pari*, equal. Every Church has its traditions, and does well to venerate them. But, in every earlier age of the Church, the appeal had always been to Scripture as the supreme and incomparable authority, to which all tradition must conform, and by which it must be tried. At the time of the Council of Trent some of the bishops asked for a clearer definition of these traditions. What are they, and where are they to be found? No clear answer was given then; no clear answer has ever been given.

This is the one central point of division between the

---

1. It is to be noted that, some years later, this statement of doctrine was reaffirmed in a Plenary Session of the Council, at which a very much larger number of bishops was present.

Church of England and the Church of Rome. Often controversy rages about problems, such as the papal supremacy, transubstantiation, and so on, which while important and interesting in themselves, are nevertheless secondary. The question of the authority, the basis, on which the Church has been built up, is primary. The Church of England stands simply and uncompromisingly for the Catholic position held by the Church through the centuries: 'Holy Scripture containeth all things necessary to salvation; so that whatsoever is not read therein, nor may be proved thereby, is not to be required of any man, that it should be believed as an article of the Faith, or be thought requisite or necessary to salvation' (Article VI). From this standard the Church of Rome has departed, with disastrous consequences to itself. The Tridentine Church is not the great historic Church of the West.

Politically, England under Henry VIII had taken its full share in international affairs and had been impoverished by the many foolish wars into which he had led it. (Henry's debasement of the coinage was one of the chief causes of the economic crisis in the next reign.) But religiously England had lived its life a little apart, and followed a path very unlike that of any other country. Now this could no longer be so. England too was to be drawn into the whirling waters of international religious change. More than a hundred years of controversy and strife were to pass before anything like stable religious peace could be re-established.

CHAPTER 3

Revolution

'WOE to thee, O land, when thy king is a child' (Eccles. x. 16). Bishop Hugh Latimer, perhaps somewhat incautiously, took this as the text of his second Lenten sermon in 1549. But this was not the general view at the time. Edward VI had been brought up as a Protestant. A sickly, over-intelligent, emotionally-starved child, there is no doubt of his sincere attachment to the cause of the Reformation, or of his devotion to Archbishop Cranmer. So the reforming party thought of Edward rather as the young Josiah, who would complete that which his father had taken in hand, and lead England into the fullness of the fellowship of the Covenant of God. This was the theme of Cranmer's sermon at the new king's coronation.

It is idle to speculate what would have happened if Edward had lived to manhood instead of dying at the age of fifteen. Harsh history records that, as a matter of fact, in his days England fell into the hands of violent men, and that in the end it was to the nation's advantage that those days were short.

First, there were the violent and unscrupulous nobles, more interested in their own enrichment than in anything else. Somerset the Protector, the first ruler in Edward's name, was an upright man, more humane than most of his contemporaries, but not strong enough to breast the desperate tides of the country's affairs in his day. His successor Warwick, later Duke of Northumberland, is one of those figures in English history for whom there is nothing whatever to be said. When under sentence of death in the reign of Mary, he wrote, 'An old proverb there is, and that most true – a living dog is better than a dead lion. Oh that it would please her good Grace to give me life – yea, the life of a dog.' And, having played in with the extreme reforming

party to serve his own ends, on the scaffold he revealed that
he had always been a Catholic at heart.

Then there were the violent among the people. In earlier
days the faith of the people had been fed on pilgrimages
and miracles and relics of the saints. Now a violent reaction
had set in against the old ways and against the power which
the priests had exercised through them. Henry VIII had set
the example, with the breaking-up and despoiling of the
shrine of St Thomas à Becket at Canterbury. In many places
people seem to have felt that they might exercise their
liberty, and express their devotion to their new-found faith,
by following his example. There are many shameful tales of
the pillaging and desecrating of churches, and of the wanton
misuse of the sacred treasures that they had contained. This
was a lamentable by-product of the Reformation.

Then there were the violent among the Reformers. Every
great movement has its iconoclasts; the English Reformation
had not a few, and, but for certain restraining influences,
they might well have carried all before them. When Edward
came to the throne, England had already become, as it has
remained ever since, the great home of the refugees. Some,
like Jan Laski (John à Lasco) the Pole (1499–1560), had
been driven out of their own lands by persecution. Some,
like the Italian humanist Peter Martyr Vermigli (1500–62),
Professor of Divinity at Oxford, and the eirenical Strassburg
theologian Martin Bucer (1491–1551), for a time Professor
of Divinity at Cambridge, may have come at Cranmer's
invitation. Some of these men would have carried the Reformation much further than Cranmer wished to see it go.
Their influence on English religious life has been a good
deal exaggerated; but they had their English followers,
and these were among Cranmer's problems.

Cranmer was still the chief architect of the English
Reformation; but, before we turn to study his work in the
greater liberty that the new reign gave him, it may be well
to take a glance at one or two of his colleagues.

Nicholas Ridley was translated from Rochester to London
in 1550. Here he set in hand a vigorous campaign against
survivals from earlier days; he pressed forward with the

removal of stone altars from the churches, and their replacement by wooden Communion tables as the order required. A man of considerable intellectual energy, he seems to have been the first to convince Cranmer that, in the light of the new knowledge of the Bible, the definition of the presence of Christ in the Holy Communion in the scholastic terms of 'accident' and 'substance' was no longer acceptable, and that a new formulation had to be found.

John Hooper (c. 1500–55) a former Cistercian monk, became in 1550 bishop of Gloucester and Worcester. He was one of the violent men, and would have liked altogether to abolish the wearing of vestments in church, so much so that a period of imprisonment in the Fleet was necessary to induce him to wear the minimum of ecclesiastical raiment, without which he could not be consecrated bishop. Yet the records of his administration reveal him as an active, warm-hearted pastoral bishop. And it is to his Visitation of his diocese that we owe the best evidence we have of the kind of material the reforming bishops had to work on in the parishes. Of 311 clergymen interviewed, nearly sixty per cent were unable to repeat the Ten Commandments, ten could not repeat the Lord's Prayer, and one of those who could expressed the opinion that it was so called because it was given by our Lord the King. It is probable that twenty years of uncertainty in religion had had some effect on the level of clerical life; it is unlikely that what Hooper encountered in his diocese was other than typical of the situation in the country at large.

Hugh Latimer (1492–1555) is perhaps every man's favourite among the Reformers. In theology he was never a clear thinker, and there were times at which it seemed that he hardly knew himself what he believed. As early as 1532 he had been in trouble for his views, and in 1539 had been compelled, or perhaps had thought it prudent, to give up his bishopric of Worcester. Under Edward VI he did not resume episcopal office, but chose instead to become preacher-in-chief of England. It is by his sermons that his name has been immortalized. While the other Reformers were deeply concerned about individual and personal holiness, Latimer had seen perhaps more clearly than any other man of his day that there is also a necessary holiness of society, and that where that is not in being God is not being glorified. In plain, straightforward English, racy, of the soil, and adorned with vivid illustrations from daily life, he set forth the righteousness of God for England. It is tempting to quote him at length. 'Schools are not maintained. . . . It will come to pass that we shall have nothing but a little English divinity that will bring the realm into a very barbarousness and utter decay of learning. It is not that, I wis, that will keep out the supremacy of the Bishop of Rome.' 'If the King's honour, as some men say, standeth in the great multitude of people, then these graziers, enclosers and rent-raisers are hinderers of the King's honour; for whereas there have been a great many householders and inhabitants, there is now but a shepherd and his dog. . . . Ye had single too much, and now ye have double too much, but let the preacher preach till his tongue be worn to the stumps, nothing is amended.'

But, when all is said and done, it was Cranmer who was the chief architect of the Reformation, and to him we must now return. When his aims are considered as a whole, it becomes clear that he had thought deeply about every aspect of the Church's life, and that his purpose was to fit it, under every aspect, to pursue its independent way in the new world. It is only in approaching the subject in this way that we become aware of the full range and power of his genius. His touch on five separate parts of the life of the Church has to be considered. It is true that in two of these his efforts led to nothing, and that in a third he was only very partially successful. This is no more than to say that his time was short and adversaries were many. That he achieved so much is memorable; that he attempted so much constitutes him one of the greatest churchmen of all times.

We may start with the most notable and permanent of his achievements – the fixing of the liturgical order of the English Church. Considerable obscurity rests upon the details of the history of the first English *Book of Common Prayer*, and upon the manner of its reception by Church and

State; but all that need concern us here is clear enough. The purpose for which the book was compiled is set forth plainly in the Preface. 'By this order, the curates shall need none other books for their public service, but this book and the Bible; by the means whereof the people shall not be at so great charge for books, as in time past they have been. And where, heretofore, there hath been great diversity in saying and singing in churches within this realm: some following Salisbury use, some Hereford use, some the use of Bangor, some of York, and some of Lincoln: Now from henceforth all the whole realm shall have but one use.' The book met with wide acceptance on the part both of those who had accepted the New Learning, and of those who clung to the Old. The latest date by which it was to be brought into use in all parishes was Whitsunday, 9 June 1549; but its use had begun in 'Paul's choir, with divers parishes in London and other places in England' not later than the beginning of Lent.

Cranmer's first concern was with the daily offices of morning and evening prayer. He was familiar with the seven daily offices of the monks, and with that form of them to be found in the Breviary, the reading of which was obligatory on the clergy of the medieval Church. But this was a long and complicated operation, and had become ever more complex with the passing of the centuries. The aims of Cranmer were threefold – simplicity, brevity, and a great increase in the amount of Holy Scripture to be read. If imitation is the sincerest flattery, it is interesting to note that each successive revision of the Roman Breviary has been a move in the direction indicated by Cranmer. For his solution was marked by real liturgical genius. The seven services of the monks were reduced to two; and by a subtle combination of the year, the month, the week, and the day, the services were given that fixed and unvarying framework which is the basis of all liturgical stability, combined with enough variety to avoid monotony.

The first point to be settled affected the reading of the Psalter. It had for centuries been the custom that the Psalms were read through in a week. This was possible for monks;

it was rather too heavy a burden for the busy priest of a parish. Cranmer decided that the Psalms should be read through once a month, from beginning to end, according to the days of the month.

For the lectionary, Cranmer followed the civil or calendar year, except that special lessons were appointed for the principal Sundays and Holy Days according to the liturgical year. His purpose was that the Old Testament should 'be read through every year once, except certain books and chapters, which be least edifying, and might best be spared', and that the New Testament 'shall be read over orderly every year thrice . . . except the Apocalypse, out of the which there be only certain lessons appointed upon diverse proper feasts'. Modern man would probably find Cranmer's lessons very long; but his lectionary continued to be used, with no radical alteration, until 1871. In that year it was considerably recast, with proper Old Testament lessons for all the Sundays of the year. But his basic principle – that the continuous reading of the Scriptures should be based on the calendar year – was not interfered with. It was only in 1922 that a lectionary was authorized in England which attempted to base both the Sunday and week-day lessons on the liturgical year. As long as we have a variable Easter, with consequent variations in the number of Sundays after Epiphany and after Trinity, no absolutely satisfactory scheme of Bible-reading in church can be devised. The unsatisfactoriness of the lectionaries at present most in use suggests that Cranmer after all was right, and that the sooner we return to his methods, adapted to a rather different understanding of the Scriptures, the better it will be for the worshippers, and particularly for the clergy, of the Church of England.

The service of Matins and Evensong began with the Lord's Prayer and ended with the Third Collect. The ancient Canticles, *Te Deum*, *Benedicite*, and so forth were retained. The ancient custom of prayer through Versicle and Response was taken over unaltered; and many of the ancient collects were retained, not so much translated as re-thought by Cranmer into chaste, athletic, melodious English prose. The English clergy were provided with an invaluable instrument of spiritual discipline and corporate worship, which has served them well to the present day. It was Cranmer's hope that the lay-people too would come to church to take part in the daily services; but here his hopes have not been fulfilled. The habit of daily worship in church has never died out, but there are still far too many Anglican parishes in which the daily offices are not said in church.[1]

Even more important than Cranmer's work on the daily offices was his handling of the central act of Christian worship, the Holy Communion.

A beginning had been made in 1548, with an Order of Communion in English, to be inserted into the Latin Mass after the Communion of the Priest. The book of 1549 contained the first complete English Communion Service, under the title 'The Supper of the Lord and the Holy Communion commonly called the Mass'.[2]

The basis of this beautiful and profoundly moving service was the Sarum Missal, the 'Use of Salisbury', and the handling of the ancient traditional material was in the main conservative, though Cranmer allowed himself considerable freedom of adaptation. The *Gloria in Excelsis* came near the beginning of the service. Then followed Collect, Epistle, Gospel, and Creed. Next came an immensely long Canon, including almost the whole of what we know as the Prayer for the Church Militant, the Prayer of Consecration and the Prayer of Oblation, followed by the Lord's Prayer. This prayer included an invocation of the Holy Spirit: 'with Thy Holy Spirit and word vouchsafe to bless and sanctify these Thy creatures of bread and wine, that they may be unto us the Body and Blood of Thy most dearly beloved Son Jesus Christ.' Before the Communion of

1. Perhaps Cranmer's ideal has come nearer to fulfilment in distant lands than in England. A census of church attendance taken in July 1943 in the diocese of Tinnevelly in South India revealed that shortened Evensong was being said daily in more than seven hundred churches and chapels in the diocese, and was being attended by, on an average, twenty thousand worshippers, fifteen per cent of the Christians in the diocese.

2. It is almost impossible to describe an order of service. The only thing is to have the text before one. The text of the two Prayer Books of the reign of Edward VI is easily available in the Everyman Edition.

Priest and People was inserted the Order of Communion of 1548 – Invitation, Confession and Absolution, Comfortable Words, and Prayer of Humble Access. The service ended with the Prayer of Thanksgiving and the Blessing, as we have them in our present order.

As was to be expected, the book met with a mixed reception. Some would not have it at all. The introduction of the book was one of the causes of the easily suppressed rebellion of the Cornishmen, who knew little or no English, and were not content with the substitution of an unknown language which they did not want for another unknown language to which they were at least accustomed.

Stephen Gardiner and other conservatives accepted the book, and in particular the Order of Holy Communion, because they believed that it was possible to find in it, or to read into it, the medieval doctrine of the Mass – a dangerous position, the assertion of which inevitably had consequences very different from those that Gardiner and his friends would have desired.

The extreme Reformers were pleased that Cranmer had gone as far as he had, and displeased that he had not gone a good deal further. Martin Bucer wrote, by invitation, an elaborate estimate of the book in twenty-eight chapters. On the whole he was able to approve of the book as Scriptural and primitive, though there were certain changes which he wished to see made. Our greatest authority, Dr F. E. Brightman, remarks that his book, 'while it is the only detailed estimate which survives, is also notable for its keen appreciation of the merits of the book and for the ability and moderation and discrimination of its criticism'.[1]

The formidable John Hooper was less well-pleased. He wrote bluntly, on 27 March 1550, 'I am so much offended with that book, and that not without abundant reason, that if it be not corrected, I neither can nor will communicate with the Church in the administration of the supper.'

It was clear that a revision would soon be called for. The question was as to the direction in which the revision would move.

1. *The English Rite*, Vol. I (1915), p. cxlii.

Critics of Cranmer have laid great stress on the influence exercised on his mind in these critical years by the foreign refugees, and have suggested that it was they who drove him further than he would otherwise have gone in the Protestant direction. Now it is certain that Cranmer had great respect for some of these men and enjoyed their company. It is equally certain that his mind moved slowly and cautiously, but that, when it came to a decision, the decision was his own. We do not know everything about the preparation of the Second Prayer Book of King Edward VI. We do know a great deal about Cranmer's mind and methods. It seems clear that he gave weight to two objections to his first effort – that it was still possible to interpret the English Communion service in terms of the medieval concepts; and that he had not gone far enough to bring that service into line with what we read in the New Testament of the institution of the Lord's Supper by our Lord Himself.

Cranmer's own temper was conservative, and he had no love for startling innovations. But he was determined to restore what he believed to be Catholic orthodoxy, and to exclude certain forms of medieval doctrine which he believed to be irreconcilable with Scripture. Two medieval errors were in his judgement particularly grave.

In the first place, he desired to exclude absolutely the idea that the presence of Christ in the Eucharist is a *local* presence. Here he was exactly in line with the classical thought of the Western Church on this subject, as it was expressed by St Thomas Aquinas: 'Christ's body is not in this sacrament as in a place. . . . Hence it remains that Christ's body is not in this sacrament as in a place, but after the manner of substance. . . . Hence in no way is Christ's body locally in this sacrament. . . .' (*Summa*, Part III, Q.76, Art. 5). 'Christ's body is at rest in heaven. Therefore it is not movably in this sacrament. . . . Hence it is clear that Christ, strictly speaking, is immovably in this sacrament' (Part III, Q.76, Art. 6). This is sound doctrine. But simple people can easily fall into misunderstandings concerning it. It is difficult for a Sicilian peasant, watching a procession of the Blessed Sacrament, not to imagine that Christ is present in the Sacrament as in a place, and that His presence is being moved from place to place. Cranmer was resolved that such simple people should not be exposed to the dangers of theological misunderstanding.

The other popular error concerned the nature of the Eucharist as a sacrifice. All sound theology agrees that the sacrifice of Christ on the Cross was full, perfect, and complete; it can never be repeated, and nothing can be added to it. On this there is not a shadow of difference between 'Catholic' and 'Protestant' theologians. But, when medieval man heard that the priest was ordained to offer propitiatory sacrifices on behalf of the living and the dead, when he saw the amount of money spent on chantries and on the multiplication of Masses, he could easily fall into the error of supposing that the priest did in some way add something to the sacrifice of Christ. Against this misunderstanding, too, Cranmer was at pains to guard.

So much for the negative side. On the positive side, studying the records in the Gospels, Cranmer saw that Christ's words of institution, 'This is my body, This is my blood' (words of distribution, some have called them), were immediately followed by reception on the part of the disciples. This was the pattern that he now determined to follow. Consecration and communion were to become a single act, separated in time by a brief moment, but not separable in thought or understanding. The moment this principle is grasped, an immense number of difficulties in Eucharistic theology simply vanish. This can be illustrated from a parallel sphere. There are two essential parts in Christian marriage – the solemn plighting of the troth in church, and the consummation of the marriage. These can be separated in time, but they are only correlative aspects of a single act; it is difficult, if not impossible, to define the significance of one out of relation to the other. So in the Communion. The Lord's Supper was given for communion; it is only in the complex act of consecration and communion together that the nature of the presence and the self-giving of Christ in His Church can be understood. This is the central principle around which Cranmer constructed his revised service; in the Order of 1552, there is no 'Amen' at the end of the Consecration Prayer; the 'Amen' is at the end of the Lord's Prayer after the Communion.

The Order of 1552 has been much criticized. Some have gone so far as to suggest that Cranmer, having decided to disrupt still further the ancient order of the Canon, arranged more or less at random the elements that he decided to retain. We must recall that Cranmer was a great liturgical genius, and that this was his considered masterpiece. This, until the year 1928 and even later, was the only form of the Communion service that the majority of Anglicans knew. It has fed and satisfied Anglican devotion for more than four centuries. We shall do well to approach it with reverence, to ask what Cranmer was trying to do, whether what he was trying to do was good, and how far he was successful in achieving it.

Part of the trouble is that hardly anyone now living has ever seen the Communion celebrated as Cranmer intended it to be celebrated. He had utterly repudiated the idea that the Communion is a form of individual and personal devotion; it is something that is *done* by priest and people together, according to the commandment of their crucified and risen Lord, and in the assurance of His presence with them. To this end he had ordered that the Communion table was to be brought into the body of the church. Churches in those days were not cluttered up with pews; clearly it was Cranmer's idea that the faithful would come and stand around the Table. This alone explains the otherwise mysterious complaint of the Cornish rebels that the Communion had been turned into 'a Christmas game'. There was indeed a startling contrast between the Mass, in which the priest at the far end of the chancel recited almost inaudibly unintelligible words, and this new form, in which every word could be heard, every gesture seen, and in which the worshipper felt that he himself was an actor.

If we are to catch the rhythm of Cranmer and to appreciate the architectonic form of the service, the whole must be read without a pause from 'Lift up your hearts' to the blessing. The communion of priest and people must follow upon the Consecration Prayer without any pause, the communicants placing themselves during the communion of the priest in readiness to receive. After the service is over, we may remain as long as we like to attend to our own devotions; during the service we are not individuals, we are part of the Body of Christ, doing together with others that which Christ has commanded us to do. In modern times, so many pauses have been introduced into the service that Cranmer's intention has been fatally obscured.[1] There has been a tendency for Holy Communion to be regarded as an occasion for private and individual devotion on the part of both priest and people, and for the corporate nature of the service as the act of the Church to be forgotten. There has been a considerable recovery in recent years of this corporate aspect of the Eucharist; but the Anglican Communion as a whole still limps a very long way behind its great Archbishop.

Once the central principle has been grasped, the reason for most of the structural changes in the Order of 1552 can be understood without difficulty. The Ten Commandments have been introduced into the preparatory part of the service. The Prayer for the Church is brought forward, and separated from the Canon – a great improvement, and evidence for the fact that Cranmer has now made up his mind as to the maximum length of time for which a congregation can be expected to maintain its attention without relief. The Preparation (Invitation, etc.) is brought forward before the beginning of the Canon, except that the Prayer of Humble Access immediately precedes the Prayer of Consecration – an illustration of another of Cranmer's liturgical principles, that each climax in the service should be followed, by way of relief, by a quieter and more meditative section. The Prayer of Oblation comes after the Communion. Then, influenced no doubt by the saying in the Gospel that 'after they had sung an hymn they went out into the mount of Olives', Cranmer by a wonderful stroke of liturgical genius brought the *Gloria in Excelsis* from the

1. No priest who has understood the form and rhythm of the service could fall into the liturgical solecism of taking the 'ablutions', the cleansing of the vessels, after the communion of the people instead of after the blessing, as the rubric orders.

beginning to the end of the service, and by so doing gave the English rite a far more glorious conclusion than that of any other Eucharistic rite in the world.[1] We open soberly with the words of the Law; we move forward, through Gethsemane and Calvary, to the glory of the resurrection and the ascension. At no moment are we allowed to sink back into our separateness as individuals; till the very end of the service we must join with our fellows in self-dedication, adoration, and praise.

So much for the liturgical aspect of the question. We must now ask what doctrine of the Communion this new service was intended to express.

In the first place, it is to be noted that the Act of Uniformity, by which the new book was enforced, nowhere states, and by implication denies, that any change of doctrine was intended in the liturgical changes that had been made. The 1549 Order had been 'agreeable to the word of God and the Primitive Church, very comfortable to all good people, desiring to live in Christian conversation, and most profitable to the state of the Realm'. The new book is set forth only for the perfecting and fuller explication of that which is already present in the first. Furthermore, when Cranmer at his trial was twitted with his constant changes of view as to the doctrine of the Communion, he maintained that he had never held but two views – the old medieval doctrine which he had held up till 1547, and that which he had tried to express in the books of 1549 and 1552. After all, only Cranmer knew, and he was an honest man. We shall probably do well to be cautious in looking for theological differences between the two books.

It may at once be admitted that Cranmer went further than was necessary in attempting to exclude the possibility of belief in a local presence of Christ in the Eucharist. The words of administration in the first book had been 'The body of our Lord Jesus Christ, which was given for thee, preserve thy body and soul unto everlasting life', 'The blood of our

---

1. I am glad to find that my opinion is shared by the late F. C. Eeles, whose churchmanship was hardly the same as my own. See *Thoughts on the Shape of the Liturgy* (Alcuin Club Tracts, xxiv) 1946, p. 49.

Lord Jesus Christ, which was shed for thee, preserve thy body and soul unto everlasting life'. These were now changed to 'Take and eat this in remembrance that Christ died for thee, and feed on him in thy heart by faith with thanksgiving', 'Drink this in remembrance that Christ's blood was shed for thee and be thankful'. In 1559, by a happy spirit of comprehensiveness, the two forms were joined together, as they stand in our Prayer Book today, and constitute a masterly summary of the whole of Eucharistic doctrine.[1]

The sacrificial aspect of the Holy Communion is not emphasized in the rite, but it is very far from being wholly absent. Apart from the offering of the bread and the wine, and the 'sacrifice of praise and thanksgiving', the sacrificial idea is present at three central points. In the prayer of consecration the sacrifice of Christ on the Cross for the sins of the world is made a present reality. In communion, as in the peace offerings of the Old Testament, the worshippers through eating and drinking are made partakers in that sacrifice and in all its benefits. Up to this point all the emphasis is on what man receives from God; in himself he has nothing that he can offer. After communion, the Church, in union with the Christ whom it has received, offers itself as a living sacrifice to God. Most Anglicans find themselves satisfied with this simple, rational, Scriptural exposition of the Eucharistic sacrifice. Some, feeling that this is too 'subjective', would wish to go further in the direction of restoring what they regard as the 'objective' character of the sacrifice.

What exactly is the Eucharistic doctrine that Cranmer desired to set forth in this rite? Although he composed a whole book on the subject, *A defence of the true and catholike doctrine of the sacrament of the body and blood of our Saviour Christ* (1550), there is still much controversy as to what view he himself held. The Reformers had given up the scholastic

---

1. When communicants are many and priests few, it is almost impossible that the whole of the words should be said to every communicant; but it is inexcusable that in some churches the second part, with its emphasis at this central point on the *Eucharistic* character of the rite, is never heard by the communicants at all.

and syllogistic way of thinking, but they had not yet found exact and scientific ways of expressing their new under-standing of the faith; this was to be the work of the following generation. It is easier to identify the errors which Cranmer desired to exclude than the truths which he desired to establish.

At one extreme, he rejected transubstantiation, the physical change of the elements in the Communion into the Body and Blood of Christ; this, as the 28th Article states (in a clause added in 1563), 'overthroweth the nature of a sacra-ment', because it *identifies* the sign given by our Lord with the gift that is given through the sacrament. At the other extreme, there is no place in the English Church for what is called Zwinglianism (though it is probable that Zwingli himself never taught any such doctrine), the view that the elements in the sacrament are *nuda signa*, signs and nothing more. The point can be made clear by asking the question what the Holy Communion *is*. Is it a pious rite carried out by men in memory of a Friend? Or is it an act of the living Christ in His Church? Cranmer, and the English Church with him, came down emphatically on the side of the second alternative. On nothing is he clearer than that it is Christ Himself who in the Holy Communion is giving Himself to us. In every celebration the true Celebrant is the risen and living Saviour.

As to the nature of the presence of Christ in the Eucharist, the discussion then as now was bedevilled by a confused and inaccurate use of terms. For this reason it is better, if possible, to avoid completely the term 'the real presence'. In terms of our current speech, it is natural to suppose that the contrary of 'real' is 'unreal' or 'imaginary'; and it is easy to imagine that, when the Reformers denied 'the real presence' of Christ in the Sacrament, they were affirming His 'real absence'. Nothing of course could be further from the truth. Cranmer asked himself the question, 'How can Christ be present in the Sacrament?' His first answer is that He cannot be present in the Eucharist as He was in the days of His incarnation, when He accepted the limitations of time and space, and was never present in more than one place at once. Secondly, He cannot be present in the Eucharist as He will be present in His 'Parousia', His second coming, since that is a future mystery, the nature of which, is not yet revealed. This is what underlies the strange and unsatisfactory language of the 'Black Rubric' at the end of the Communion Service, 'And as concerning the natural body and blood of our Saviour Christ, they are in heaven and not here: for it is against the truth of Christ's true natural body to be in more places than in one at one time.'[1] If these two possibilities are removed, what is left?

Nothing is more striking in the Reformation than the recovery of the almost forgotten doctrine of the Holy Spirit. Of Cranmer's special interest in this doctrine the Prayer Book itself, with its constant references to the Holy Spirit, is evidence. It seems that we have here the clue that Cranmer himself followed, as he wrestled with the problem of the Eucharistic presence. If we put the question in the form, 'In what epoch of revelation are we now living?' the only possible answer is, 'In the epoch of the Holy Spirit.' How then can Christ be present among us in these days? The answer is, as He promised, through the Holy Spirit, whom He has sent to us in His name. We are now on the way to understanding what Cranmer and his colleagues meant when they spoke of a spiritual presence of Christ in the Holy Communion.

Many people feel that a spiritual presence is somehow less real, objective, valid, than a physical presence; that a spiritual presence in some way depends upon or is created by our thoughts and imagination. But this was not the thought of Cranmer. Perhaps we should do him greater justice if we were to use a capital, and write 'Spiritual presence'. To Cranmer, the presence of Christ in the Holy Spirit was richer, fuller, more glorious, than that which his adversaries affirmed in their doctrine of 'the real presence', just as His risen life is richer and more glorious than His life in the time of the incarnation. If we find this hard to

---

1. The 'Black Rubric' was added at the last moment to the Prayer Book of 1552 without the authority of the Act of Uniformity, was removed in 1559, but restored in a slightly revised form in 1662.

grasp, may that not be because we do not take the doctrine
of the Holy Spirit as seriously as the Reformers?[1] It is to be
regretted that Cranmer's busy life as Archbishop did not
give him time to work out his idea in full theological detail.
But, even if he had done so, it is likely that he would have
been anxious to avoid too exact definition, and to leave a
mystery at the heart of what to our limited intelligence must
ever remain a mystery.[2]

The three years between 1549 and 1552 had already revealed what was, and was to remain, the saddest failure of
Cranmer's attempts to restore Catholic orthodoxy in faith
and practice. It had been his hope that, when the Lord's
Supper was offered to the people in their own tongue as an
act of the Church in which they could participate, they
would come in crowds not as spectators of the Mass but as
communicants. But the people had been brought up in the
late medieval tradition, which made communion a rarity.
They had become accustomed to communicate only once a
year, at Easter; and from this custom neither invitation nor
exhortation was able to move them. In order to make plain
the corporate character of the service, the Prayer Book had
laid down that 'there shall be no celebration of the Lord's
Supper, except there be some [1552 'a good number'; 1662
'a convenient number'] to communicate with the priest'.
The result was that in many parishes on many Sundays,
contrary to Cranmer's godly intent, the service broke off
before the Canon of the Mass, and there was no communion.

It was perhaps for this reason, in part also to meet the
criticism of some of the extreme Reformers, that the penitential introduction to Morning and Evening Prayer, including the General Confession and Absolution, was
introduced. These splendid specimens of liturgical English

1. I am glad to find this, which has long been my own conviction,
confirmed by the latest authoritative book on Cranmer, *Thomas Cranmer,
Theologian*, by G. W. Bromiley (1956) pp. 69 ff.

2. Bishop Andrewes showed himself a true follower of Cranmer, when
in his answer to Cardinal Bellarmine (1610) he wrote: 'Praesentiam
credimus non minus quam vos *veram*; de modo praesentiae nil temere
definimus.' 'No less than you we believe in a *true* presence; we make no
rash definition as to the nature of the presence.'

manifest in the highest degree one characteristic of the Prayer Book to which attention has not yet been drawn – its extraordinary theological density. A whole course of sermons can be preached on the General Confession; and the more deeply it is considered, the profounder is seen to be its understanding of the nature of sin, of repentance, and of the grace of God. But at the same time this introduction manifests one of the less attractive features of the English liturgy (and one which probably we owe mainly to Bucer) – its tendency to bring in homily and exhortation at every possible point. The exhortations are in almost every case admirable in style and theological content; but they tend to interrupt the movement of worship, and it is good that nowadays they are read less frequently than used to be the case.

The Prayer Book plays so important a part in the life of the whole Anglican Communion that it has been necessary to deal with it at some length. Cranmer's other services to his Church can be handled much more briefly.

Amid the burning controversies that raged in the sixteenth century, it was necessary that a Church should know where it stood, and should make its position plain both to friends and enemies. Cranmer had a principal hand in drawing up the Forty-two Articles, which were put forth in 1553 as 'Articles agreed on by the Bishops and other learned men in the last Convocation at London . . . for the avoiding of controversy . . . in certain matters of religion'.[1] These Articles are not an attempt to provide a complete confession of faith; their purpose is to make clear the position of a Church which sets before itself the aim of being Catholic, avoiding on the one hand the late medieval traditions of Rome, and on the other the excesses of the Anabaptists. Cranmer's Articles of 1553 formed the basis of the Thirty-nine Articles, which were finally accepted in 1571, and are still of authority in the Church of England. In the meantime a thorough revision had taken place; seven articles had been

1. This was a misstatement. The Articles had not been submitted to Convocation and were in fact published on the sole authority of the King. Cranmer protested to the Council against this mishandling of his work.

suppressed, four added, seven shortened, and five rewritten. But the manner and temper of Cranmer are still evident in the document, as it finally emerged from this long process of revision.[1]

In the main the Articles are an admirable body of divinity admirably expressed. But to this general commendation some qualifications must be added. Some of them deal with controversies which are no longer of concern to the Church. True to the Anglican principle of comprehension, some of them are plainly marked by Cranmer's purpose not to define anything that God has not defined, and make the Anglican position less clear than could be desired by friends or critics in other Churches. The sixteenth century was midway between the great classical period of Western theology and the beginning of modern scientific thought and expression, and to some phrases in the Articles exception can rightly be taken. But in most cases it is clear that, though the appropriate forms of expression were lacking, the Articles are trying to express something true and deeply significant. Article XIII, *Of Works before Justification*, has perhaps been criticized more sharply than any other. It reads, 'Works done before the grace of Christ, and the inspiration of his Spirit, are not pleasant to God, forasmuch as they spring not of faith in Jesus Christ . . . we doubt not but they have the nature of sin.' This has found an unexpected defender in William Temple, who in his exposition of the words of Jesus, 'Apart from me ye can do nothing' (John xv. 6), wrote: 'Article XIII states the matter in the unsympathetic tone born of theological controversy; but what it says is true. "The nature of sin" is self-centredness – the putting of self in the centre where God alone should be. We are all born doing this; that is Original Sin. . . . We cannot too harshly drive this truth into our souls, however

1. The changes are excellently set out in *The Prayer Book Dictionary* (1912), Art. 'Articles of Religion'. Canon Dixon's tribute to Cranmer's work is notable: 'The broad soft touch of Cranmer lay upon them when they came from the furnace. . . . Nearly half of them are such as are common to all Christians; but even in these the brevity of statement and avoidance of controversy is to be admired.' *History of the Church*, iii. p. 520.

eager we may be to trace "the grace of Jesus Christ" in others, even in atheists. Apart from Him, I can do nothing.'[1]

A word must be added as to the status of the Articles in the Anglican Communion. For almost three centuries, all holding office in any capacity in the Church of England had to affirm unconditionally their assent to the Thirty-nine Articles. Since 1865 subscription in England has been in the form, 'I assent to the Thirty-nine Articles and to the Book of Common Prayer and of ordering of Bishops, Priests and Deacons. I believe the doctrine of the Church of England as therein set forth to be agreeable to the Word of God.' In some Provinces of the Anglican Communion the Articles have been retained, and in others not. For instance, in the Church of India, Pakistan, Burma, and Ceylon, the Articles may be cited as a valuable historical document; but otherwise in that Church they have no authority, and subscription to them is not required of anyone.

In the next of his great schemes Cranmer was not to attain any success at all. A Church must have worship and articles of faith. It must also have a law – no Church in the world has ever been able to dispense entirely with rules and regulations. The law of the Church, the Canon Law, at the end of the Middle Ages, was a vast confused body of legislation. After the Reformation it was uncertain what parts of the Canon Law were still in force in England. The Act of 1533 declared that such canons as were not 'contrarient to the laws, customs and statutes of this realm, nor to the damage and hurt of the King's prerogative royal', were to remain in force; but it proved almost impossible to put an accurate legal interpretation on these phrases. With this the precise and accurate Cranmer was very ill-satisfied. It was not till October 1551, however, that he was able to secure the appointment of thirty-two commissioners to produce a new code. Most of the work, as usual, fell to the Archbishop, acting with advisers. Under his direction the project went rapidly forward, and the commissioners were successful in producing what has been described as 'an ordered, coherent, and intelligible body of ecclesiastical law'.

1. *Readings in St. John's Gospel*, Vol. II (1942), pp. 260-1.

The *Reformatio Legum Ecclesiasticarum*, which was printed for the first time in 1571, is interesting for the combination of conservatism and liberalism which it displays. In many respects Cranmer was four centuries ahead of his time. His plan included the revival of diocesan synods, in which he would have permitted the participation of laymen. As grounds for divorce he was prepared to recognize not only adultery, but also desertion, long absence, and cruelty. No action was taken in the time of Elizabeth to follow up Cranmer's plans and proposals; and no Anglican Church has ever again made so thorough an attempt to re-think and to re-state the laws under which its children should live.

In 1604 a set of 141 Canons was passed by the Convocation of Canterbury (by that of York in 1606) and issued on the authority of Letters Patent from the King. These are not a systematic statement of Church Law, but rather a collection of rules and regulations governing a number of details in the Church's life. And it is important to note that in 1605 Parliament passed an act laying down that 'No Canon or Constitution ecclesiastical made in the last ten years, or thereafter to be made, should be of force to impeach or hurt any person in his life, liberty, lands or goods', unless it had first been confirmed by Parliament. Since the Canons were passed only by Convocation, which is an assembly of clergymen, the courts of law have always held fast to the principle that these Canons are not binding on laymen unless confirmed by the legislature.[1] The independent Anglican Churches have not found insuperable difficulties in the way of providing themselves with suitable sets of Canons. In England the difficulty is greater by reason of the state connexion. At the time of writing the English Canons were in process of revision, and this has proved to be a long, tedious and controversial business.

Cranmer's next concern was the formation of a devout, educated, and preaching ministry. Few of the parish priests of the sixteenth century had ever preached, that work in the

---

1. They are, however, binding on laymen who have expressly or implicitly agreed to be bound by them, such as churchwardens, or lay rectors, who are responsible in England for repairs to chancels.

main being left to visiting friars and revivalists. As the returns of Bishop Hooper's Visitation show, many of them were incapable of doing so. And of those who were qualified not all were licensed to preach. In the days before news-papers, the pulpit was one of the most effective instruments of propaganda. It could be used unwisely for the stirring up of passion over controversial religious issues; it was difficult to ensure that preachers would refrain from comment on current political problems. Not unnaturally the rulers both in Church and State felt it better that licences to preach should be issued only to those whom they regarded as 'safe men'; and some recent events may incline us to think that Tudor sovereigns had better reason for being cautious in this matter than has been always recognized. But Cranmer never regarded this as more than a temporary muzzling of the freedom of the word of God.

One of his most interesting and far-sighted proposals was that a part of the revenues of cathedrals should be used to create centres for the training of candidates for ordination. The establishments of the cathedrals were still very large; with the reduction in the number of Masses and the aboli-tion of the chantries, there was very little for many of the cathedral priests to do. Unhappily nothing came of this proposal, and centuries were to pass before the Church of England took seriously the preparation of all those who were to serve it in the churches and parishes of the land.

A more immediately practical step was the revival of an ancient Catholic practice in the provision of homilies to be read in church by those who were unable or unauthorized to preach their own sermons. A book of twelve homilies was ready in 1543, but was not published until 31 July 1547. In this book, four (possibly five) of the homilies were from the pen of Cranmer; these included the highly important sermons on 'The Salvation of All Mankind' and 'Of the true and lively Faith'. A second book, with twenty-one titles, was provided in 1571.

Hearers today would probably regard the homilies as long, tedious, complex, and unintelligible. But we note once again Cranmer's confidence in his fellow-countrymen's ability to understand and to respond to a reasoned exposition of the Christian faith. In those days everyone had to go to church, whether he wished or not; in many churches no sermon was ever provided other than the reading of homily. There is little doubt that, through constant hearing, the substance of many of these discourses, like that of the Bible and of the Prayer Book itself, passed into the consciousness of the hearers, and helped to fashion the characteristically Anglican attitude towards the Christian faith.

Finally, we must note Cranmer's efforts to promote unity and common action between the Churches which had accepted the Scriptures as their ultimate standard in all matters of faith and conduct. More clear-sighted than almost any other ecclesiastical statesman of his time, he foresaw the dangers that must arise for all the Churches of the Reformation from the Council of Trent and the Counter-reformation. He was constantly in correspondence with Melanchthon in Germany, with Calvin in Geneva, and with other leaders elsewhere, urging them to come to England, where the calm and safety of the English situation would be more conducive to quiet and patient thought than the troubled and turbulent atmosphere of the continental countries. He was convinced, perhaps naïvely, that if the leaders could meet and open their hearts to one another, they would find that their agreements were far more extensive than they imagined. Nothing came of the project. The continental leaders lacked Cranmer's wider vision, and nothing was done. But, as the scholar who has most deeply studied these efforts has written, 'The frustration of magnificent hopes does not necessarily brand them as fatuous. . . . Cranmer is not to be judged merely on his record as the chief of the English Reformers. His espousal of a cause far wider than national reform – the integration of the severed Reformation Churches of Europe – is evidence of a sincere ecumenicity and of a certain noble grandeur of design challenging to later generations.'[1] If Cranmer had had his

---

1. John T. McNeill, in *A History of the Ecumenical Movement 1517–1948* (1954), p. 58. See also the fuller treatment in the same writer's *Unitive Protestantism* (1930), Chap. vi.

way, the World Council of Churches would have come into being in 1548, and not in 1948.

But the time of Cranmer and his colleagues was running short. They were sowing for a larger and later harvest; but the time of sowing was coming to an end. On 6 July 1553 the boy king died. After the futile episode of the attempt to set Lady Jane Grey on the throne – an attempt favoured by the dying Edward VI[1] – Mary Tudor came to the throne. For the moment the religious revolution had come to an end; it was certain that it would be followed by counter-revolution.

1. On Cranmer's very unwilling acceptance of the King's desire, see A. F. Pollard: *Thomas Cranmer* (1904), pp. 297–9.

CHAPTER 4

Counter-Revolution

WHEN Mary Tudor ascended the throne, she was thirty-seven years of age. She had had a dreadful time of it for more than half her life. In 1533 she had been pronounced illegitimate, and had been constrained to sign a paper declaring that her father's union with her mother had been incestuous. Various plans for her marriage had been made, and all had fallen to the ground. At times her life had been in danger. She was prematurely aged, and throughout her brief reign was a sick and anxious woman.

Mary had inherited a great deal of the Tudor intellectual ability. She was virtuous, a tireless worker, and not without gleams of compassion. But the two dominant features in her character were her devotion to the memory of her mother, and her devotion to the Roman Catholic Church. Through all her trials she had clung to the faith of Catherine of Aragon, and that had sustained her through her darkest days. Now she believed herself to have been providentially ordained to restore the true faith in England. Her many merits were counterbalanced – and here she differed markedly from her father – by a curious inability to understand the thoughts and feelings of her fellow-countrymen, and a grave incapacity to handle them aright. When she became Queen, she was enthusiastically welcomed by her people; when she died five years later, hardly a tear was shed for her.

It was certain that many changes in religion would take place. But Mary was soon to discover that she could not have her own way in everything, and that she would not be able to go as fast and as far as she would have wished. At every stage of his reformation Henry VIII had secured the consent of Parliament, and what a statute had declared to be the law of the land only a statute could alter. Mary wished to put

the clock back beyond the beginning of the changes introduced by her father; she was to find that most of her advisers, while desiring the re-establishment of many things
that the reign of Edward VI had destroyed, were unwilling
to go beyond the state of religion that had existed in the
second half of the reign of Henry VIII. At two points in
particular she found herself checked. Parliament insisted
that she must assume the title of Supreme Head of the
Church, which was hers by law and which she detested; and
that there could be no question of the resumption of Church
lands which were now in the hands of lay owners. The
Imperial ambassador Simon Renard wrote to his master
that more lands were in the possession of Catholics than of
heretics, and that in no circumstances were they prepared
to countenance a new religious settlement that would result
in their own impoverishment.

Nevertheless the work of restoration went rapidly forward.
Seven reforming bishops were removed, and replaced by
those favourable to the policies of the Queen. Services were
held according to the ancient service books. A large number
of priests, probably between a fifth and a quarter of the
parish priests of England, were deprived, most of them for
having committed the sin of matrimony. When Henry VIII
dissolved the monasteries, pensions were settled on all
monks and nuns who had to leave the cloister, and, with
some irregularities of payment, these were continued through
all the troubles of the times.[1] No such provision was made
for those turned out under the Marian regime, and considerable suffering was the result. After a year of preparation, on St Andrew's Day 1554 the realm was solemnly
absolved by the papal legate from the sin of schism,
and brought back into the communion of the Roman
Church.

At three points Mary's errors and miscalculations had
already begun to threaten the success of her plans.

The first was her choice of a successor for Cranmer as
Archbishop of Canterbury. Reginald Pole had royal blood
in his veins, and it had even been suggested that he might

1. One survivor was still drawing his pension in 1602!

marry Mary.[1] He was not a bad man; in fact he had been one of the best of the cardinals at the Council of Trent, and had been favourable to a moderate measure of reform. But he had lived out of England, as an attainted traitor, for nearly twenty years, and had little understanding of the changes in the national life and temper that had taken place in that stormy period.

An even graver error was Mary's marriage. Foreigners were not popular in England, and Spaniards were the least popular of foreigners. But Mary was determined to have a Spaniard, and to her own undoing she got what she wanted. Philip II of Spain was the heir to a tremendous inheritance. Like his wife he was a bigoted Catholic. But he was ten years younger than Mary. The marriage, which was to have brought the Queen happiness and an heir, proved barren, and brought instead frustration, sorrow, and the bitterness of desertion. From the start the marriage was unpopular in England; the French ambassador reported that, when it was announced, rebellion was in the air. But Mary was a Tudor, and she would have her way; and, though unwillingly, her subjects acquiesced in her choice.

Mary's third and gravest miscalculation was her belief that the minds of Englishmen could be changed by persecution.

Although no burning of heretics took place until 4 February 1555, eighteen months after the beginning of the reign, persecution had been in the air from the very first day. Cranmer knew well what was coming, and within a month or two of the death of Edward VI had advised his friends to fly for their lives.[2] Nicholas Ridley wrote: 'If thou, O man of God, do purpose to abide in this realm, prepare and arm thyself to die; for both by Antichrist's accustomable laws and these prophecies, there is no appearance of any other thing – except thou wilt deny thy

---

1. He had been Cardinal of the Roman Church since 1536, but was only in deacon's orders, and it was thought that the necessary dispensations would easily be obtained. It is typical of late medieval practice that in 1527, although still a layman, he was appointed Dean of Exeter.

2. Cranmer's *Works*, Vol. II, pp. 44–5: To Mrs Wilkinson.

master Christ.' Most of the foreign divines – Peter Martyr' John à Lasco, and others – who had come to England at Cranmer's invitation, departed to the continent with many members of their flocks. Four bishops, five deans, and other less conspicuous Anglicans followed their example.

It has been suggested that this emigration was planned and carried out as part of a deliberate policy – to preserve and develop the leadership through which at a later date the Reformation could be re-established and carried further in England. The hand of the prudent William Cecil, later Elizabeth's great minister, has been seen as active in these arrangements.[1] It is hardly necessary to attribute to careful planning what would seem to be a very natural reaction of men and women in peril of their lives. It is certain that many Englishmen went abroad. It is certain that they lived and worshipped and quarrelled famously in many foreign cities. It is certain that many of them returned to serve, and to distress, the Church in the reign of Queen Elizabeth. It is not necessary to suppose that they foresaw all this in the days of their exile, though they may have hoped for it. So many kingdoms suffered from violent changes and rapid reversals of policy in the sixteenth century that to accept exile in the hope of return was as common a practice then as it has become in the twentieth century. What has made the Marian exiles conspicuous was that they were privileged to succeed, where others, such as the recusant exiles under Elizabeth, were doomed to failure.

Once the fires had been lighted, they burned with dreadful frequency for the rest of the reign. The exact number of the martyrs cannot be established with certainty, but there is no reason to question the general accuracy of the reckoning which puts it at about three hundred. Five bishops suffered at the stake. About a hundred of the victims were priests. Sixty of the lay folk were women. The great majority of the martyrs died in London, or in other parts of

1. This is the main thesis of Miss C. M. Garrett's *Marian Exiles* (1938), which is valuable for the details which it gives of many of the exiles and their life abroad, even though its main thesis may not be entirely acceptable.

south-east England; this was the region in which the doctrines of the Reformation had taken deepest root, the conservative north and south-west being very much less affected.
The towns provided far more martyrs than the countryside,
and this too corresponded to the lines along which the new
ideas had spread. Many of those who died were humble folk
of the artisan class. Some of those arraigned recanted; most
of them held firm to their profession, and went cheerfully
and courageously to their deaths. There are very few indi-
cations of fanaticism, still fewer of hysteria, in the records of
their trials and of their deaths. These were people who knew
what they believed, as they had found it written in the
Scriptures, and were prepared to die for what they affirmed
to be the true and uncorrupted understanding of the
Christian faith.[1]

Already before the end of 1553 Ridley, Latimer, and
Cranmer were lodged in the Tower. The nature of their
treatment there can be inferred from a remark of Latimer
to the Lieutenant of the Tower: 'You look, I think, that I
should burn; but except ye let me have some fire, I am like
to deceive your expectation, for I am like here to starve for
cold.' Later they were transferred to the Bocardo at Oxford,
described by Bishop Coverdale as 'a stinking and filthy
prison for drunkards, whores and harlots, and the vilest sort
of people'. It seems that during his long ordeal here Latimer
was for a time out of his mind. 'Master Latimer was crazed,'

1. The story of the English martyrs has lived on in the racial memory
of the British people and in the vivid pages of Foxe's *Book of Martyrs*, the
reading of which was sedulously encouraged by the Elizabethan government, and has continued in certain circles until the present day. Foxe's
accuracy and reliability have been virulently assailed, and the impression has been spread abroad that the traditional understanding of the
Marian persecution must be modified in many respects. It is good that
the whole matter has been carefully investigated by a scholar, Dr J. F.
Mozley, who by his own admission 'approached him with suspicion'; his
conclusion is that Foxe's inaccuracies have been greatly exaggerated:
'He will lead us to historic truth, help us to plant our feet upon the solid
rock. . . . We respect him for his zeal and earnestness, for his enormous
painstaking. We thank him for the historic knowledge, which he has
preserved from destruction.' (*John Foxe and His Book* (1940), pp. 237,
239.)

wrote Ridley, 'but I hear now, thanks be to God, that he amendeth again.' We sometimes forget, in the heroic end of the martyrs, the awful valley of the shadow through which almost all of them had to pass. 'Pardon me, and pray for me,' wrote Latimer while still in the Tower: 'pray for me, I say, pray for me, I say. For I am sometimes so fearful, that I would creep into a mousehole; sometimes God doth visit me again with His comfort.'

It was plain from the start that the main issue on which the prisoners would be judged was their belief as to the nature of the presence of Christ in the Sacrament. Here is Latimer's own account of his self preparation to meet the challenge: 'I have read over of late the New Testament three or four times deliberately; yet can I not find there neither the popish consecration, nor yet their transubstantiation, nor their oblation, nor their adoration, which be the very sinews and marrow-bones of the mass.' Later, in public debate, he affirmed that he and his companions in trial had read the New Testament seven times over in search for the doctrine to which they were required to subscribe, and had not found it. Here was the very heart of the dispute. On what are we to depend for our faith – on the Holy Scriptures in the Church, or on the authority of the Church unsupported by the Scriptures? This is a living issue today. The Anglican objection to the development of Roman Catholic dogma since the Reformation is that the additions which that Church has made to the Catholic faith rest only on the authority of the Western Church without any demonstration from Scripture of their truth. The principle maintained by all the Anglican Churches today is precisely that for which the martyrs in the Marian persecution died.

The tactics of the judges followed the well-known formula, 'If you do not believe A, then you must believe B. If you do not accept our doctrine of transubstantiation, then you do not believe in any effective presence of Christ in the Sacrament at all.' Here again, the words of Latimer are significant, precisely because he was not such a good scholar as Ridley or Cranmer, and reveal something of the perplexity to which many were reduced by the difficulty of finding the right terms in which to express their meaning: 'I say there is none other presence of Christ required than a spiritual presence; and this presence is sufficient for a Christian man, as the presence by which we both abide in Christ, and Christ in us, to the obtaining of eternal life, if we persevere in his true Gospel. And the same presence may be called a real presence, because to the faithful believer there is the real or spiritual body of Christ: which thing I here rehearse, lest some sycophant or scorner should suppose me, with the anabaptist, to make nothing else of the sacrament but a bare and naked sign.' The sense of Ridley's answer was the same, though more elaborately expressed.

Cranmer's ordeal was the hardest, just because he was more sensitive than any of the others, and because he was blessed with the kind of mind that can always see both sides of every question. He had carried almost to its extreme the sixteenth-century doctrine of the authority of the Christian prince, the prince appointed by God Himself, in the Church. His undying regard for Henry VIII and his deep affection for Edward VI combined with his historical studies and his reading of the Old Testament to produce an almost mystical veneration for the commands of the prince. The situation had changed; his basic conviction had not. After his brief and unwilling consent to the affair of Lady Jane Grey, Cranmer had accepted Philip and Mary as his lawful sovereigns. But, if these lawful sovereigns bade him deny what he had taught and teach what he had denied, who was he to resist them? His submission to their authority included, in large measure, submission to the standard of doctrine which they wished to enforce in the English Church.

But this was not enough for his persecutors. They desired his complete humiliation, and with it the humiliation of the whole cause of the Reformation. Under the long stress of his imprisonment and of the alternation of fear and hope, Cranmer's nerve broke. He signed one recantation after another, each more degrading than the last. His enemies thought that they had attained their aim. But this was not the last word in the story.

Before Cranmer was led out to die, on 21 March 1556, there was one last ceremony in St Mary's Church at Oxford. After the customary sermon, Cranmer stood to make his dying declaration. Some preliminaries ended, he came to the crucial passage in what he had to say: 'And now I come to the great thing that so troubleth my conscience, more than any other thing that I said or did in my life: and that is my setting abroad of writings contrary to the truth, which here now I renounce and refuse, as things written with my hand contrary to the truth which I thought in my heart, and written for fear of death, and to save my life, if it might be. . . . And forasmuch as my hand offended in writing contrary to my heart, it shall be first burned. And as for the Pope, I refuse him as Christ's enemy and Antichrist, with all his false doctrine. And as for the Sacrament . . .' At the winning-post his enemies had been defeated. In the clamour he got no further. 'Play the Christian man,' said Lord Williams, 'remember your recantations and do not dissemble.' 'Alas, my Lord,' was Cranmer's answer, 'I have been a man that all my life loved plainness, and never dissembled till now against the truth; which I am most sorry for.' So he went to the place of martyrdom, walking so fast that those who accompanied him had difficulty in keeping pace with him. During the agony he was not seen to speak or move, except that he wiped his forehead with his hand.

We have seen something of the immense services that Cranmer had rendered to his Church, and indeed to the whole of Christendom. The greatest service of all was rendered in his death. These things were not forgotten; it will be an evil day for England, if they should ever come to be forgotten.

There can be no doubt that the main responsibility for the persecution rests on the Queen, and, when every possible excuse has been made for her, the burden of her guilt is heavy. The Marian bishops were happy to see the restoration of the ancient laws against heresy. Bonner of London earned an unenviable notoriety through his vigour as a persecutor, and through the disgusting glee which he manifested at the degradation of Cranmer. Pole was second only to Bonner. But the Church could not itself burn anybody; it could do no more than hand heretics over to the secular arm. It was the custom of the age to pass savage statutes, and to maintain great flexibility in the application of them. The secular arm would act or remain passive according to the directives received from the sovereign. It was the will of Mary to deliver hundreds of her subjects over to a horrible death; the secular arm moved only in response to the impulse of her will.

It is often stated that in the sixteenth century no one was tolerant. This is inaccurate; the kingdom of Poland in that century made memorable experiments in the peaceful co-existence of rival confessions.[1] From the inaccurate premiss the misleading conclusion is drawn that, in the matter of persecution, there was really no difference between Catholic and Protestant. This is not so. The Churches of the Reformation had inherited from the medieval Church the evil doctrine that heresy is a crime which may rightly be punished by death, and nearly a century was to pass before they had finally rid themselves of this inherited poison. But Cranmer himself had protested earnestly against the view that a faith professed under compulsion can have any value; and Calvin's partial responsibility for the burning of Servetus in Geneva in 1553 called forth a storm of protest in all the Protestant Churches, precisely on the ground that Calvin had been unfaithful to that new understanding of human liberty which was perhaps the greatest contribution of the Reformation to the welfare of mankind. In the whole history of England, there had never been anything the least like the Marian persecution; nothing in the least like it was ever to occur again. The numbers of the victims, if nothing else, distinguish it clearly from every other persecution.

During the fifteenth century burnings of Lollards had gone forward at an average rate of one a year. In the thirty-eight years of the reign of Henry VIII sixty persons suffered

---

1. The Consensus of Sendomir (1571) is one of the great landmarks in the history of religious toleration.

death as heretics; this number has to be rather more than doubled if we include those who in Henry's eyes, though not in their own, were political criminals. In forty-five years, under Elizabeth I, rather less than two hundred Roman Catholics were executed, but the Government put to death only those whom it regarded, rightly or wrongly, as members of a subversive fifth column. In little more than three years of the reign of Mary three hundred 'heretics' were burned at the stake. Here there was no question of any political consideration; these people were loyal subjects and prayed for the Queen, though their prayers doubtless included the hope and the wish that she might come to a better mind. If ever there was a religious persecution, in the strictest acceptation of the term, it was this one. Mary's victims were far fewer than those who perished in the holocausts by which the Catholic powers of the Continent were disgraced in this period. But they were too many for the minds and consciences of the English people. It had been Mary's aim to re-establish that form of the Christian faith in which she herself believed; her achievement was to make it certain that Roman Catholicism would never again be the faith of the English nation.

The reign ended in unrelieved gloom. In 1555 the aged Cardinal Caraffa became Pope under the title Paul IV. In 1557 this angry and venomous old man vented ancient spite on Pole by depriving him of his office of papal legate, and summoning him to Rome to answer a charge of heresy – a summons which Pole, already a dying man, was unable to obey. Calais had been ignominiously lost in war against the French. Mary was more deeply in debt than any English sovereign before her. The Crown and the nobility were poor, captains and soldiers not to be found, the people insubordinate and divided, injustice rampant and prices high. England was pinched between the French in Calais and the French in Edinburgh. The future seemed dark indeed.

The struggles between the rival faiths in England were tense and dramatic. An old world was dying, a new world was striving to be born; in that elemental strife all men, whether they would or no, were in a measure caught up.

'The doctrinal revolution enforced by Cranmer under Edward VI, and the Counter-Revolution of Gardiner, Pole, and their assistants under Mary, exposed our agitated Islanders in one single decade, to a frightful oscillation. Here were the citizens, the peasants, the whole mass of living beings who composed the nation, ordered in the name of King Edward VI to march along one path to salvation, and under Queen Mary to march back again in the opposite direction; and all who would not move on the first order or turn about on the second must prove their convictions, if necessary, at the gibbet or the stake. Thus was New England imposed on Old England; thus did Old England in terrible counter-stroke resume a fleeting sway; and from all this agony there was to emerge under Queen Elizabeth a compromise between Old and New which, though it did not abate their warfare, so far confined its fury that it could not prove mortal to the unity and continuity of national society.'[1]

1. Winston S. Churchill: *A History of the English-speaking Peoples* (1956), Vol. II, p. 81.

CHAPTER 5

Reformation

Like her sister Mary, Elizabeth I had had a frightful girlhood. Like her, she had been declared a bastard. Again and again she had nearly been the victim of the schemes of scheming men. Many times her life had been in danger. When she came to the throne at the age of twenty-five, she had learned to keep her own counsel, to think courageously but when necessary to act tortuously, and above all to call no man her master.

Elizabeth had all the intellectual power of her father. She was well read in Latin and Greek, and could speak French, German, and Italian fluently. She had Henry's wonderful flair for understanding her subjects and for retaining their loyalty and devotion, whatever she might do and however badly she might treat them. Her manner of speaking was in general complex and involved; but at crises it could rise to really splendid heights of eloquence, as in the famous speech at Tilbury in 1588: 'Let tyrants fear; I have always so behaved myself that, under God, I have placed my chiefest strength and safeguard in the loyal hearts and goodwill of my subjects, and therefore I am come amongst you, as you see, at this time, not for my recreation and disport, but being resolved in the midst and heat of the battle, to live or die amongst you all, to lay down for my God, and for my kingdoms, and for my people, my honour and my blood even in the dust. I know I have the body but of a weak and feeble woman; but I have the heart and stomach of a king and of a King of England too, and think foul scorn that Parma or Spain, or any prince of Europe should dare to invade the borders of my realm; to which rather than any dishonour shall grow by me, I myself will take up arms, I myself will be your general, judge, and rewarder of every one of your virtues in the field.'

When this remarkable young woman came to the throne of a weak, impoverished, and distracted country, she was threatened on every side by innumerable enemies at home and abroad. For more than thirty years her life was in constant danger. What in 1558 neither friend nor foe could foresee was that she would outlive all the other Tudors, that she would remain unmarried, and that she would so dexterously play off her enemies one against another that England would emerge from her reign at the pinnacle of power, fame, and artistic achievement.[1]

Elizabeth's first and most urgent task was to make up her mind as to the course that she would pursue in religion.

Much mystery has been made as to the Queen's own personal convictions, but perhaps the mystery was not of her creating. She was not given to wearing her heart upon her sleeve. She disliked religious controversy and found long sermons tedious. But she spoke often, plainly and consistently of her religious position, as when she told Parliament on 24 November 1586, 'Although I may not justify, but may justly condemn, my sundry faults and sins to God, yet for my care in this government let me acquaint you with my intents. When first I took the sceptre, my title made me not forget the giver, and therefore began as it became me, with such religion as both I was born in, bred in, and, I trust, shall die in; although I was not so simple as not to know what danger and peril so great an alteration might procure me – how many great Princes of the contrary opinion would attempt all they might against me, and generally what enmity I should thereby breed unto myself. Which all I regarded not, knowing that He, for whose sake I did it, might and would defend me. Rather marvel that I am, than muse that I should not be if it were not God's holy hand that continueth me beyond all other expectations.' Unless

1. A pleasant picture of the feelings of her people is to be found in the opening words of Holinshed's Chronicle of the reign: 'After the stormy, tempestuous, and blustering windy weather of Queen Mary was overblown, the darksome clouds of discomfort dispersed . . . it pleased God to send England a calm and quiet season, a clear and lovely sunshine, a quitset from former broils of a turbulent estate, and a world of blessings by good Queen Elizabeth.'

Elizabeth was a consummate hypocrite, these words have
the unmistakable ring of sincerity and truth.

Elizabeth had been brought up from birth in the faith of
Cranmer and the Reformers. It is almost certain that she
would have liked to retain more colour and ceremonial than
the austere mind of the Edwardian Reformation permitted.
There was the dreadful scandal of the crucifix and the lighted
tapers in the Queen's chapel, when such papistical follies
had been forbidden by the law of the land – 'the enormities
yet in the Queen's closet retained'. Archbishop Parker
managed to get this cause of scandal removed, but ten years
later it slipped back when he was not looking. Elizabeth
would have preferred a celibate clergy, and was capable of
expressing herself very forcibly on the subject, so much so
that the gentle Archbishop noted on one occasion: 'I was in
a horror to hear such words to come from her mild [*sic!!*]
nature and Christianly learned conscience, as she spoke
concerning God's holy ordinance and institution of matrimony. I marvelled that our states in that behalf cannot
please her highness.' But further back than the state of
religion at the end of the reign of Henry VIII the Queen
would not go; and, as it turned out, the state of religion in
the country would not let her go so far.

The new and uncertain factor in the situation were the
refugees, the gentlemen who had lived abroad during the
period of the persecutions. Now they all came flocking back.
Preaching at Queen Mary's funeral on 14 December 1558,
Bishop White of Winchester remarked that 'the wolves be
coming out of Geneva and other places of Germany and
have sent their books before, full of pestilent doctrines, blas-
phemy, and heresy to infect the people'. Once returned, the
exiles with astonishing rapidity organized themselves, made
common cause with like-minded brethren who had stayed
at home, and tried to carry through a radical reformation
on the lines of the doctrine and discipline with which they
had become acquainted and with which they had fallen in
love in their residence abroad. 'In the beginning of Her
Majesty's reign a number of worthy men . . . desired such a
book and such orders for the discipline of the Church as

they had seen in the best-reformed Churches abroad.' Such was their strength that they gained a good deal more than the Queen could have wished; that they did not gain more was due to the resolute and obstinate resistance which Elizabeth maintained over forty years to all those who desired to turn her Church into a dependency of the best-reformed Churches of Geneva and of Zürich.

Elizabeth would have liked to go slowly, and if possible to win to her cause some of the more moderate among the Marian bishops. Faced by the recalcitrancy of what would later have been called the Puritan element, she found it prudent to agree to the settlement of all the main issues of religion within five months of her accession to the throne.

The first of the crucial acts was the Act of Supremacy. In this, Elizabeth was declared to be not, like her father Henry VIII, 'the only supreme head in earth of the Church of England', but 'the only supreme governor *of this realm*, as well in all spiritual and ecclesiastical things or causes as temporal'. It has often been suggested that the change from 'head' to 'governor' marked a distinction without a difference. But, in fact, the greater precision of the Elizabethan definition was very important indeed. Late medieval theory distinguished very clearly between *potestas ordinis*, the spiritual or sacramental authority mediated to the bishop or priest in his ordination but derived directly from Christ, and the *potestas jurisdictionis*, the administrative authority by which the Church was ordered. Critics of the English Church had maintained, and sometimes still imagine, that the secular rulers of England have claimed priestly powers within the Church. Elizabeth was rightly concerned to make it plain that this was not so; though it has to be admitted that she allowed to herself a characteristically Tudor liberty in the interpretation of the limits of the *potestas jurisdictionis*.

The second Act was that of Uniformity. The Prayer Book of 1552 was restored in its entirety with some small but very significant changes. The 'Black Rubric' was omitted. In the words of administration at the Holy Communion, as has been already mentioned, the form of 1549 was combined with that of 1552. The petition in the Litany to be delivered 'from the Bishop of Rome and all his detestable enormities' was suppressed. An 'Ornaments Rubric' was inserted and passed apparently almost without comment; as one of the leaders later expressed it, 'as we were unable to prevail, either with the Queen or the Parliament, we judged it best, after a consultation on the subject, not to desert our churches for the sake of a few ceremonies'. Peace was not so easily to be restored, but later conflicts were still happily hidden within the womb of time; it is probable that Elizabeth had attained to as wide a range of agreement among her subjects as was possible at that time.[1]

The next move was to fill up the vacant bishoprics and to restore the organization of the Church. In her choice of an Archbishop of Canterbury Elizabeth was extraordinarily fortunate. Matthew Parker (1504–75), like Cranmer, was a quiet, shy, Cambridge scholar. He had been one of the company that had read Lutheran books at the White Hart Tavern, and had early embraced Reformation principles. He had a great veneration for Cranmer, and once expressed his desire to recover the 'great notable written books of my predecessor Dr Cranmer. . . . I would as much rejoice while I am in the country to win them, as I would to restore an old chancel to reparation.' A man of simple piety, he was entirely free from fanaticism and bigotry. He would have preferred above all things to be left at peace with his books in the University, to which he had rendered such notable services. When his name was put forward for the Archbishopric, he was horrified, and besought the Queen not to lay this cross upon him. But the Queen was firm, and rightly so; in all England a better man for the position could not have been found.

In the consecration of Matthew Parker the greatest care was taken to maintain continuity with the past, and above all to ensure that the succession of episcopal consecration was

1. It is to be noted that once again the religious settlement was a *Parliamentary* settlement; the Convocations of the clergy had no hand in it, though later they played a very active part in the religious history of the times.

unbroken. Four bishops performed the consecration according to the form in the Edwardian Ordinal, and of these, two
had been consecrated in the reign of Henry VIII under the
old order.

Passionate attempts have been made from the Roman
Catholic side to prove that Parker was not validly consecrated, and that the Church of England has no valid succession
of episcopal order and ministry from the days of the
apostles. All kinds of improbable legends have been circulated; each in turn has been exploded by patient historical
research. In recent years, the Roman attack has tended to
concentrate on the question of 'intention'. What did the
consecrators of Parker intend to do? Did they intend to
consecrate a bishop to stand in the great and unbroken
succession of bishops from the earliest days of the Church?
Or did they intend to create a new office and order,
distinct from that which had been in existence before the
Reformation? A conclusive answer to this question is provided by the *Preface* to the Order of Service which they used:
'It is evident unto all men, diligently reading holy Scripture, and ancient authors, that from the Apostles' time there
hath been these Orders of Ministers in Christ's Church:
Bishops, Priests and Deacons, which Offices were evermore
had in such reverent estimation, that no man by his own
private authority might presume to execute any of them.
. . . And therefore, to the intent that these orders should be
continued, and reverently used, and esteemed, in this
Church of England, it is requisite, etc.' In many things the
Church of England may be accused of ambiguity; these
sentences are marked by a superb lucidity, and leave no
doubt at all that the intention of their authors, and of those
who used this service, was to continue in the Church of
England those orders of bishop, priest, and deacon which
had existed in the Church since the time of the apostles, *and
no others*. Whether they were successful in carrying out their
intention or not is another question, and on this the critics
have a right to say their say.

It is well known that on 15 September 1896 Pope Leo
XIII closed the question from the Roman side in the Bull

Apostolicae Curae, in which he declared Anglican Orders to be wholly and absolutely null and void. The Bull is a very curious document, and Roman Catholic scholars have had considerable difficulty in interpreting it. This is a problem for Roman Catholics and not for Anglicans. From the Anglican point of view, all that the Bull means is that an Anglican priest knows in advance that, if he joins the Church of Rome and wishes to become a priest, he will have to be unconditionally re-ordained. Roman Catholic priests who become ministers of Free Churches are sometimes re-ordained, on the ground that their previous ordination has not qualified them to preach the Word and minister the Sacraments in the true and biblical understanding of them. Roman Catholic priests who join an Anglican Church are not re-ordained; they are required to repudiate Roman errors and to undergo a period of probation, after which they are permitted to exercise their ministry, if licensed to do so by a bishop. It is to be regretted, however, that Leo XIII by his Bull added one more to the already long list of difficulties which stand, and which seem to stand unalterably, in the way of the union of all Christian people in one Church.

The headship being settled, it remained to proceed with the organization of the body. All the clergy, justices, mayors, royal officials, and several other classes of people were required to take the Oath of Supremacy. It was first tendered to the bishops. Ten sees were vacant at the time; to their credit the survivors, with one exception, refused to take the Oath, and were consequently deprived. Bonner the persecutor died in the Marshalsea prison; but not one of the Marian bishops was put to death. The others were allowed to live under conditions which varied from comfortable and dignified sequestration to some hardship and discomfort. None played any further part of importance in the life of the country. Of the new bishops the majority had been refugees abroad, and therefore belonged to the extreme and Calvinistical section of the reforming party. This was probably not as the Queen would have wished it to be; but among these men were the best, the most learned, the most devoted servants of the Church. Elizabeth kept the notable

firebrands out of the episcopate; otherwise she had to use
such material as was available to her hand.

The mass of the parochial clergy, then as earlier, seems to
have been willing to follow any directions given by the
Government. Some, however, refused to take the Oath and
were deprived. Estimates of the numbers of these nonconformists vary wildly, between two hundred and two thous-
and; on the evidence so far gathered it is impossible to reach
any certain conclusion, but the true figure is probably much
nearer to the lower than to the higher of these estimates.
But the clergy were gravely demoralized by all the changes
of the preceding thirty years. Many were openly hoping for
another change. Some were no more than sullen conformists,
and there are many complaints in the episcopal visitations
of priests who read the English service inaudibly or in a
slovenly manner. It appears that some incumbents said the
new services openly in church and continued to say the
Roman Mass in secret. The new bishops had a plentiful
supply of problems on their hands.

It is even more difficult to estimate the general attitude
of the people as a whole. There is no doubt that among
them was found every shade of opinion, from the staunch
papalist, who could not bow down at any price in the house
of Rimmon, through the outward conformist and the indifferent, to the passionate supporter of the new ways. The
influence of the Queen was steadfastly exercised in favour of
moderation and against any kind of persecution. She
affirmed that she had no desire 'to make windows into
men's souls', and there is no doubt that she spoke the truth;
if men would promise loyalty to her and to her Government,
she would not inquire too closely into the exact nature of
their inner and personal beliefs. Elizabeth held as firmly as
the Pope or Philip of Spain that nation and Church should
be co-extensive; but she desired that as far as possible her
people should be brought to this unity in faith and worship by
persuasion and not by the sword. Convinced as she and her
ministers were of the superiority of the wares they had to
offer, they believed that, if only the people could be brought
to Church to worship in English and to hear the true doctrine,

they would come in time to buy willingly in the Anglican market. Continuous pressure was maintained, and the process of persuasion was no doubt helped by the fine of 12*d*. a Sunday which could be imposed for non-attendance at church. The position of the recusant may have been uncomfortable; but it is noteworthy that in the first twelve years of the reign no Roman Catholic suffered death for his faith.

This comparatively peaceful state of affairs was brought to an end not by the Queen but by Pope Pius V. On 25 February 1570 he issued the Bull *Regnans in Excelsis*, in which he declared Elizabeth to be guilty of heresy, and of encouraging heresy, to have incurred excommunication, and therefore to have forfeited her 'pretended right' to the English crown; her subjects were no longer bound by any oath of loyalty to her, and *under pain of excommunication* could no longer yield her obedience. Pius himself frequently explained that it was his intention to encourage the English Roman Catholics to rise in rebellion against their lawful sovereign, and to set at rest the consciences of those who had hesitated so to rise before Elizabeth had been declared a heretic and deposed by the papal power. By a stroke of the pen the Pope of Rome had made every English Roman Catholic, on pain of excommunication, a potential traitor to his country and his Queen.

The lengths to which the authorities of the Western Church were prepared to carry their hatred of Elizabeth is painfully revealed in a letter written by the Cardinal Secretary of State to the papal nuncio in Madrid on 12 December 1580. Certain gentlemen of England had approached the nuncio to inform him that they had entered into a pact to assassinate Elizabeth, and wished for an assurance that, if they were successful in an attempt which might cost them their lives, they would not incur sin. The answer must be quoted at some length: 'Since that guilty woman of England rules over two such noble kingdoms of Christendom and is the cause of so much injury to the catholic faith, and loss of so many million souls, there is no doubt that whosoever sends her out of the world with the pious intention of doing

God service, not only does not sin but gains merit, especially having regard to the sentence pronounced against her by Pius V of holy memory. And so, if those English gentlemen decide actually to take in hand so glorious a work, your Lordship can assure them that they do not commit any sin. We trust also in God that they will escape danger.'[1]

The excommunication must be judged to have been, from the Roman Catholic point of view, a grave error, for which the English Roman Catholics were to pay dear. 'No event in English history, not even the Gunpowder Plot, produced so deep and enduring an effect on England's attitude to the catholic church as the bull of Pius V. Englishmen never forgot their queen's excommunication. . . . The story of the excommunication, and of the pope who freed men from their oaths, and subjects from their allegiances, was a weapon that kept its edge for centuries and effectively put a stop to every thought of toleration for the papists.'[2] Two hundred and fifty-nine years were to pass before the co-religionists of Pius V were to be judged worthy to take a share in the government of their country. It may be doubted whether the evil effects of the Bull of 1570 have yet entirely passed away.

It is fear that makes men cruel. For many years Elizabeth and her ministers lived in fear. The dangers they had to face were by no means imaginary. In 1572 hatred of Protestants in France broke out in the notable crime of the massacre of St Bartholomew's Eve; when the news reached Rome, the Pope ordered a *Te Deum* to be sung, and commanded a procession to be held to thank God for the grace bestowed on Christendom. On 9 July 1584 William the Silent, Prince of Orange,[3] fell by the hand of an assassin. He was the same age as Queen Elizabeth. Everyone in England knew well what chaos would have followed if the plots

1. The Italian text of the letters can be found in A. O. Meyer: *England and the Catholic Church under Queen Elizabeth* (E. Tr. 1916), pp. 490–1.

2. A. O. Meyer, op. cit., p. 85.

3. It was of him that the American historian J. L. Motley wrote the splendid words: 'As long as he lived, he was the guiding-star of a brave nation, and when he died the little children cried in the streets.' *Rise of the Dutch Republic*, pt. vi, ch. vii.

against her life had been successful. Some of these plots were imaginary; far too many of them were all too real.

But the gravest danger lay elsewhere. From the beginning of the reign, a number of young men, some of them of the best families, had been fleeing from the country and betaking themselves abroad.[1] There they were being trained as missionaries for the spiritual reconquest of England. William Allen, later Cardinal (1587), had founded in 1568 the famous Seminary at Douai (at Rheims 1578–93); there these young men were trained in the full doctrine and discipline of the Counter-reformation, and hardened to face what must needs be a perilous mission to England. Herein lay the tragedy of the situation; to the Seminary priests their mission presented itself in terms of pure devotion to the cause of Christ; from the standpoint of the Government they were traitors from the moment they set foot on the shores of their native land. Religion and politics had become hopelessly confused.[2]

Fear makes men cruel; and the Pope had put into the hands of Elizabeth and her ministers the perfect weapon for their purposes. It was necessary to ask only two questions of the suspected person: (1) Did he acknowledge Elizabeth as his lawful Queen? (2) Did he believe that the Pope had power and authority to excommunicate and depose the Queen? If a strict Catholic answered according to his conscience, he was sentencing himself to death. If he refused to answer, he could be cruelly put to death for contumacy. The campaign of repression developed along abominable lines – the employment of informers, the regular use of

1. Many of them were from the University of Oxford. Cambridge, always firmer in its hold on Anglican principles, yielded a much smaller number: 'From Cambridge there was never more than an occasional recruit', says Fr Hughes: *The Reformation in England*. Vol. III, p. 340, n. 1.

2. It hardly helped the situation that, on the eve of the Armada, Allen issued an 'Admonition to the Nobility and People of England', in which Elizabeth was described as 'a most unjust usurper and injurer of all nations, an infamous, depraved, accursed, excommunicate heretic; the very shame of her sex and princely name, the chief spectacle of sin and abomination in this our age; and the only poison, calamity and destruction of our noble Church and Country'.

inhuman tortures, the perversion of justice to secure conviction where evidence was inadequate. As Burleigh's information service improved, priests who were little more than
boys were taken from the ships on which they had sailed,
and carried off directly to prison and the gallows.

It must not be supposed that all Roman Catholics were
traitors. Many of them managed to avoid the dilemma into
which the Pope had forced them, and to square their consciences with a genuine and unquestioning loyalty to their
Queen and country. There were in fact two types of Catholicism in the country, and through the changes and chances
of time these have in a measure persisted in England almost
until our own day. There was the old easy-going Catholicism of Warham and Wolsey. Now there was the new
Catholicism of the Council of Trent and the Counter-
reformation. The dangers through which the Church had
passed had fired its defenders with new zeal, passion, and
devotion, with a new willingness to suffer and to die. Above
all the crisis of the sixteenth century had provided the
Roman Church with a trained clergy. Among the greatest
achievements of the Council of Trent was its creation of the
seminary. When the Protestants spoke bitterly of the returning emissaries of the Roman Church as 'Seminary priests',
the name hit the mark; these young men, trained at Douai
or elsewhere, had learned discipline and devotion; they had
been trained in the methods of controversy. The training
might be narrow and scholastic, but it was calculated to
make them more than a match for anything that they were
likely to meet in England. Therein lay the danger. Most of
these priests were insignificant people. But in such men as
the pure and guileless Edmund Campion (1540–81), birth,
breeding, intellectual gifts, and social charm combined with
deep and almost apostolic Christian faith to produce a
character of singular attractiveness. Such men deserved a
better fate than a traitor's death at the hands of their fellow-
countrymen.

As long as the danger lasted, that is to say from 1570 to
1590, the repression of the recusants continued in unremitting
severity. During that period executions took place at the

average rate of nine a year. The Queen herself was weary of
the slaughter, and the people themselves revolted against
the sickening spectacle of living men being disembowelled
before the public gaze. There was no violent revulsion in
favour of the old religion. The Catholics were weakened by
dissensions in their own ranks. During the last years of the
reign, the severities decreased and died away. But the
danger had passed; the recusants were no longer in a position to threaten the stability or the security of the realm.

The second danger, that which arose from the people who
came to be called Puritans, was perhaps even greater
because more intimate than that which threatened the
Church from the side of the recusants.

The term 'Puritan', like the term 'Christian' itself, seems
to have been originally a term of reproach invented by the
enemies of the movement. (The first instance of the use of
the term in print is from the year 1564.) The word in itself
has nothing to do with a particularly grave or grim attitude
to the pleasures of this world, though some Puritans held to
rather more austere ethical standards than the majority of
their contemporaries. A puritan, to cite the excellent definition given in the *Oxford Dictionary*, was 'a member of that
party of English Protestants who regarded the reformation
of the Church under Elizabeth as incomplete, and called for
its further "purification" from what they considered to be
unscriptural forms and ceremonies retained from the unreformed Church'. These were the men who had come
under the influence of Calvin at Geneva or Bullinger at
Zürich, who had seen there the living examples of 'the best
reformed Churches', and believed that the discipline that
had been good for France and Scotland would be good for
England too.

It is difficult to give an exact account in brief compass of a
movement that to a large extent worked underground, and
had its own dissensions and divisions. But the aims of the Puritans may perhaps be grouped under four main headings.

The first aim, and the most continually stressed, was the
provision of a skilled and preaching ministry. This was
urgently necessary and unexceptionable, if it could have

been separated from the revolutionary and destructive tendencies of the movement as a whole. To assist one another
in this godly work, groups of ministers would meet from time
to time for counsel and mutual edification in the form of
'exercises'. One of the brethren would expound a passage
of Scripture, previously chosen, and the others would then
discuss and criticize. Laymen were permitted to be present,
but were not allowed to speak. This served as good practical
training in the study and exposition of the Scriptures.

The second aim was the reduction of ceremonies to that
bare minimum which was regarded as justifiable in the light
of the Scriptures. The signing with the Cross in baptism
was a superstitious ceremony. Much vilification was directed against the harmless, decent surplice. It might have been
thought that sensible men would not engage in violent controversy over things so indifferent in themselves. The
trouble is that very few of us can ever admit that anything
that concerns our religious faith and its expression is indifferent. A flag is only a square of cloth; but, if it has come
to be the symbol of the pride and dignity of a great country,
what it signifies bears very little relation to the stuff of
which it is made. So the wearing or the not wearing of a few
yards of white linen can present itself as a matter on which
the fate of empires and churches may depend.

Thirdly, the Puritans wished to introduce into the Church
of England the whole Presbyterian system as they had known
it in Geneva, the only system which they believed to be
agreeable to the Word of God. This would involve the
abolition of the office of bishop (in the traditional sense of
the term), equality of ministers, the organization of the
Church in presbyteries, the formation of consistories in
every parish for the exercise of discipline, and participation
of laymen in the government of the Church. Parts of this
programme have been incorporated into the Anglican
system in many areas of the world, with great benefit to the
Provinces that have adopted them. But, if the Puritans had
had their way, the universal Church would have been the
poorer through being robbed of the peculiarly Anglican
witness to the truth of Christ.

The extremer Puritans went to astonishing lengths in the attempt to carry out their programme. In 1568 Archbishop Grindal wrote to Bullinger of some of the extremists among them in the following terms: 'Some London citizens of the lowest order, together with four or five ministers, remarkable neither for their judgement nor learning, have openly separated from us, and sometimes in private houses, sometimes in the fields, and occasionally even in ships, they have held their meetings and administered the Sacraments. Besides this, they have ordained ministers, elders and deacons, after their own way, and have even excommunicated some who had seceded from their Church.'

But even those who were not extremists were prepared, while remaining within the national Church, to go far towards subverting its order and discipline. In 1572 a number of the leaders decided secretly to form a presbytery. This was done at Wandsworth; and, in the words of the Church historian Fuller, 'This was the first-born of all presbyteries in England, and *secundum usum Wandsworth* as much honoured by some as *secundum usum Sarum* by others.' Presbyteries were formed with great rapidity in other regions. One of the aims of these bodies was, in effect, to take the control of ordination out of the hands of the bishops. It was laid down that no one should *offer* himself for the ministry; he must be called by a Church. If duly called, he should impart the matter to the *classis* or conference of which he was a member; and, only if the call was approved by that body, should he present himself to the bishop for ordination.

Ere long the presbyteries had a national, though still underground, organization. The chief artificer of this widespread conspiracy seems to have been a remarkable man, John Field, 'a great and chief man amongst the brethren of London, and one to whom the managing of the discipline ... was specially by the rest committed'. At least once, and probably more often, the godly brotherhood was able to hold a national synod in London, the aim being 'to ask a full reformation and to accept of none, if they had not all'.

The Puritans were masters of secret organization. They had also a pretty taste in controversy. There was much in the English Church against which criticism could be legitimately directed; in other points the almost childish narrowness of the Puritan temperament is painfully evident. Let one specimen of Puritan rhetoric, drawn from the 'Request of all true Christians to the most honourable High Court of Parliament' (1580), suffice: 'Let cathedral churches be utterly destroyed . . . very dens of thieves, where the time and place of God's service, preaching and prayer, is most filthily abused; in piping with organs, in singing, ringing ond trolling of the Psalms from one side of the choir to another, with squealing of chanting choristers. . . . Dumb dogs, unskilful, sacrificing priests, destroying drones, or rather, caterpillars of the Word. . . . Dens of lazy, loitering lubbards.'

The reference to the High Court of Parliament leads us to the fourth of the Puritan aims. They would have had a State Church in the full sense of the term, completely under the control of Parliament, and robbed of the last traces of spiritual independence. It was only through Parliament that Elizabeth had been able to carry out her moderate measures of reform. The Puritans had many friends in Parliament. It was through Parliament that they planned to capture the Church and to transform it into that which they sincerely believed it ought to be.

The thorough, indeed violent, Reformation which the Puritans desired to carry out by law would have completely changed the character of the English Church. On 27 February 1587 Anthony Cope introduced into the House of Commons a Bill, which after a long preamble, ended in two brief clauses for the better government of the Church. The first authorized a new Prayer Book, a revision of the form that had been in use among the English exiles in Geneva, in which a Presbyterian form of Church constitution was incorporated. The second was to declare 'utterly void and df none effect' all 'laws, customs, statutes, ordinances and constitutions' which related to the Church and its government. The Puritans would first make a desert, and then introduce a Presbyterian kind of peace. Professor Neale does not exaggerate when he says that 'through the plotting of the godly brotherhood and their organized group of parliamentary agents, Queen Elizabeth was menaced with revolution in both Church and State'.[1]

The modern reader may feel impelled to ask, Why, if these men so gravely disapproved of the Church of England as it was, did they not leave it? We are so familiar with the existence of Nonconformists, with the coexistence of several Christian bodies in one realm, that that would seem to be the obvious solution. But not for the Puritans; that was not in the least the way in which they thought. They were not Separatists or Sectaries. There were Separatists in the reign of Elizabeth, such as Henry Barrow and John Penry, who were hanged for treason in 1593. These men represented the attitude of 'tarrying for no man', as contrasted with the Puritan attitude of 'tarrying for the magistrate' in the carrying out of reform; they repudiated Elizabeth's supremacy, and denied to the Church of England the name of a true Church. The Puritans had no intention of departing from the national Church; it was their intention to capture it, and put it to rights – a task for which they regarded themselves as much better fitted than anyone else. That was what made them dangerous; they were enemies within the gate, who had convinced themselves that they were the Church's true friends.

What had the threatened Church to oppose to this inner subversion and gnawing away of its vitals?

The answer, initially, must be – the Queen, and not very much else. Having cooperated with Parliament in 1559, the Queen thereafter affirmed, and continued to affirm in no uncertain tones, that Parliament had nothing whatever to do with matters of religion; that belonged to her prerogative, and she would see to it; she had her bishops, and, if anything needed to be reformed, they would attend to it under her direction. On 28 February 1585 the Speaker reported the Queen's views to the Commons; she had repeatedly forbidden them to 'meddle with matters of the Church, neither in reformation of religion or of discipline'. 'As she knew the doctrine preached in the Church of England to be as sincere

1. J. E. Neale: *Elizabeth and her Parliaments 1584-1601* (1957), p. 156.

as might be possible, so she knew the discipline thereof not so perfect as might be. . . . Yet to redress it in such open manner as we sought, whereby the whole state ecclesiastical might be overturned, or at the least defaced, that she most disliked. For as she found it at her first coming in, and so hath maintained it these twenty-seven years, she meant in like state, by God's grace, to continue it and leave it behind her.' The Commons did not like all this, but in those days personal rule was a very real thing, and they submitted. This may not be the ideal way of governing a Church; but, if Elizabeth had been less stubborn, there might be no Anglican Church today. And the royal prerogative was in fact the bulwark of what little spiritual freedom the Church retained, just as episcopacy, when bishops know their business, is the great safeguard against other and much less endurable forms of authority.

She had her bishops. They would do the work, and endure the unpopularity, and she would remain Gloriana, undisturbed and serene.

In 1566 the gentle Matthew Parker issued his *Advertisements* 'by virtue of the Queen's Majesty's letters commanding the same', and fixed the minimum conformity in matters of clerical dress which the clergy were required to observe. They were to wear a surplice at all times of their ministrations, and a cope was to be worn at celebrations of the Holy Communion in cathedrals and collegiate churches. The measure was not popular, and Parker never knew whether the Queen would support him. 'I was well chidden at my prince's hand; but with one ear I heard her hard words, and with the other and in my heart, I heard God. And yet her Highness being never so much incensed to be offended with me, the next day coming by Lambeth bridge into the fields, and I according to duty meeting her on the bridge, she gave me her very good looks, and spake secretly in my ear, that she must needs countenance mine authority before the people to the credit of my service.'

Parker's successor Grindal (1576) was inclined to be somewhat favourable to the 'exercises' of the Puritans. When the Queen issued strict instructions to him to put an end to them, Grindal replied, 'I am forced, with all humility, and yet plainly, to profess that I cannot with safe conscience and without offence to the majesty of God, give my assent to the suppressing of the said exercises; much less can I send out any injunction for the utter and universal subversion of the same.' He urged the Queen, 'When you deal in matters of faith and religion, or matters that touch the Church of Christ, which is his spouse . . . you would not use to pronounce so resolutely and peremptorily, *quasi ex auctoritate*, as ye may do in civil and extern matters.' This was not the tone in which Harry's daughter expected to be addressed. Grindal was confined to his house for six months, and sequestered from the exercise of his office.

It was in Grindal's successor John Whitgift (1530–1604) that Elizabeth found her 'little black husband', the ideal instrument for her will.

This singularly unlovable man was the perfect type of academic archbishop. Narrow, diligent, incorruptible, excellent in administration, cold and unimaginative, yet not without compassion, he seemed to be the very man that those uncomfortable times demanded. Like most of his opponents he was a Calvinist in his theological views; but in Church affairs he had adopted honestly and conscientiously the moderate position which Elizabeth was determined to maintain.

Whitgift, as Master of Trinity and Vice-Chancellor of the University of Cambridge, had already had experience of the troubles that Puritans could cause. His chief opponent was Thomas Cartwright (1536–1603), universally recognized to be 'the head and most learned of that sect of dissenters then called puritans'. Such was the influence of his teaching that Cambridge in 1565 was astonished to behold the fellows and scholars of Trinity and St John's worshipping in chapel without their surplices. Such was the commotion created by this shocking occurrence that Cartwright found it prudent to retire to Ireland, where his friend Adam Loftus, Archbishop of Dublin, hoped that he might be promoted to the Archbishopric of Armagh. Ireland escaped this fate, and in 1569 Cartwright was back in Cambridge as Lady Margaret

Professor of Divinity. From his Chair he attacked the godly order of the Church of Elizabeth, and put forward proposals that would have led to its complete transformation on the Presbyterian model. It is not surprising that in December 1570 he was deprived of his professorship, and in the following year of his fellowship of Trinity.

The ten years between 1583 and 1593 were years of steady, patient, persistent, and unsympathetic repression of the Puritans. The Puritans themselves were much to blame. The bishops could not be pleased to be described, as they were described in the first of the Martin Marprelate tracts, as 'profane, proud, paltry, popish, pestilent, pernicious, presumptuous Prelates'. On the other hand it is hard to feel anything but distaste for Whitgift's methods. His chief instrument was the Court of High Commission, a prerogative court, the methods of which were entirely different from those of the courts of the common law. In 1584 Whitgift secured for the Court the right to use the oath *ex officio*, an oath which the judge could require the accused person to take, and, having taken which, he was then required on oath to give evidence against himself. Whitgift next set forth Twenty-four Articles, couched in such terms that no honest Puritan could escape deprivation by accepting them. A heavy premium was placed on perjury; the Puritans saw with dismay many of the best and most earnest ministers in the kingdom deprived of their livelihood and of the opportunity of exercising their ministry.

Naturally such proceedings aroused opposition, not only from the side of the Puritans, but among those who were concerned with the good fame of the realm. Lord Burleigh wrote anxiously to Whitgift, 'I think the inquisitors of Spain use not so many questions to comprehend and to trap their preys. . . . I favour no sensual and wilful recusants. But I conclude that, according to my simple judgement, this kind of proceeding is too much savouring of the Romish inquisition, and is rather a device to seek for offenders than to reform any.' It could not be held unreasonable that ministers who absolutely refused to accept the order and the discipline of the Church which they had undertaken to serve should be deprived. But was it right that an ecclesiastical court should have, and exercise, the right to impose such civil penalties as fine and imprisonment?[1] Cartwright himself, having refused to take the oath *ex officio*, was committed to the Fleet, and spent nearly two years in prison. Many who were not Puritans disliked such methods and held them to be less than Christian.

The only thing that can be said in favour of Whitgift's methods is that they were successful. As the older leaders died, their places were not taken by younger men of equal calibre. By 1593 Puritanism within the Church of England was a dwindling force; it no longer had power seriously to threaten the Elizabethan establishment.

So, by 1593, the Church of England had shown plainly that it would not walk in the ways either of Geneva or of Rome. This is the origin of the famous *Via Media*, the middle way, of the Church of England. But a 'middle way' which means 'neither this nor that' seems a rather negative road. And a middle way which is no more than a perpetual compromise, an attempt to reconcile the irreconcilable, is not likely to inspire anyone to heroism or to sanctity. Such is the caricature of the Anglican position which is the current coin of controversialists, and nothing could be further from the truth. Anglicanism is a very positive form of Christian belief; it affirms that it teaches the whole of Catholic faith, free from the distortions, the exaggerations, the over-definitions both of the Protestant left wing and of the right wing of Tridentine Catholicism. Its challenge can be summed up in the phrases, 'Show us anything clearly set forth in Holy Scripture that we do not teach, and we will teach it; show us anything in our teaching and practice that is plainly contrary to Holy Scripture, and we will abandon it.' It was time that this positive nature of Anglicanism should be made plain to the world. It was the good fortune of the Elizabethan Church that it produced the two greatest of the positive controversialists of English ecclesiastical history.

1. It is to be noted that, until 1813, a sentence of excommunication passed by an ecclesiastical court in England carried with it a number of civil disabilities.

John Jewel (1522–71) had been a refugee under the Marian persecution. On his return, he at once made his mark as one of the ablest defenders of the reformed position, and was consecrated Bishop of Salisbury on 21 January 1560. At Oxford he had been a pupil and protégé of Peter Martyr, whom he regarded as his spiritual father, and had been inclined to the left wing of the Reformation. Gradually he receded from this position, and realized that the first task of the Church of England must be to defend its own position against the renewed assaults of the Church of Rome. The result of his meditations on this subject was his *Apologia pro Ecclesia Anglicana*. Published in Latin in 1562, the book at once became widely known everywhere in Europe. In the brilliant English translation by Anne, Lady Bacon, it has become a minor English classic,[1]

Jewel's book is short and clear, and compares favourably with the vast and dreary tomes in which most sixteenth-century controversy is buried. His aim is to deal with one single question: 'Who are the innovators?' He takes his stand on the Scriptures and the primitive Church of the first six centuries – the still undivided Church of East and West. To fix this limit is not to deny the possibility of development in understanding of the Christian faith beyond the position reached in A.D. 600; it is to affirm that by that date the clear outline of Christian doctrine had been drawn, and that all further developments will be found within that outline and not beyond it. The real innovators are the Popes and their supporters. This being so, Jewel's method is to 'shew it plain that God's holy Gospel, the ancient bishops, and the primitive Church do make on our side, and that we have not without just cause left these men, or rather have returned to the apostles and old catholic fathers'. There was no evidence in early times for the supremacy of the Pope, as later times had held it, and to reject that supremacy did not of itself constitute a lapse into heresy. Hopes of reform

1. Professor C. S. Lewis remarks on this translation, 'If quality without bulk were enough, Lady Bacon might be put forward as the best of all sixteenth-century translators.' *English Literature in the Sixteenth Century, Excluding Drama* (1954), p. 307.

through a general council had had to be abandoned; this being so, national Churches had liberty to carry out for themselves the urgently needed reform.

So much for the defence against the Church of Rome. The defence against the Puritans fell into the hands of an even greater man than Jewel. Like Jewel, Richard Hooker (*c.* 1554–1600) was an Oxford man; unlike him, he had been a small child when Elizabeth came to the throne, and had known nothing of the bitterness of the earlier tempests. For a short time he was Master of the Temple in London. He held a number of quiet country cures, and in these he had leisure to acquire his immense learning, slowly to ripen his thought, and to master the handling of his lapidary style. Four books of his *Laws of Ecclesiastical Polity* appeared in 1593, when he was about forty years old, a fifth in 1597. Before his early death in 1600 he had finished but not revised three other books; the first complete edition did not appear till 1662. Into this tremendous work Hooker poured all the resources of a capacious mind – a wonderful knowledge of Scripture, of the Fathers, of the schoolmen, of classical antiquity, of almost everything of profit that had been written in the Christian world.

The first thing that strikes the reader of Hooker is, of course, his style, a complex, flexible instrument, in which no concessions are made to mere rhetoric, but in which majesty of expression can be naturally wedded to the intensity and splendour of the thought.' Now if nature should intermit her course, and leave altogether, though it were but for a while, the observation of her own Laws; if those principal and Mother Elements of the World whereof all things in this lower World are made, should lose the qualities which now they have; if the frame of that Heavenly Arch over our heads should loosen and dissolve itself; if Celestial Spheres should forget their wonted Motions and by irregular volubility turn themselves any way as it might happen; if the Prince of the Lights of Heaven, which now as a Giant doth run his unwearied course, should as it were through a languishing faintness begin to stand and rest himself; if the Moon should wander from her beaten way, the times and

seasons of the year blend themselves by disordered and confused mixture, the Winds breathe out their last gasp, the
Clouds yield no Rain, the Earth pine away as Children at
the withered breasts of their Mother, no longer able to
yield them relief; what would become of Man himself
whom these things now do all serve?'

Another marked characteristic of Hooker is what Professor C. S. Lewis calls 'sequaciousness'. His thought, and
therefore the structure of his book, is architectonic; it
moves steadily forward to its object, taking in and considering all kinds of subsidiary objects on the way, but never
widely or seriously deflected, and with a singularly deft
linking together of all its parts and members. It is this that
makes it difficult to quote Hooker; he has to be taken in
large draughts to be enjoyed. A further consequence of this
way of handling the material is, once again to quote Professor Lewis, that 'The *Polity* marks a revolution in the art of
controversy. Hitherto, in England, that art had involved
only tactics; Hooker added strategy. Long before the close
fighting in Book III begins, the puritan position has been
rendered desperate by the great flanking movements in
Books I and II. . . . Thus the refutation of the enemy comes
in the end to seem a very small thing, a by-product.'[1] It is
this largeness of view that makes possible for Hooker his
beautiful tolerance and fairness; he always says the best that
can be said for an opponent's position before refuting it,
sometimes stating it better than the opponent would have
been able to state it for himself.

What, then, is the subject of the great work? It is not a
general defence of the Christian faith; it is not a book of
doctrine. It is a discussion of the Order of the Church, of the
principles on which that Order should be based, and of the
freedom that regional Churches ought to enjoy within the
fellowship of the one Church. For Hooker, as for the Puritans whose views he is to refute, the basis of all things is the
Word of God, and that Word is supremely to be found in the
Holy Scriptures. But this is not the only Word of God to

1. C. S. Lewis, op cit., p. 459. The whole section, pp. 451–63, is most
satisfying.

man, and to all His other words also we ought to be attentive. This leads on to that characteristically Anglican thing,
a defence of Reason.[1] 'Wherefore the natural measure
whereby to judge our doings, is the sentence of Reason, de-
termining and setting down what is good to be done'
(1. x. 8). Reason is not the affirmation of the arrogant autonomy of man, fashioning a universe according to his own
ideas. It is that faculty in man which makes it possible for
him to receive the revelation of God, to receive revelation
in the form of the Word of God. But, to receive it, he must
be humble, and ready to listen to God, whenever and however He speaks.

So Hooker goes on his tranquil way. He will not reject
anything simply because the Roman Catholic Church has
used it in the days of darkness. All things are to be proved,
and joyfully accepted as good, provided they tend to edification, even though there be no express commandment concerning them in Holy Scripture. In nothing is this eirenical
temper more remarkably shown than in what Hooker has to
say about the Sacrament of the Holy Communion: 'Take
therefore that wherein all agree, and then consider by
itself what cause why the rest in question should not rather
be left as superfluous than urged as necessary . . . the sacrament being of itself but a corruptible and earthly creature
must needs be thought an unlikely instrument to work so
admirable effects in man, we are therefore to rest ourselves
altogether upon *the strength of his glorious power* who is able
and will bring to pass that the bread and cup which he
giveth us shall be truly the thing he promiseth.' 'What these
elements are in themselves it skilleth not, it is enough that
to me which take them they are the body and blood of
Christ, his promise in witness hereof sufficeth, his word he
knoweth which way to accomplish; why should any cogita-
tion possess the mind of a faithful communicant but this,

1. I once horrified a German colleague at the World Council of
Churches by saying, 'You will not find a single Anglican theologian who
does not believe in reason' – horrified, because what an Anglican means
by 'reason' is not the least the same as what a German means by
'*Vernunft*'.

O my God thou art true, O my soul thou art happy!'
(v. lxvii. 7, 12).

'The real presence of Christ's most blessed body and
blood is not therefore to be sought for in this sacrament,
but in the worthy receiver of the sacrament.' These words
have caused perplexity to some readers, who have thought
that Hooker was here putting forward a minimal or 'recep-
tionist' view of the Eucharistic presence, as though it was the
faith of the believer or of the Church that in some way
created that presence.[1] But this is far from Hooker's meaning. He is here simply showing himself a good disciple of
Cranmer. He is maintaining that the Sacrament cannot be
understood, if it is broken up into separate and isolated
parts. It can be understood only if it is considered in its
complex totality as oblation, consecration, and communion.
We cannot theologically think of the Gift, unless we at the
same time think of the Giver, and of those who devoutly
and thankfully receive.

We must come down from the heights of theology to much
more humdrum affairs. The Church lives not in its great
men but in its little men, not in the eloquence of famous
preachers but in the patient faithfulness of obscure pastors
and unknown Christians. In the past the Elizabethan
Church has had many critics and few admirers; it is only in
the last few years that better justice has been done to its
solid achievements. Its task was that of consolidation, and
that is less attractive than adventure. No one knew this
better than Hooker: 'He that goeth about to persuade a
multitude that they are not so well governed as they ought
to be, shall never want attentive and favourable hearers. . . .
Whereas on the other side, if we maintain things that are
established, we have not only to strive with a number of
heavy prejudices deeply rooted in the hearts of men . . . but
also to bear such exceptions as minds averted beforehand
usually take against that which they are loth should be
poured into them.' It is true that the greater part of that

1. In any case the term 'receptionism' is to be avoided. It is not one of
the classical terms of theology, and the earliest citation for it in the
Oxford English Dictionary is from the year 1867.

which was high and heroic was to be found among Puritans or recusants; but perhaps in the end the fustian of Anglican piety has worn better than more immediately attractive forms of clothing.

The first, stark fact about the Elizabethan Church was that it was very poor. Elizabeth was always in need of money; Parliament was more generous to her than might have been expected, but the idea that 'the King should live of his own' persisted long and died hard. Like her father, the Queen cast her eyes upon the revenues of the Church, and few men managed to become bishops without first making over part of the revenues of the see to the Crown. Some modern authors have written of this as the 'spoliation of the Church', but this involves a misunderstanding of the contemporary point of view. Elizabeth still held that all dominion belonged to her, and that there was no other absolute ownership in the country. Her ancestors had made the bishops great nobles, and had enriched them with lands in order that they might maintain a state consonant with their great position. Elizabeth had no need of such bishops. Nobles they should remain, but of the minor nobility; and for that status they would have need of much less wealth than the bishops of earlier centuries. Whitgift reckoned that what remained in his hand as Archbishop of Canterbury, after all pensions and annuities had been paid, was £1,500; this was a large sum in those days, but hardly enough to meet all the obligations that rested even upon an Elizabethan Archbishop.[1]

The example set by the Queen on the highest level was sedulously followed by those of her subjects who had control of the parishes. The reign was marked by a steady process of the diversion of ecclesiastical revenues into lay pockets. When country gentlemen demanded a better educated and

1. Not everyone thought this a bad thing. On 2 November 1559, John Jewel wrote to Josiah Simler: 'For we required our bishops to be pastors, labourers and watchmen. And that this may be the more readily brought to pass, the wealth of the bishops is now diminished and reduced to a reasonable amount, to the end that, being relieved from that royal pomp and courtly bustle, they may with greater ease and diligence employ their leisure in attending to the flock of Christ.'

more virtuous ministry, bishops and others replied with some natural heat that such ministers could hardly be found, if the patrons of the livings provided less for the maintenance of the clergy than for the support of their own domestic servants.

Hardly anyone has a good word to say of the bishops of Elizabeth's time. It may be that justice has not been done them. The considered judgement of Mr A. L. Rowse is this: 'I do not know a single Elizabethan bishop who was a bad man. Some were failures, some muddled. . . . The great majority of them were conscientious hard-working men struggling in difficult circumstances with a heavy burden of administrative toil.'[1] These carefully chosen phrases sum up both the success and the failure of these bishops. It was in large measure due to their efforts that the Church of England survived; but in that Church as they made it one essential ingredient was lacking. It has been the tragedy of the Church of England that it has rejected so much that it might well have kept and kept so much that had better been rejected. It clipped the bishops' wings; it did not recover the true episcopate, in which the bishop is a Father in God, first to his clergy, and through them to all the lay folk of his diocese. These bishops, like their medieval predecessors, were judges, administrators, civil servants on the highest level; only in rare cases were they pastors, fathers, or friends.

Of the clergy at the beginning of the reign not much good can be said. There were many gaps in the ranks, and these had somehow or another to be filled. In one month after his consecration Bishop Grindal ordained a hundred candidates; Matthew Parker is recorded as having ordained a hundred and fifty at one time. This could not continue. Archbishop Parker, writing to the Bishop of London on 15 August 1560, admitted that, in ordaining 'artificers, unlearned, and some even of base occupations', the bishops had made a grave mistake. There is no reason to suppose that these ministers were more ignorant than their predecessors of medieval times; the trouble was that now a great deal more was expected of them – they were to be ministers of the

1. A. L. Rowse, *The England of Elizabeth* (1591), p. 389.

Word and not only of the Sacraments. But what is remarkable is not the poverty of the beginnings, but the immense improvement in the course of the reign.

The Church of England has never lacked for learned and distinguished servants, and such were present in the time of Elizabeth as in all other times. The best known of those who did not become bishops is without doubt Bernard Gilpin (1517–83), the Apostle of the North, a delightful man, who endured the suspicion of being a heretic in the reign of Mary, and of being a crypto-Papist in the reign of Elizabeth. Mr Rowse deserves our gratitude for having called our attention to the exemplary ministry of Dr John Favour of Halifax, also in the backward and largely recusant North.[1] Dr Favour had Puritan sympathies, but he managed to hold his living for thirty years from 1593, and to exercise a rich and varied ministry, 'preaching every Sabbath day, lecturing every day in the week, exercising justice in the commonwealth, practising of physic and chirurgery in the great penury and necessity thereof in the country where I live, and that not only for God's sake, which will easily multiply both clients and patients'. We would gladly know more of men like Parson Latham of Barnwell in Northamptonshire (d. 1620, after fifty-one years as Vicar of Barnwell), who out of the meagre stipend of an Elizabethan incumbent left charities by which his name is still remembered; and John Orton, 'first warden of Parson Latham's hospital', who died in 1607 aged 101 (1506–1607, a goodly span of varied time).[2] For all the vitriolic criticisms of the Puritans, such men were neither so few nor so uninfluential as later generations have sometimes supposed.

As the reign advanced, the proportion of clergy who had been at the universities steadily increased, and in 1600 nearly half of them had a licence to preach. But those who had not had such advantages were not to be left quiescent in their ignorance. In the diocese of Lincoln, Bishop Cooper ordered that 'ministers bend themselves diligently to the study of the Holy Scriptures', and added that they should

1. Op. cit., p. 432.
2. *Victoria County Histories*, Northamptonshire, Vol. II. p. 75.

acquire in English or in Latin the *Decades* of Henry Bullinger of Zürich. Whitgift as Archbishop issued in 1586 similar instructions, with the precision of a pedant and the severity of a schoolmaster. All clergy (if not Masters of Arts or licensed to preach) were required to possess themselves of a Bible, the *Decades*, and a note-book. One chapter of the Bible was to be ready daily, and one sermon from the *Decades* weekly. More learned ministers were appointed as tutors to supervise the studies of their less learned brethren.

There were many methods of enforcing discipline in Elizabethan England, and there is no doubt that these measures had their effect. Nevertheless it remains true that the Church of England, alone among the great Churches of Christendom, has never yet taken seriously the provision of adequate training for the ministry. The Reformation swept through the greater part of Switzerland because the Roman priests could not stand in controversy before the young men trained in the academies of Geneva and Zürich. We have already seen the power and effectiveness of the seminary training instituted by the Council of Trent. England alone had nothing comparable to offer. Some English priests have been so learned as to earn for their order the admiring title *Stupor mundi*[1]; the general level has been such that the same words might be applied in a very different sense. Four centuries have passed since the accession of Queen Elizabeth; in all these years the problem has never been radically dealt with, and the ignorance of the average Anglican theological student today is the astonishment of his opposite number in every country on the continent of Europe.

If it is hard to pass any judgement on the general body of the clergy, still harder is it to estimate the standards and achievements of the general body of the laity. But here too there can be little doubt that immense progress was made in the course of the reign. Among the many dislocations of the

---

1. The phrase seems to be first found in *An Account of the Life and Death of John Hacket*, by Thomas Plume (reprinted 1865), p. 33. There is no full history of training for the ministry in the Church of England, but valuable materials in F. W. B. Bullock, *A History of Training for the Ministry of the Church of England 1800-1874* (1955), Introduction, pp. 1-27.

revolutionary period must be included the dislocation of education; and it may be that the level of literacy was lower at the end of the reign of Queen Mary than it had been thirty years before. But Protestantism in all its forms has always been a champion of popular education; the time of Elizabeth was the time of the founding or re-founding of many schools. With the spread of literacy, the reading of the Bible received new impetus. The second half of the reign of Elizabeth was the period in which, in the famous phrase of J. R. Green, the English people became the people of a book, and that book the Bible.[1] All modern research combines to show that Green's statement was no exaggeration.

As we have seen, in the days of Henry VIII the people had enjoyed their first honeymoon with the English Bible. But Bibles had been rare, and were generally to be found in churches. Now the Bible was in the hands and in the homes of ordinary people. The form in which the Word now came to them was principally the Geneva Bible, a revision carried out by the exiles on the Continent in the reign of Mary. Here for the first time the Bible text was divided into verses. Tyndale remained the basis, but the revisers corrected many errors and added many felicities which remain in our Bible today. Above all, they packed their margins with comments and explanations, in which the reader was introduced to an extreme Protestant and Calvinist understanding of the faith and of the Church. In this the Geneva men were not unique – very few translators in those days trusted the Word of God to do its own work without their help – and, as is the nature of Calvinists, the Geneva men were more thorough and consistent than others. It was the vigour of these notes that led the bishops to prepare their own revision, the Bishops' Bible of 1568 and 1572. The bishops are warmly to be commended for their decision 'to make no bitter notes upon any text', less warmly for the policy that 'all such words as soundeth in the Old Translation to any offence of lightness or obscenity be expressed with more convenient terms and phrases'. On the whole this was an unsatisfactory

1. *A Short History of the English People* (edn. of 1876), p. 447. The whole section should be read.

and backward-looking piece of work that has left very few traces on the forward march of the English Bible.[1] It was the Geneva version that was the Bible of Shakespeare and the Bible of the people. The number and size of the editions published after 1580 is astonishing. And, as always happens when the Bible becomes part of the national heritage of a people, there was widespread improvement in morals, a great rise in the sense of human responsibility. The Court was still marked and marred by a number of the more unpleasant features of the Renaissance; Anglo-Saxon Christianity as a whole was beginning to take on that profoundly ethical character which it has never wholly lost.[2]

Next to the Bible we must reckon the influence of the Prayer Book. In the sixteenth century men did not live as long as they do today. By 1603 there were very few men living, apart from known recusants, who had ever heard or seen the Roman Mass. All that they knew was the Prayer Book of 1559; and what they knew they had rightly come to love. They had to go to church every Sunday on pain of a fine; by constant hearing, the words and the rhythms of the Prayer Book knit themselves about their hearts. Anglicans who go to church regularly find that in time they have come to know large parts of the service by heart, and so are equipped with priceless treasures for use also in their private devotions. Bible and Prayer Book together are the marrow and the sinews of English Church life.

The miracle had happened. Attacked and threatened from every side, contrary to all expectation, that lovely thing the English Church had survived, and Christendom was for ever the richer by its survival. Much remained to be done. The future course was to be far from quiet or peaceful – few Churches in Christendom have had to struggle so hard to maintain themselves, or have been so beset by

---

1. We have not retained the bishops' understanding of Ophir in Psalm XLV. 9: 'Ophir is thought to be the Island in the West coast, of late found by Christopher Columbo: from where at this day is brought most fine gold.'

2. Some details are compactly given in P. Hughes: *The Reformation in England*, Vol. III, pp. 228–31.

ruthless enemies. But the candle has never been removed, and the Anglican Churches today are, in all essentials, what the Elizabethan settlement made them.

At the end of this long survey of the storms of the sixteenth century, it may be well to pause and ask what the English Church had done, and what it had not done, in the course of its Reformation.

It had *maintained* the Catholic faith, as that is set forth in the Scriptures, the Creeds, and the doctrinal decisions of the first four General Councils.

It had *restored* the Catholic doctrine of the supremacy of Holy Scripture in all matters of doctrine and conduct.

It had *restored* Catholic practice in the provision of worship in a language understanded of the people.

It had *restored* Catholic practice in the encouragement of Bible-reading by the laity.

In the Holy Communion, it had *restored* Catholic order by giving the Communion to the laity in both kinds, both the Bread and the Wine, instead of only in one kind, as was the practice of the medieval Church.

In Confirmation and Ordination, it had *restored* Catholic order by making the laying on of hands by the Bishop the essential in the rite.

It *aimed at restoring* the Catholic practice of regular Communion by all the faithful.

It had *retained* the three-fold Order of the ministry: bishops, priests, and deacons.

It had most carefully *retained* the succession of the bishops from the days of the Apostles.

It had *retained* the liturgical order of the Christian year, though in a considerably modified and simplified form.

It had *repudiated* the supremacy of the Pope, as that had developed since the days of Gregory VII.

It *denied* that the Pope had authority to interfere in the civil affairs of States and to depose princes.

It *claimed* liberty for national Churches, within the fellowship of Christ's Holy Catholic Church, 'to decree Rites or Ceremonies' (Article xx).

It *rejected* the scholastic philosophy, and the late medieval definitions, especially of transubstantiation, which had been based on it.

It *rejected* late medieval ideas of purgatory, indulgences, and the merits of the saints.

It *retained*, unfortunately, the medieval ideas of property, of jurisdiction, and of ecclesiastical administration.

It *maintained* continuity of administration, most of the episcopal registers showing that the work of the Church was carried on through all the troubles without the intermission of a single day.

It *claimed* to be a living part of the world-wide Church of Christ.

Was this, then, a new Church or the old Church reformed and restored? If it is necessary, as Boniface VIII affirmed, for the salvation of every human creature that he should be subject to the Bishop of Rome, then the position of the Church of England is indefensible. But, if we look to Scripture, to the life of the early Church, to the great central tradition of the Church through all the centuries, we may come to think that the Church of England in the reign of Elizabeth was right in claiming, as it claims today, that it and no other body is the Catholic Church in England. Certainly a large section of Elizabethans would have been behind their Queen, when she wrote to the Emperor Ferdinand in 1563: 'We and our subjects, God be praised, are not following any new or foreign religions, but that very religion which Christ commands, which the primitive and Catholic Church sanctions, which the mind and voice of the most ancient Fathers with one consent approve.'

CHAPTER 6

The Seventeenth Century

ELIZABETH died, and James VI of Scotland became James
I of England. Everyone knew that this marked a revolution,
but no one could tell in advance what kind of a revolution it
would be.

James was not a very nice man or a very good man; but
he was a genuinely learned man, and he had a considerable
interest in religion. The coming of a king from the land of
Presbyterianism fanned for a moment the hopes of the Puritans; but James manifested all of a sudden and at once an
intense appreciation of the English Church, and this attitude
on the part of the new king left a deep impress on the future
history of that Church.

Almost the first thing that happened in the new reign was
that James was confronted by the Millenary Petition, a
document signed by many ministers of Puritan inclina-
tions, in which certain moderate requests for changes in the
order of the Church were moderately set forth. In January
1604 a Conference was held at Hampton Court, at which
James made all too plain his feeling about Puritans and
Presbyterians. 'I will tell you', he said, 'I have lived among
this sort of men ever since I was ten years old, but I may say
of myself, as Christ did of Himself, though I lived amongst
them, yet since I had ability to judge, I was never of them.'
If the Puritans came to power, 'I know what would become
of my supremacy. No bishop, no king. When I mean to live
under a presbytery, I will go into Scotland again, but while
I am in England, I will have bishops to govern the Church.'
'If this is all they have to say, I will make them conform
themselves or I will harry them out of this land or else do
worse.' All this was a grave mistake. At that moment the
Puritans were a group which might have been reconciled by
gentle handling; they were turned into an opposition whose

smouldering resentment was to have grave consequences, not in the days of James, but in those of his successor. If the Church of England has suffered much at the hands of its enemies, it has often itself taken the initiative in turning into enemies those who might have been its friends.

Little came of the Conference. Some minute changes were made in the Prayer Book. The one important consequence was the decision to undertake a new translation of the Bible, a project in which the King himself was considerably interested. The result of this decision was the publication in 1611 of the world's best-seller, the Authorized Version of the Bible.[1]

Six companies, of about fifty members in all, were formed, among them the most learned men in the kingdom, not excluding even a number of the more moderate Puritans. The Bishops' Bible was taken as the English basis, though the translators consulted all the existing English translations and the modern translations in other European languages. All the best resources of Greek and Hebrew scholarship, as these existed at that date, were brought to bear on the work.[2] The revision, as accomplished by the companies, was sent in to a smaller group in London, in which Bishop Lancelot Andrewes of Chichester (from 1609, of Ely) seems to have played the leading part. 'They went daily to Stationers Hall, and in three quarters of a year [apparently 1607] finished their task. All which time they had from the Company of Stationers XXX^s per week, duly paid them; though they had nothing before but the self-rewarding, ingenious industry.'

The great translation is not without its defects. The translators were almost too conscientious in their adherence to

1. An excellent account of the Authorized Version and the work that went into it is to be found in *Ancient and English Versions of the Bible*, ed. H. Wheeler Robinson (1940), pp. 196–227. Although the term 'Authorized Version' has become so widely current, it does not appear that the book was ever 'authorized', though it is 'appointed to be read in churches'.

2. 'As a translation of the Massoretic text of the O.T. as available to the translators in the rabbinical Bibles and the polyglots, A.V. is on the whole as accurate a rendering as the combined Christian scholarship of Europe would at that time have been able to produce.' D. Daiches: *The King James Version of the English Bible* (1941), p. 208; interesting and valuable as written from the Jewish point of view.

the traditional Hebrew text, and in certain passages wrote
what they must have known to be nonsense. They were over-
particular in rendering Hebrew idioms at the expense of
ordinary English style. They were not invariably happy in
their sense of rhythm. The style was unnecessarily archaic
as compared even with the style current at the time when
they were working; this gives a solemn and hieratic tone to
the translation, but at the expense of ready understanding.
They rightly eschewed the principle of uniformity in rendering the Greek and Hebrew words, but allowed themselves a
wholly unnecessary latitude in the matter. Among other
criticisms, Mgr R. A. Knox has remarked that 'Certainly it
has impoverished our language; or why is Shakespeare such
a mass of obsolete words?'[1]

But, when every criticism has been made, and every allowance for the imperfection of scholarship in the early days
of the seventeenth century, it remains the fact that it is this
version that has worked its way into every corner of the life
of the English-speaking world, and into the hearts of uncounted Christians of the British race. It has enlightened
them in life and comforted them in death. No other version
has taken its place or even challenged its supremacy. Surely
F. W. Faber, who himself left the Church of England for that
of Rome, is right in saying that one of the hardships endured
by the 'convert' is the exchange of the glories of the Authorized Version for the lack of glory of the version of Douai.[2]

The not very interesting reign of James I had its glories, and
among them the two most notable churchmen of the period.

Lancelot Andrewes (1555–1626), bishop successively of
Chichester, Ely, and Winchester, was reckoned among the
most learned men and the most famous preachers of his age.
Faithful to the line which Hooker had marked out, Andrewes
carried it a stage further. 'Hooker had vindicated on its

1. *On Englishing the Bible* (1949), p. 47.
2. Faber wrote: 'It lives on the ear like a music that can never be
forgotten, which the convert hardly knows how he can forego. Its
felicities often seem to be almost things rather than mere words. . . . In the
length and breadth of the land there is not a Protestant, with one spark
of religiousness about him, whose spiritual biography is not the Saxon
Bible.'

behalf the rights of Christian and religious *reason*, that reason which is a reflection of the mind of God. Andrewes vindicated on its behalf the rights of Christian *history*. Hooker had maintained the claims of reason, against a slavish bondage to narrow and arbitrary interpretations of the letter of Scripture. Andrewes claimed for the English Church its full interest and membership in the Church universal from which Puritan and Romanist alike would cut off the island Church by a gulf as deep as the sea.'[1] Yet it is to be noted that, though Andrewes was a strong defender of episcopacy as an integral part of the Church's life, he absolutely refused to unchurch those Christian bodies which in the turbulent days of the Reformation had lost the episcopate: 'Even if our order be admitted to be of divine authority, it does not follow that without it there can be no salvation, or that without it a Church cannot stand. Only a blind man could fail to see Churches standing without it. Only a man of iron could deny that salvation is to be found within them.' This is a latitude of judgement that Anglicans in later times have not always maintained.

Great as were the services that Andrewes rendered in his life, the greatest was rendered after his death. In 1648 Richard Drake published a little book of devotions, drawn from a manuscript which Andrewes had prepared for his own use. 'Had you seen the original manuscript,' writes Drake, 'happy in the glorious deformity thereof, being slubbered with his pious hands, and watered with his penitential tears, you would have been forced to confess that book belonged to no other than pure and primitive devotion.' What Cranmer was to the public liturgy of the Church, Andrewes was to the world of private devotion. The *Preces Privatae* have perhaps been more widely used in the Anglican world than any book of prayers outside the Prayer Book, and deservedly. 'The piercing and rapid energy of Andrewes' devotions, their ordinary severe conciseness, their nobleness and manliness, their felicitous adaptations, their free and varied range, the way in which they call up before the mind the

1. R. W. Church: *Pascal and Other Sermons* (1895), p. 90. The whole of the admirable sermon on Andrewes should be read.

whole of the living realities of God's creations and God's revelations, and, in the portions devoted to praise, their rhythmical flow and music . . . all this is in the strongest contrast to anything that I know of in the private devotions of the time.'[1]

The one serious blot on the fair fame of Andrewes is his acquiescence in the burning of the Arian heretic Legate on 18 March 1612, and of one Wightman on 11 April of the same year. The gravest blame for these executions must rest, however, on the King, who at that time was anxious to make a demonstration before the world of his orthodoxy. In discussion with Legate, the King asked him whether he did not pray. When Legate replied that he had not prayed for seven years, the King cried out in anger, 'Away, base fellow. It shall never be said that one stayed in my presence that hath never prayed to our Saviour for seven whole years together.' This was the last occasion in England on which the flames were kindled for heretics. It is deplorable that the Church of England was so slow to recognize the wickedness of such persecution; but at last the lesson had been learned, and far earlier than in many other European countries. It has to be noted that James regarded himself as an authority on demonology, on which he had written a book; and in his reign fifty poor crazed creatures were put to death as witches; against this no one in either Church or State seems to have entered any protest.

The other great ornament of the reign of James I was John Donne, Dean of St Paul's (1573–1631). Donne had been a Roman Catholic, and his earlier life had been marked by profligacy. As a result of earnest study and sincere conviction, he accepted the Anglican position, and was ordained in 1615. Almost immediately upon his ordination he became known as one of the best preachers in England. When he reached the zenith of his powers, the English pulpit attained a height of dignity, a splendour of passionate eloquence, such as has rarely been equalled, and certainly never surpassed.

It is often imagined that Donne was the type and original

1. R. W. Church, op. cit. pp. 89–90.

of the gloomy Dean; even Dr S. C. Carpenter writes of his sermons as 'packed with melancholy, recondite thought, constantly obsessed by the ideas of sin and death'.[1] But this is what comes of reading Donne in anthologies and not at large. It is true that, when he thinks of the eternal issues that lie before the soul of man, his language rises to memorable heights, and it is such passages that catch the eye of the anthologist: 'When it comes to this height . . . that mine enemy is not an imaginary enemy, fortune, nor a transitory enemy, malice, but a real, and an irresistible, and an inexorable, and an everlasting enemy, the Lord of Hosts Himself . . . we are weighed down, we are swallowed up, irreparably, irrevocably, irrecoverably, irremediably.' Yet the keynote of the very sermon in which these solemn words occur is to be found in the following passage: 'I would always raise your hearts, and dilate your hearts, to a holy joy, to a joy in the Holy Ghost. . . . God hath accompanied and complicated almost all our bodily diseases of these times, with an extraordinary sadness, a predominant melancholy and faintness of heart, a cheerlessness, a joylessness of spirit, and therefore I return often to this endeavour of raising your hearts, dilating your hearts with a holy joy. Joy in the Holy Ghost, for under the shadow of His wings, you may, you should rejoice.'[2] Here speaks the authentic Donne. His preaching is a true preaching of the Gospel.

James I loved the English Church, yet he did it greater harm than perhaps any other English monarch. 'King James regarded positions in Church and State not as offices of trust but as sinecures or perquisites to be given away or sold to the highest bidder. The bureaucracy thus became a vast system, or market, of patronage; and this patronage, moreover, was not operated by the King himself; it was operated in his name by royal favourites.'[3] The King was

1. *The Church in England, 597–1688* (1954), p. 359.

2. Sermon Preached at St. Paul's, 29 January 1625/6, on Psalm LXIII. 7; eds. Simpson and Potter, Vol. VII, pp. 56–7, 68–9.

3. H. Trevor Roper: 'King James and his Bishops', in *History Today*, September 1955, pp. 571–81. This is an exceedingly important article which I would most gladly see expanded into a book.

the fountain of honour, but, wrote Dr Hacket, 'There was one pre-eminent pipe through which all graces flowing from him were derived – George Villiers, Duke of Buckingham.' Any Anglican Churchman who desired advancement must needs become a courtier; even the austere William Laud confided to his diary 'I dreamed that the Duke of Buckingham came into bed with me and showed me great affection.' Only so could Laud begin to climb. 'From this time he prospered at the rate of his own wishes and ... was left, as was said before, by that omnipotent favourite in that great trust with the King, who was sufficiently indisposed towards the person or the principles of Mr Calvin's disciples' (Clarendon).

So it came about that, when Whitgift's lieutenant and successor the efficient Bancroft died, the primacy went not, as had been expected, to Andrewes, but to the insignificant George Abbot, of whom the best that can be said is that he was no worse than 'indifferent, negligent, and secular'. Another of the type was the ambitious Welshman, John Williams, Bishop of Lincoln, perhaps the nearest thing that the Church of England has produced to a Renaissance Cardinal. Williams kept great state at Buckden. He was a patron of poets, scholars, and churchmen, among them the most noted figures of the ecumenical interest in the seventeenth century, Samuel Hartlib, John Dury, and the Czech John Amos Comenius.[1] The trouble was not so much that Williams and his like were positively vicious, as that they had no tincture of the Gospel; they do not seem to have had any idea of what the Gospel is. Good men groaned under these things. The Church was laying up for itself a goodly store of troubles. When the Puritans, like Milton, inveighed against 'swan-eating, canary-sucking bishops', they were moved by something other than atrabilious malice.[2]

1. On these men and their activities, see *A History of the Ecumenical Movement 1517–1948* (1954), pp. 88 ff., 97 ff., 134 ff.

2. James is reported to have defended himself by saying that 'No good men would take the office on them'. This was, of course, nonsense; and it is not quite true that 'it was not until Charles II's reign, when bishops had been divested of most of their political functions, that saints could be appointed to episcopal sees'. C. Hill: *Economic Problems of the Church* (1956), pp. 28–9.

The years between 1628 (The Bill of Rights) and 1688 ('The Glorious Revolution') were almost as full of movement, of tragedy and heroism as those between 1528 and 1588; and almost as full of change and destiny for the English Church. Before we come on to consider them, it may be well to pause for a moment, and to set out briefly the spectrum of English religious life at that time. Popular history thinks in terms of Cavaliers and Roundheads; the reality was a good deal more complicated than this.

We may begin on the right with the recusants, those who still remained faithful to the Church of Rome. Under Charles I their lot was happier than it had been for more than half a century. The penal laws were still in force against them, but they were not strictly put into execution. Roman Catholics, in Clarendon's phrase, 'were grown only a part of the revenue without any probable danger of being made a sacrifice to the law'. Queen Henrietta Maria was a Roman Catholic, and had privileges of worship; it was fashionable to attend Mass in her chapel. During most of the reign the Pope was represented in London, first by the intelligent Scot George Con, with whom Charles I was specially friendly; later by the Florentine noble Rossetti.[1] Some Anglicans left their own Church to join the Church of Rome, notable among them being the minor poet Richard Crashaw (1612–49). All this was pleasant for the recusants, but it was highly dangerous for the King, since 'No Popery' is one of the few unchanging constituents of what the average Englishman calls his thoughts.

Next we come to those who would in a later age have been called high churchmen, but were then more generally known as Arminians. Here we cannot but touch for a moment on the endless debate between predestination and free-will, though it is hard for us today to understand how strongly men felt in times past about such inscrutable mysteries of the Faith. Calvin's thought was logical, and moved steadily forward from its original premiss, the absolute

1. These men were not diplomatic agents accredited to the Court; they were the personal representatives of the Pope to the Queen of England.

sovereignty of God. No man can be saved, unless God has willed that he should be saved. But, if we affirm that Christ died for all men, and yet all men are evidently not being saved, we land ourselves in a logical contradiction. We must, then, hold that Christ died only for those whom God had already in His secret council predestined to salvation. So our salvation depends on God's will that our name should be found in the Book of Life and on nothing else. The Arminians refused to accept the further horrible corollary that God had brought the vast majority of mankind into existence only to condemn them to eternal torment for failing to accept a salvation which in any case it was not within their power to accept. Clearly a place must be left for the freedom of the human will, and for man's responsibility to choose the good, as it is given him to see it. But, replies the Calvinist, that is to attribute to the human will a freedom which we know quite well that it does not possess. And so the debate went on.

Calvinism and Arminianism are both distortions of the truth. We can do justice to the Gospel only if we hold firmly both to the absolute sovereignty of God and to the moral freedom of man, recognizing that it is only in the depths of the divine wisdom that what appears to us as a logical contradiction is resolved. What is important for our story is that William Laud was one of the first in England to maintain 'Arminian' principles. He and his followers stood in the succession of Hooker and Andrewes, though without the breadth of their understanding or their eirenic temper.

Next came the Anglican Calvinists, who were quite happy with the episcopal organization of the Church as it had emerged from the Elizabethan settlement. Typical of these men was Abbot, James's Archbishop of Canterbury. They stood in the succession of Grindal and Whitgift, though without the gentleness of the one or the efficiency of the other.

To the left of the episcopalian Calvinists came the Calvinistic Calvinists, the men who still looked for their model to the best-reformed Churches of the Continent, and believed that the Church of England would never prosper, until the opportunity of which they had been frustrated by the wiles of Elizabeth was seized by men of sterner temper and more resolute will.

Still further to the left were the somewhat indeterminate groups called sectaries, separatists, or independents. What made their fortune was that eventually they could include Oliver Cromwell among their supporters; but throughout the first half of the century they had been quietly increasing their numbers and their influence. This period had been astonishingly productive in devotional books: 'reading these manuals made men feel independent of bishops and priests. . . . While the middle classes were united in opposition to episcopacy, they became divided about all else. Whereas the upper-middle-class man tended to adopt an Erastian form of presbyterianism, the lower-middle-class man often became a separatist.'[1] This sociological note is deserving of careful attention. Perhaps because of their social background, independents were likely to be more favourable to democracy than most others in that highly undemocratic age; it was one of them, Rainborow, who let fall the ever-memorable remark, 'I think that the poorest he that is in England hath a life to live as the greatest he' – an argument in favour of universal suffrage that was naturally to fall on deaf ears at that time. Up to a certain point independents believed in religious liberty, and could have made their own the noble words of John Milton: 'Give me the liberty to know, to utter, and to argue freely according to conscience, above all liberties.'

Finally, we come to the lunatic fringe – the Fifth Monarchy men, the Seekers, the Levellers, and so forth. These combined vivid apocalyptic expectation with a variety of revolutionary views on social and political affairs that made it hard for any government to tolerate them. Common opinion classed the Quakers with these others. Looking back in the light of history and of the wonderful life of George Fox (1624–91), we can see that this was cruel and unfair. But in so disturbed a time men's judgements are hardly rational, and George Fox himself, with his constant

1. G. Davies: *The Early Stuarts 1603–1660* (1937), p. 193.

inveighing against 'steeple-houses' and the lifelessness of all established forms of religion, was not exactly a conciliatory figure.

The tragedy of these times was that Laud and the Puritans could not understand one another. For, whatever we may think in detail about these men and their doings, it is impossible to understand them at all, unless we realize that what they cared about above all else was holiness. This was the serious part of the nation. Laudians and Puritans were united in reaction against the frivolity and shallowness of the Jacobean bishops. Laud was in love with the beauty of holiness, with the priestly holiness of the sanctuary. It mattered to him very much that all things should be done decently and in order. The wearing of the surplice was the symbol of regard for seemliness and of obedience. The Puritan was concerned for the prophetic ideal of holiness – an exacting ethical code, and an adoring submission before the majesty of God. To him the wearing of the surplice was a bowing down in the house of Rimmon, an unworthy compromise with 'the mother of harlots'. We can see why Laud and the Puritans could not understand one another; we must lament the blindness that set them at one another's throats.

It has to be admitted that the Puritans could be extremely irritating. They had inherited from their Elizabethan predecessors a remarkable flair for scurrilous propaganda, and showed less than discretion in the use they made of it. The bishops could not be expected to be pleased when Dr Henry Burton, rector of St Matthew's, Friday Street in the City of London, referred to them all in a sermon as 'upstart mush-rumps'[1]; when one worthy lady pegged up her washing in the chancel of her church, saying that, if the parson brought his old linen into the church, she would do so too; when

---

1. Compare Milton's 'This impertinent yoke of prelatry, under whose inquisitorious duncery no free and splendid wit can flourish'. Bastwick's contribution (in his *Litany* of 1637) was to call the bishops 'littel toes of Anti-christ ... those Fishmongers that have bought and sold Christ's best fishes and made them the mundungus and garbage both of sea and land'.

another marched into Lichfield cathedral, accompanied by the town clerk and his wife, and ruined the altar-hangings with a bucketful of pitch.

In 1633 Laud became Archbishop of Canterbury. From that time on, his were the mind and the will that moved the Church of England. Laud was a learned, virtuous, narrow-minded, choleric, unlovable man. The most attractive thing about him was his unfailing devotion to the University of Oxford, and the encouragement that he gave to learning, particularly to oriental learning. Because of his high church tendencies, he was suspected of leanings towards popery; it was even rumoured that he had been offered a cardinal's hat as the reward for submission to Rome. But Laud wrote the truth when he wrote, 'Something dwelt within that would not suffer that, till Rome was otherwise than it was at the present time.'[1] He was honestly and sincerely devoted to the English Church, and desired it to be a glorious Church according to his own ideals. Yet it remains also true, in the brilliant phrase of Miss C. V. Wedgwood, that 'Laud was more of a tidier-up and setter-in-order than a true reformer'.[2]

Laud would start with his own brother bishops. No more easygoing ways were to be tolerated. They were to be harried into holding visitations of their dioceses and suitably harrying their clergy. Laudian agents were set on to watch and report on them. Joseph Hall of Exeter, a satirist and poet of some distinction, complained that men were 'set over me for my espials; my ways were curiously observed and scanned'. The bishops did not like it very much.

Then Laud would proceed to deal with notable offenders, without regard to quality or reputation. 'He intended the discipline of the Church . . . should be applied,' says Clarendon, 'to the greatest and most splendid transgressors, as well as to the punishment of smaller offences and meaner offenders. . . . Persons of honour and great quality, of the court, and of the country, were every day cited into the high-commission court, upon the fame of their incontinence, or other scandal in their lives, and were there prosecuted to

1. *Works*, III, p. 219.
2. *The King's Peace* (1955), p. 93.

their shame and punishment; and . . . the shame (which they called an insolent triumph upon their degree and quality, and levelling them with the common people) was never forgotten, but watched for revenge.' It was good that great offenders should not feel that they were above the law, and Laud's proceedings may have done some good. But in truth this was no way to attempt to reform the morals of a country; this kind of confusion between sin and crime has always done more harm than good.

Next we must take into account the violent persecution of the Puritans. One or two examples will suffice. In 1630 Alexander Leighton, the father of Robert the future Archbishop of Glasgow, wrote a book called *Zion's Plea against Prelacy*, a violent controversial tract. He was sentenced to be fined £10,000, to be degraded and imprisoned for life; and 'for further punishment and example to others, to be brought into the pillory at Westminster (the Court sitting) and there whipped, and after his whipping be set upon the pillory for some convenient space, and have one of his ears cut off, and his nose slit, and be branded on the face with a double S.S. for a sower of sedition; and shall then be carried to the prison of the Fleet, and, at some other convenient time afterwards, shall be carried into the pillory at Cheapside upon a market day, and there be likewise whipped, and then be set up on a pillory and have his other ear cut off'.[1] Leighton was sixty years old and a Doctor of Divinity; even in that callous age it was felt unseemly that such penalties should be inflicted on such men.

On 14 June 1637, the three chief libellers, the acid lawyer Prynne, the impudent priest Henry Burton, and the swashbuckling doctor Bastwick, were brought to trial. This was a Star Chamber matter, and Laud did not speak until after sentence had been pronounced, when he delivered a long oration in defence of the bishops. All three men were sentenced to a fine of £5,000, to stand in the pillory and lose their ears, and to perpetual imprisonment. In addition Prynne was to be branded on the cheek with the letters S.L.

1. It is not certain whether the whole of this savage sentence was carried out.

for 'seditous libeller', but interpreted by Prynne as 'Stigmata Laudis'. Laud could never learn that it does not do to provide the opposition with martyrs. The martyrdom of the three men was transformed into a triumph. Bastwick, characteristically, added a pretty touch by producing a surgeon's knife and instructing the executioner in the art of cropping ears.

Laud seems to have been singularly unaware of the opposition that he was arousing; but it is the sober judgement of a modern historian that, in ten years, he had succeeded in making the bishops the most bitterly hated body of men in the country. And he had succeeded in uniting against himself and his policy a large number of people who were divided in everything except their hatred of the Church of England, as Laud would have made it. The heart of the trouble was that Laud, in his thinking, was almost wholly medieval; his plans for the dignity, the independence, and the authority of the Church looked back to a day that had passed for ever, and broke on the solid opposition of the lay folk, including many who were heartily loyal to the Church as Elizabeth had left it.

Many were offended by the rites and ceremonies which Laud was anxious to introduce. These things were harmless in themselves; many of them have become common practice in the Church of England, not always to its advantage. But, in days when tempers were so greatly inflamed, insistence on the details of the Church's order could well lead to the destruction of the whole.

Laud's policy of introducing churchmen into offices of great weight and dignity in the State seemed to the laymen to threaten both their prerogatives and the stability of the country: 'The unseasonable accumulation of so many honours upon them, to which their functions did not entitle them (no bishop having been so much as a Privy Councillor in very many years), exposed them to the universal envy of the whole nobility, many whereof wished them well as to all their ecclesiastical qualifications, but could not endure to see them possessed of those offices and employments, which they looked upon as naturally belonging to them . . .

and some of them, by want of temper or want of breeding, did not behave themselves with that decency in their debates towards the greatest men of the kingdom as in discretion they ought to have done.'[1]

Laud's desire was for an independent clergy, dutifully submissive to their bishops and to centralized control. But the clergy could not be spiritually independent if they were dependent on laymen for their daily bread. The great trouble of the Church was that it was poor. Godly laymen were prepared to augment stipends, to create lectureships and so on, provided they could have the kind of ministers they wanted. But this did not at all suit Laud's book. 'To be politically useful to the government, Laud and Charles would have agreed, the establishment must enjoy a position of dignity and independence. . . . Dignity, in Laud's view, necessitated a well-paid clergy, though paid on a graded scale; independence must also be given an economic basis by setting the clergy free from the financial pressure of laymen, whether they were patrons, impropriators, town corporations nominating and paying lecturers without reference to a bishop, members of congregations augmenting stipends, or the Feoffees for Impropriation.'[2] One way to secure the independence of the establishment would be to recover some of the Church lands and revenues that had been lost to laymen. Wentworth had had some success with this policy in Ireland. The sacredness of property is the first article in the Englishman's creed: 'Your lands in danger' added fuel to the already blazing flame of the layman's dislike of Laud and all his ways.

Among Laud's most dangerous opponents must be numbered the common lawyers. We have already seen with what impatience Elizabethan England had endured the Court of High Commission, and the enforcement of ecclesiastical discipline by civil penalties. Here was Laud using the same methods, vigorously and, from the point of view of his enemies, cruelly and unscrupulously. But history had

1. Clarendon: *History*, 1, 117. Laud himself was described by Mrs Hutchinson as 'a fellow of mean extraction and arrogant pride'.
2. C. Hill: *Economic Problems of the Church* (1956), p. 307.

marched on since the days of Elizabeth. Sir Edward Coke
(1554–1632), the greatest common lawyer of all time, had
educated men's minds as to the supremacy of the common
law. Naturally one of the chief objects of his animus were the
prerogative courts, and in the reign of James I, as Chief
Justice of the Court of Common Pleas, he had aroused the
opposition of the ecclesiastical administrators by issuing
writs of prohibition against the Court of High Commission. In the reign of Charles I these things had not been
forgotten.

In view of what followed, it has been necessary to enlarge
a little on the causes and forms of strife and controversy. But
the reader must not form the impression that Christians of
three centuries ago did nothing but bite and devour one
another. In point of fact, a great many of the best men and
women withdrew themselves from turmoil to live in godly
quietude. In considering an age so rich in outstanding
personalities, the only difficulty is that of selection.

The most famous examples of this godly life in retirement
belong to what we have called anachronistically the high
church wing of the Church.

Nicholas Ferrar (1592–1637) was the founder of what
came to be known as the 'Arminian Monastery' at Little
Gidding in Huntingdonshire. Here, with members of his
family, he lived a life of extreme simplicity, devotion, and
practical service. 'Their apparel had nothing in it of
fashion, but that which was common; yet plain; and much of
it for linen and woollen spun at home. . . . They gave no
entertainment but to the poor, whom they instructed first,
and then relieved, not with fragments, but with the best
they had . . . the devil had the less power to tempt them,
that he never found them idle. . . . Four times every day they
offered up their supplications to God, twice in the words of
the Common Prayer in the Church; twice in their family,
with several petitions for their own needs, and for such as
desired, upon some special occasions, to be remembered by
them to God. . . . By night they kept watch in the house of
the Lord, and two by turns did supply the Office for the
rest, from whence they departed not till the morning.

Their scope was to be ready like wise virgins with oil in their lamps, when the Bridegroom came.'[1]

Among the closest friends of the Ferrars was George Herbert (1593–1633), the brilliant Fellow of Trinity College, Cambridge, and Public Orator of the University. Presented in 1630 to the cure of the village of Bemerton near Salisbury, Herbert settled down to be that ideal parish priest whom he depicted in *A Priest to the Temple*. In humility and simplicity he showed what the life of an Anglican country parson can be. Daily he read the Morning and Evening Prayers of the Church 'at the canonical hours of ten and four', and had to pray with him 'most of his parishioners and many gentlemen in the neighbourhood, while some of the meaner sort would let their plough rest, whenever Mr Herbert's saint's-bell rang to prayers'. Herbert is *par excellence* the poet of Anglicanism. Some of his poems are marred by the conceits that were fashionable in his day; most of them express, in simple and moving terms, devotion to God, love for the Saviour, and humble conformity to the rule of that Church of which he was a minister. One who wishes to know what Anglicanism is and has not much time for study cannot do better than to pay attention to the life, the poems, and the prose of George Herbert.

Ferrar and Herbert were clergymen. As a type of the godly layman of the day we may take Lucius Cary, Viscount Falkland (1610–43), the influence of whose goodness and purity long outlived him. His house at Tew near Oxford became a place of resort for 'the most polite and accurate men of that university', who came there, not so much for repose as study; and 'to examine and refine those grosser propositions which laziness and consent made current in vulgar conversation'. Falkland's mother was a recusant, and tried to win him over to her way; but 'having diligently studied the controversies and exactly read all, or the

<hr>

1. John Hacket: *Scrinia Reserata: The Life of Archbishop Williams* (1639). The account of Little Gidding in J. H. Shorthouse's *John Inglesant*, though marred by some inaccuracies, does convey wonderfully well the atmosphere of the place.

choicest of the Greek and Latin fathers, and having a memory so stupendous, that he remembered, on all occasions, whatsoever he read', he remained steadfast in the Church of his origin. But 'he was so great an enemy to that passion and uncharitableness, which he saw produced, by differences of opinion, in matters of religion, that in all those disputations with priests, and others of the Roman Church, he affected to manifest all possible civility to their persons, and estimation of their parts'. After his death at the battle of Newbury in 1643, his friend Hyde (later Clarendon) wrote sadly of his prodigious parts of learning and knowledge, his inimitable sweetness and delight in conversation, his flowing and obliging humanity and goodness to mankind, and his primitive simplicity and integrity of life.

Among simpler people we meet with Alice Thornton, who was of the opinion that the Church of England was that 'excellent, pure and glorious Church then established, which for soundness in faith and doctrine none could parallel since the Apostles' time'. Those who think that strict and devout piety was found only among the Puritans would do well to note the rule in Alice Thornton's household, that the whole family was called to prayers by a little bell at six in the morning, at two in the afternoon, and again at nine at night.

The opposition to Laud had become so strong that it was clear that something must break somewhere. The breaking-point proved to be not in England but in Scotland.

As early as 1600 James VI had appointed three titular bishops in Scotland. In 1610 a further step was taken by the consecration of three 'superintendent moderators' as bishops for Scotland. Archbishop Bancroft decided that, though these three men had previously had only presbyterian ordination, they need not first be ordained presbyters according to the Anglican rite. This consecration *per saltum* served as one of the precedents for what was done at the inauguration of the Church of South India in 1947. It was made clear that the Church of Scotland, though now episcopal, was in no way subordinate to the Church of England.

Charles I, with his superb devotion to the Church of England, was naturally anxious to introduce into his northern kingdom that order which he found so satisfactory in the southern. He moved to his aim with all that ineptitude which marred almost all the projects of the Stuart kings. He took to using Scottish bishops in high office of state. A proposal was on foot to restore titular abbots, with seats in the Scottish Parliament. In 1628 an Act of Revocation had been put forward, with the aim of resuming all Church lands which had fallen into the hands of laymen since 1542. This was not put through in its entirety, but it was quite enough to alarm a very large number of people. 'Enough was already done to alarm all that were possessed of the Church lands, and they, to engage the whole country in their quarrel, took care to infuse it into all people, but chiefly into the preachers, that all was done to make way for popery.'[1] The spark that touched off the conflagration was the Liturgy of 1637; but so small a spark would not have produced so great an effect, unless the conflagration had been carefully prepared in advance.

Scotland had been for many years the scene of considerable liturgical confusion. The Scottish Reformers had not altogether abandoned the liturgical principle, and had put out their *Book of Common Order*; but this was widely disregarded by the ministers. On the other hand, the English Prayer Book was used in some places, and was fairly widely known. From 1634 onwards the question of a regular Prayer Book for Scotland was being actively debated, and, though a good deal still remains obscure, we can trace fairly accurately what happened.[2] Charles I and Laud would have liked to introduce the English Book without change; but this was not to the mind of the Scottish bishops, who, for a number of reasons, national, liturgical, and anti-Presbyterian, wished for considerable changes. As Laud himself wrote:

1. G. Burnet: *History of my Own Times*, 1, p. 34.
2. Perhaps we should put the date rather earlier; Laud writes: 'When I was first Bishop of London [1628], his Majesty expressed a great desire which he had to settle a liturgy in the Church of Scotland, and this continued in agitation many years.' *Works*, III, p. 278.

'I laboured to have the English liturgy sent them, without any omission or addition at all, this or any other; so that the public Divine service might, in all his Majesty's dominions, have been one and the same. But some of the Scottish Bishops prevailed herein against me; and some alterations they would have from the Book of England.'[1] The remark of James Wedderburn, Bishop of Dunblane, that the second part of the words of administration in the Communion service 'may seem to relish somewhat of the Zwinglian tenet, That the Sacrament is a bare sign taken in remembrance of Christ's passion', may be taken as an indication of the direction in which they wished to go.

The Scottish liturgy of 1637, at its first introduction, produced so great a tumult that it was hardly ever used. But, as the venerable ancestress of a distinguished progeny, it is of the highest importance in Anglican liturgical history, and demands fuller treatment than the bare mention of its immediate failure. Other changes need not concern us here. Those made in the Communion service are of lasting interest.[2] The service follows the English order up to the *Sursum Corda* ('Lift up your hearts') except for the addition to the Prayer for the Church Militant of a commemoration of the departed, and a thanksgiving for the saints. The Prayer of Consecration is immediately followed by an extended Prayer of Oblation, and this, in turn, by the Lord's Prayer and the Prayer of Humble Access. Only the first half of the words of administration is retained. The 'Collect of Thanksgiving' (so called here) is always to be said, not as an alternative to the Prayer cf Oblation. In brief, the 1637 service goes back a good distance in the direction of 1549, without entirely disrupting the order of 1552.

This order is the basis of many subsequent efforts at liturgical revision in the Anglican Churches. It underlies, naturally, the various forms of the Scottish liturgy, including that which is authorized in the Episcopal Church in

---

1. *Works*, III, p. 356.
2. All this is set out in great detail and on the basis of very careful research in Gordon Donaldson's *The Making of the Scottish Prayer Book of 1637* (1954).

Scotland today. From Scotland it travelled to America with
Bishop Seabury, and survives, with certain changes, in the
American Prayer Book. Its influence is clearly identifiable
in the South African liturgy and in the abortive English revision of 1928. The liturgies adopted by the Canadian and
Indian Churches also follow the general lines of 1637. It is
widely held that this form of the service, with the longer
Canon, is a great improvement on, by some even that it is
more 'Catholic' than, the service of 1552.

Not all students of liturgies would agree. As we have seen,
the central principle of 1552 is that communion should immediately follow our Lord's words of institution or distribution. The service of 1552 is so carefully drafted, so unified in
its structure, so balanced in its parts, that, if any major
alteration is made in its structure, it is no longer the service
as Cranmer planned it, and it is hard to find any logical
reason why any particular part of it should be in one place
rather than another. The services of 1549 and of 1552 are
both the product of liturgical genius. Either of them can be
used as a basis for revision. It seems hardly likely that a
conflation of two different structures will ever be completely
successful.[1]

When things began to break, they broke fast. Passion
raged among the Scottish people. 'I think our people,' wrote
the moderate Robert Baillie, 'possessed with a bloody devil,
far above anything that ever I could have imagined, as
though the Mass in Latin had been presented. The ministers
who had the command of their mind, does disavow their
unchristian humour, but are no ways so zealous against the
devil of their fury, as they are against the seducing spirit of
the bishops.' On 28 February 1638 the Scottish leaders met

1. It has been my privilege to celebrate the Holy Communion according to almost all the rites which are authorized in the Anglican Communion, except the lately adopted Japanese order, and frequently
according to the American Prayer Book. But I always come back with
immense thankfulness to the stately measured order and the strong
rhythms of 1662 (which is 1552 slightly revised and amended). If we are
to have an alternative order, it seems to me that it would be a great
deal better to return to the order of 1549, which, with some minor
revisions, would give a satisfactory service for use today.

at the Greyfriars Church in Edinburgh and set their hands to
the National Covenant, of which the last paragraph read:
'We promise and swear by the great name of the Lord our
God to continue in the profession and obedience of the aforesaid religion; that we shall defend the same and resist all
those contrary errors and corruptions according to our
vocation, and to the utmost of that power that God hath
put into our hands, all the days of our life.' In November of
the same year, the Scots held at Glasgow an Assembly of the
Church, the legality of which was hotly contested, in the
course of which they abolished episcopacy, deposed and excommunicated by name all the bishops in Scotland, and
appointed a permanent commission to maintain what they
regarded as a godly discipline in the Church. This could not
mean anything but war.

In 1640 Charles in his distress convened Parliament, and
almost immediately dissolved it. With astonishing blindness he allowed Convocation to continue, with doubtful
legality, since in general Convocation sits only when Parliament is in session. The Laudian majority, which seemed
incapable of learning anything, chose this disastrous moment to pass a set of highly provocative Canons. It was laid
down that 'For subjects to bear arms against their kings,
offensive or defensive, upon any pretence whatsoever is at
least to resist the powers which are ordained of God; and . . .
they shall receive to themselves damnation'. The clergy were
to expound this doctrine to their parishioners at least once
a quarter. Members of the learned professions were to take
an oath that they would never wittingly subvert 'the government of the Church by Archbishops, Bishops, deans and
archdeacons, &c. as it stands now established'. This
'Etcetera Oath', as it came to be called, 'the curled lock of
Antichrist', roused Puritan fury to its height.

Before the end of the year Laud had been arrested at
Lambeth, conveyed to the Tower, and told that, like his
friend Strafford, he would be charged with high treason. He
remained in prison for four years, and at last was brought to
trial in 1644, when he was seventy-one years old. His trial
was a mockery of justice. If he had done all the things of

which he was accused, and done them ten times over, all together they would not have amounted to a grain of treason. But the rebellious Commons were determined that their enemy should die. He was beheaded on 10 January 1645. Laud could not learn that it does not do to make martyrs; his enemies had not learnt it either. His execution was a wicked and indefensible action, and such actions are liable to bring a nemesis with them. Laud had caused the Puritans a good deal of trouble during his life; he has caused them a great deal more since his death. For more than three hundred years the ghost of William Laud has stood between the English Nonconformists, the successors of the Puritans, and their brethren of the Church of England.[1]

After the Church of Scotland came the turn of the Church of England. Here too events moved rapidly, but through a great deal of confusion. Almost all were agreed that there must be a limitation of the powers of the episcopate, but how far this should go was still a matter of debate. That wise, learned, and temperate man, James Ussher, Archbishop of Armagh (1581–1656), put forward a plan for the 'reduction of Episcopacy unto the form of synodical government' (1640), an ingenious combination of presbytery with episcopacy, such as might have been listened to in quieter times, but then had little chance of success, since it made no provision for lay participation in the government of the Church. The current was setting in a more radical direction, and on 27 May 1641 the House of Commons gave a first reading to the 'Root and Branch Bill', 'An Act for the utter abolishing and taking away of all archbishops, bishops, their chancellors, commissaries, deans and chapters, archdeacons, prebendaries, chanters, and canons, and all other their under officers'. Though there was much support for this Bill, Parliament wisely saw that it would not do to advance too rapidly with the pulling down of the old, until some thought had been given to the form that the new should take. Recourse was had to consultation with an assembly of divines.

1. The same is true, though in a rather different way, of the execution of King Charles the Martyr on 30 January 1649.

In July 1643 the Westminster Assembly began its deliberations. Of the 121 divines nominated to membership (to be reinforced by thirty lay assessors), almost all had had episcopal ordination, and many varieties of opinion were held among them.[1] The eminent Puritan Richard Baxter (1615–91) wrote, in perhaps pardonable enthusiasm, of this Assembly that 'the divines there congregate were men of eminent learning and godliness, and ministerial abilities and fidelity . . . as far as I am able to judge by the information of all history of that kind, and by any other evidences left us, the Christian world since the days of the Apostles had never a synod of more excellent divines (taking one thing with another) than this Synod and the Synod of Dort were'. Certainly the Synod was diligent in business and productive. In four years, it put forth a *Confession of Faith*, a *Larger* and a *Shorter Catechism*, and a *Directory for the Public Worship of God*, all notable documents of the Faith. It made plans for the government of the Church on a non-episcopal basis, with much control in the hands of lay commissioners.

But, while the divines deliberated, history did not stand still. The Civil War was not going well for Parliament. Needing the help of the Scots, on 22 September 1643 both Houses swore to the 'Solemn League and Covenant', the National Covenant of Scotland with some modifications introduced by Parliament. For the moment the triumph of Presbyterianism seemed to be complete. In January 1645 episcopacy was abolished, and the use of the Book of Common Prayer forbidden. On 23 May 1646, Parliament ordered the adoption of the presbyterian system of Church government.[2]

In the meantime Parliament had been busy getting rid of malignant, delinquent, and scandalous ministers. Parliamentary committees in the different areas were authorized to make inquiry, and to sequester the livings of those who

1. On the first day sixty-nine divines were present, and the average attendance was about sixty.

2. The victory of Presbyterianism was never complete. It was mitigated by the influence of the Independents, and by the determination of Parliament not to abolish one form of clerical autocracy only to submit to another.

did not meet their requirements. It is impossible to determine exactly the number of those affected. Careful research in modern times has arrived at the figure of 2,425 livings sequestered, about thirty per cent of the total; but, if to these are added members of the universities, schoolmasters, and others, about 3,600 ministers were deprived of their means of livelihood. In a number of cases, a fifth of the income of the living was granted to the families of those dispossessed; but, when we consider the poverty of the majority of the parishes, it is clear that immense suffering must have been caused to great numbers of people whose only crime was their loyalty to their king and their Church. The situation was made worse by Cromwell's order of 24 November 1655, under which no sequestered minister might serve as a private chaplain or schoolmaster.

It is to be noted that seventy per cent of the parishes remained unaffected. Some of those who remained may have been mere trimmers, prepared to adapt themselves to every theological wind that blew. Others were moderate Anglicans, who held on in the hopes of better times coming; and, apart from the deprivation that they could not make use of the Book of Common Prayer in public worship, kept things going much as they had always been – 'moderate conformists that were for the old Episcopacy' is Baxter's description of them. In the places of those ejected the national Committee of Triers intruded a mixed multitude of Presbyterians, Independents, and Baptists; but the general order of a national Church and of its parochial system was not seriously altered.

Under the Commonwealth, specifically Anglican life persisted to a greater extent than has often been supposed. In the heart of London John Evelyn could find at least one church in which the Prayer Book liturgy was still in regular use. Bishop John Warner of Rochester (1581–1666) claimed that, throughout the period of the Commonwealth, 'While I lived in my own house . . . I read the liturgy morning and evening; weekly I preached privately or publicly; monthly I administered the Sacrament; and I confirmed such as came to me, or I went and confirmed them in orthodox congregations.'[1] Most striking of all is the fact that so many men, in this dark period, received episcopal ordination. The bishops were living very quietly, and in many cases in poverty. They showed few signs of leadership, and none at all of any capacity for acting together. But they had not entirely forgotten their episcopal obligations. It is clear that at least five English bishops held ordinations under the Commonwealth; and an even larger number of candidates appear to have been ordained by three Irish bishops at that time resident in England.

A serious question that agitated the Church throughout this period was that of the maintenance of the episcopal succession. No consecrations had taken place since 1644; by the end of 1659 only nine English and Welsh bishops survived. Both Puritans and Papists were said to be looking eagerly for the extinction of the Church of England by the demise of its last bishop. It may be asked why the bishops did not meet from time to time and confer the episcopate secretly on suitable persons. The answer to this question is most revealing of the seventeenth-century attitude towards episcopacy. The episcopate was not regarded as being, so to say, the personal possession of an individual; a bishop is to be consecrated by, in, and for a church; only if the necessary conditions are fulfilled can his episcopacy be regarded as canonical and regular. In the appointment of an English bishop three things were necessary – the *Congé d' élire* from the King; election by the Dean and Chapter; and consecration by at least three bishops. Of these, the first must be somewhat doubtful, when the King was in exile; the second was impossible. Some bishops indeed declared that rather than suffer 'through their default that necessary function to fail, they would either consecrate bishops without titles; or else assign them to small bishoprics themselves'; but this was not generally approved of, as both being in itself irregular and involving an infringement of the royal prerogative.

1. Quoted in R. S. Bosher's *The Making of the Restoration Settlement 1649–1662* (1951), p. 27. It was of Warner that the Church historian Fuller remarked that 'in him dying episcopacy gave its last groan in the House of Lords'.

In the end nothing was done. It seems to have been felt that to effect a consecration in England was too dangerous, in view of the increasingly close surveillance of the bishops by the government; and that to arrange for a consecration abroad presented too many problems, in view of the age of the bishops and the difficulties of travel. Happily for the Church, the exile of Charles II did not last as long as at one time seemed probable; and a sufficient number of the bishops survived to ensure the continuity of the succession at the Restoration.[1]

Under the Commonwealth England plunged from experiment to experiment both in Church and State. Cromwell's rule was efficient though oppressive. But, with his death in 1658, the whole fabric began to fall in pieces. Men began increasingly to regret the stability that in the past had been provided by the Crown and by the Church, and to cast their eyes beyond the seas to Holland where Charles II was living in exile. The idea of a restoration came to seem less improbable. And, inevitably, one of the questions that was foremost in men's minds concerned the effects that such a restoration would have on the fortunes of the Church.

Before we attempt to record what actually happened, we must take note of the development within the Church of two divergent attitudes and points of view.

Even under the Commonwealth, there were men who managed to avoid becoming absorbed in the religious contentions of the times, and to devote themselves quietly to study and to a deeper understanding of the faith. Among men of this type the remarkable group known as the Cambridge Platonists were conspicuous.[2] Their so-called Platonism was not so much a definite philosophy as an

1. It is interesting that all parties seem to have felt it essential that the English succession should be maintained by English bishops. The suggestion was made that the King should nominate a number of clerics to Irish bishoprics where his right of nomination was absolute, with a view to their later translation to English sees; but this met with little favour.

2. The best-known leaders of what was a group and not an organization were Benjamin Whichcote (1609–83), Ralph Cudworth (1617–88) and Henry More (1614–87).

inclination to a mysticism of the Neoplatonic type, a quiet confidence that, even in his earthly condition, man is capable of, and called to, commerce with the eternal verities and with the changeless world in which they dwell. The Cambridge Platonists stood in the line of Hooker, believing that human reason is the point at which God makes contact with man – their favourite text was 'The spirit of man is the candle of the Lord'. Above all they believed that it was possible for men to live together in amity, even though they disagreed, and that a modesty which was prepared to leave some of God's secrets in His own keeping was more to be commended than the arrogance which sets out to understand and to define all things in heaven and earth.

We can best understand these men by listening to the superb account given of them by Gilbert Burnet (1643–1715), later Bishop of Salisbury, who had known them well: 'Whichcote . . . being disgusted with the dry systematical way of those times . . . studied to raise those who conversed with him to a nobler set of thoughts and to consider religion as a seed of deiform nature (to use one of his own phrases). . . . All these, and those who were formed under them, studied to examine farther into the nature of things than had been done formerly. They declared against superstition on the one hand, and enthusiasm on the other. They loved the constitution of the Church, and the liturgy, and could well live under them; but they did not think it unlawful to live under another form. They wished that things might have been carried with more moderation. And they continued to keep a good correspondence with those who had differed from them in opinion, and allowed a great freedom both in philosophy and divinity; from whence they were called men of latitude. And upon this men of narrower thoughts and fiercer tempers fastened upon them the name of Latitudinarians.'[1]

A single quotation from Ralph Cudworth will indicate the modesty and humility with which the men of the second Cambridge movement approached the deepest questions of

1. This term was later used with an implication of doctrinal unorthodoxy, but not when it was first bestowed on these good and tolerant men.

theology: 'God's everlasting decree is too dazzling and bright an object for us at first to set our eye upon. It is far easier and safer for us to look upon His goodness and holiness as they are reflected in our own hearts, and there to read the mild and gentle character of God's love to us in our love to Him, and our hearty compliance with His heavenly will. . . . Let us not therefore make this our first attempt towards God and religion, to persuade ourselves strongly of those everlasting decrees; for if at our first height we aim so high, we shall haply but scorch our wings and be struck back by lightning as those giants of old were, that would needs attempt to invade and assault heaven.'

What these men were in themselves was less important than the influence they exercised on others; they were the authors of a tradition which has lasted on in the English Church. Greatest of their followers was Jeremy Taylor (1613–67; Bishop of Down and Connor 1660–67), who in 1647 published *The Liberty of Prophesying*, a plea for reason in religious belief and for toleration in practice. In 1659 Edward Stillingfleet (1635–99), later Bishop of Worcester, put forth his *Irenicon*, an attempt to find common ground between Presbyterians and Episcopalians, such as would make it possible to unite them in one Church. To these should be added Burnet himself (Bishop of Salisbury 1689–1715) and John Tillotson (Archbishop of Canterbury 1691–5). In addition to their other services, these men made a revolution in English preaching. Up to that time, again to quote Burnet, it had been 'overrun with pedantry'. The Latitude men adopted a style which was 'clear, plain and short . . . they . . . applied themselves to the matter, in which they opened the nature and reasons of things so fully, and with that simplicity, that their hearers felt an instruction of another sort than had commonly been observed before'.

But not all eminent English churchmen were of this type. There was another school of which we must not lose sight. A number of the outstanding Laudians had fled from the country; these men, too, were important, not so much for what they then were as for what they became. Some of them were bishops already; no less than eighteen of them became bishops after the Restoration, and among them were such stalwart upholders of the Laudian position as John Cosin (Bishop of Durham 1660–74) and George Morley (Bishop of Worcester 1660–62, of Winchester 1662–84).[1] Political impotence was good for the Laudians; it deepened both their religious convictions and their understanding of the nature of the Church. But present impotence was not without its balance in ambition for the future; these men knew very well what kind of a religious settlement they wanted to see in England at the Restoration; they were near to the King in his exile: above all they had the ear of Edward Hyde, soon to be all-powerful as Earl of Clarendon and Chancellor of the Kingdom. The judgement of Professor Sykes is incontrovertible: 'There can be little doubt that the issue of the Church settlement at the Restoration was predetermined by agreement between the returned Anglican exiles and Clarendon. The form and content of the religious settlement was decided, not by those Anglicans who had stayed at home and some of whom had conformed during the Commonwealth, but by the exiled divines, whose experience abroad had sharpened their antipathy to both Rome and Geneva, and whose resolve was to tread straitly the via media, be it never so narrow and arduous a path.'[2]

At first, there seemed good hope of a policy of reconciliation and comprehension. Charles II, in the 'Declaration of Breda', declared that 'because the passion and uncharitableness of the time have produced several opinions in religion, by which men are engaged in parties and animosities against each other, which, when they shall hereafter unite

1. Not all the eighteen were appointed to *English* bishoprics. Many other leading Laudians became deans. Dr Bosher (op cit., pp. 284–94) gives a fascinating list of them. Note how many passed some time as chaplains to the Levant Company; also how many became Roman Catholics.

2. In *A History of the Ecumenical Movement 1517–1948* (1954), p. 142. Just how their plans were carried through is not altogether clear; much light has been shed on the subject by Dr R. S. Bosher in the valuable book already twice cited.

in a freedom of conversation, will be composed or better understood, we do declare a liberty to tender consciences, and that no man shall be disquieted or called in question for differences of opinion in matters of religion, which do not disturb the peace of the Kingdom; and that we shall be ready to consent to such an Act of Parliament as, upon mature deliberation, shall be offered to us, for the full granting that indulgence'. This was too vague to serve as a basis of agreement; but there were not lacking signs of a willingness on both the Presbyterian and the Anglican sides to seek a moderate solution, such as would have kept within the Church the best of the Puritan elements. These hopes were to founder on the violent monarchical and episcopalian reactions in the country, and on a measure of blindness and folly on the Puritan side.

On 25 March 1661 the King appointed twelve bishops and twelve Puritan divines to meet at the Savoy or elsewhere, 'to advise upon and review' the Book of Common Prayer. The bishops insisted that the Puritans should draw up the list of their objections to the Anglican form of worship. A long list was drawn up, the *Exceptions against the Book of Common Prayer*, and Richard Baxter, one of the most learned of the Puritans and one of the most admirable Christians of the seventeenth century, devoted himself to the rapid production of an alternative liturgy on the Genevan model. This was a tactical mistake. There was no possibility of agreement on a Genevan liturgy; the bishops rejected most of the Exceptions, making concessions only on seventeen points, and most of these not of major importance.[1]

If the bishops were under some pressure from the Puritan side, they had also to face and deal with the desires and proposals of the leaders of the Laudian school. The Convocations which assembled in 1661 were overwhelmingly of that school. Bishop Wren of Ely wrote that 'Never could there have been an opportunity so offenceless on the Church's part for amending the Book of Common Prayer as now, when it hath been so long disused that not one of five hundred is so

1. A complete list is in F. E. Brightman: *The English Rite*, Vol. 1 (1915), p. cxcv.

perfect in it as to observe alteration'. In fact, there was before the Church, in the so-called Durham Book, the outline of a revision, which included most of the changes made in the Scottish Book of 1637, and 'if adopted, would have given the English Church a form of Common Prayer not far removed from the original Book of 1549'.[1] But the more prudent of the leaders seem to have realized that this was going too far. In the Durham Book, at the end of the proposed Order of Consecration in the Holy Communion, the major point of difference, there is a note in the handwriting of Dr Sancroft (1616–93; later Archbishop of Canterbury): 'My L.L. the B.B. at Ely House ordered all in the old method, thus . . .', and then follows a note of the order of the parts of the service, as they stand in the Prayer Books generally in use today.

There are a great many changes in detail in the Prayer Book of 1662, but they are mostly such as would not strike the ordinary worshipper. A new Preface was added, in which the aims of the revisers were stated to be the making clearer of the way in which the services were to be carried out, the elimination of that which was obscure and archaic, and the provision that the Scriptures should be read in a better version. A form for the Baptism of Adults was added, which 'by the growth of Anabaptism, through the licentiousness of the late times crept in among us, is now become necessary, and may be always useful for the baptizing of Natives in our plantations, and others converted to the faith'. Those glories of the Book, the Prayer for all Sorts and Conditions of Men, and the General Thanksgiving, were added. In the Prayer for the Church Militant, a commemoration of the faithful departed (though not intercession for them) was added – a notable improvement. In the Baptism service, the words 'sanctify this water to the mystical washing away of sin' were added.

This was a moderate, sensible, and practical revision, well calculated to secure the assent of all except such as were wedded to the Geneva way. The revision was approved by the Synod of the Church on 20 December 1661, and accepted

1. R. S. Bosher, op. cit., p. 245.

by the Houses of Commons and Lords on 24 February and 17 March 1662 respectively. It was to come into force on St Bartholomew's Day, 24 August of that year, and by the Act of Uniformity all ministers who were not in episcopal orders by that date were to 'be utterly disabled, and *ipso facto* deprived'.

Now came the crisis of conscience for the Puritan ministers. Many found it possible to reconcile their consciences with the order which had now been accepted by the Church. But about 1,760 incumbents were deprived for refusal to conform to the requirements of the Prayer Book and the ordinal, and went out into poverty and hardship. It is possible that, if the settlement had been even more moderate, if, for instance, as happened in Scotland, the requirement of episcopal ordination had not been made retrospective but had been insisted on only for the new entrants, more of the Presbyterians might have found it possible to stay within the Establishment. But this is by no means certain. Men have convictions, and they have consciences. It is possible at the same time to regret the departure of so many good men from the Church, and to believe that they served God better by acting on their convictions than they would have done by making a compromise which their consciences would have disallowed.[1]

If the Laudians had been somewhat moderate in the ecclesiastical settlement, in political affairs they set the Church of England on a course of sinning, of which it has never adequately repented, and the effects of which still darken the relationships between Anglicans and Nonconformists in England. Even before the date of crisis, according to Baxter, 'our calamities began to be much greater than before . . . we were represented in the common talk of those who thought it in their interest to be our adversaries, as the most seditious people, unworthy . . . to enjoy our common liberty among them'. It came to be taken almost for granted that religious nonconformity was synonymous

1. Anabaptists and other 'sectaries' were, of course, already excluded from ministering. The crisis affected only those who can be called Presbyterians or Puritans.

with political disaffection. Under the 'Clarendon Code' of
the Corporation Act, the Act of Uniformity, the Conventicle
Act, and the Five Mile Act, Protestant Nonconformists
including Presbyterians were constituted a second nation
within the nation, a nation that stood permanently at a disadvantage, that was denied equal privileges with others and
the right of taking part in its own government. More than
two centuries were to pass before this iniquity was done
away.

Comprehension or exclusion? Then, as now, this was one
of the major problems of English religion. In spite of the
unfavourable climate created by the Clarendon Code,
several attempts were made during the reign of Charles II
to bring back the Presbyterians to the Church, and to
secure greater toleration for the sects; in all of these the
admirable Baxter was deeply concerned. In the attempt of
1667 a leading part was played by John Wilkins, brother-
in-law of Oliver Cromwell and now Bishop of Chester. A
compromise on ordination was proposed, and Baxter was
prepared to go so far as to accept *legal* authorization to
minister in the Church of England, 'and if it were by a
Bishop, we declared that we should take it from him but as
from the King's Minister'. This plan broke down before a
Bill could be introduced into Parliament. A second attempt
in 1680 led to no better result.

When inner charity was in danger of failing, pressure
from without tended to bring Anglicans and Nonconformists
nearer together. Charles II was received into the Roman
Catholic Church on his deathbed; he had probably been
a crypto-papist for years before that time, and in any case
had had to keep on good terms with his paymaster Louis
XIV of France. James II was openly and fanatically a
Roman Catholic, and, short of completely subverting the
national Church, was prepared to use his prerogative to the
extent of straining the constitution in order to secure better
terms for his co-religionists, and incidentally for the Protestant dissenters.

In 1686 the King secured the appointment of a commission for ecclesiastical causes, suspiciously like that court

of High Commission which had been abolished by Parliament in 1641. In 1687 he expelled the Fellows of Magdalen College, Oxford, for failing to select an avowed papist as President. On 7 May 1688 he issued the second Declaration of Indulgence, and ordered that it should be read in all the churches in the kingdom. This led to the famous episode of the Trial of the Seven Bishops.

Archbishop Sancroft of Canterbury and six of his colleagues, being uncertain whether the King really possessed the dispensing power involved in the declaration, courteously presented a petition to him, requesting the withdrawal of the order. The petition was printed and circulated. The King was furious, and ordered the trial of the bishops for 'publishing a seditious libel against his majesty and his government'. As we have seen, in the days of Charles I the bishops had been the best-hated men in the kingdom; now, suddenly and overnight, they had become the idols of the people, of whatever religious affiliation (except, of course, Roman Catholics), and the declared champions of the liberty of the subject. When the verdict of 'not guilty' was declared, the nation went mad with delight.[1]

After his release, the Archbishop issued to his colleagues in the Province of Canterbury (27 July 1688) what must be one of the most remarkable eirenical documents in the whole history of the English Church, bidding them admonish their clergy 'that they have a very tender regard to our brethren, the protestant dissenters, that upon occasion offered they visit them at their own houses, and receive them kindly at their own . . . persuading them (if it may be) to a full compliance with our Church, or at least, that whereto we have already attained, we may all walk by the same rule, and mind the same thing. . . . And . . . that they warmly and affectionately exhort them to join with us in daily fervent prayer to the God of peace, for an universal blessed union of all reformed Churches, both at home and abroad, against our common enemies; that all they that do confess

1. An episode like this was perfectly suited to the genius of Macaulay. His account of the whole incident should be read as an admirable specimen of impassioned historical narrative.

the holy name of our dear Lord, and do agree in the truth of his holy word, may also meet in one holy communion, and live in perfect unity and godly love.' The Church of England has not always managed to live on so high a level as this.

The crisis came and passed. James had gone too far, and could not withdraw. On 5 November 1688 William of Orange landed at Torbay. Before the new year came, James had taken flight for France, and with him went an old England that was never to come again. The 'Glorious Revolution' was in countless ways the beginning of the history of modern England.

CHAPTER 7

The Eighteenth Century

THE Revolution of 1688 was the end and the beginning of many things.

It marked the end of the old personal rule of monarchs. William and Mary came to England at the invitation of the English people; their title to rule could not be other than a parliamentary title. For more than a century the ruler would still play a predominant part in English affairs; but the old doctrine of the divine right, 'the right divine of kings to govern wrong', was an anachronism. The new age was to be an age, not of mystery and awe, but of lucidity and understanding. No two epochs can be, as it were, surgically divided from one another; yet, if one date is to be chosen as marking more than any other the passage from the medieval to the modern world, that date must be 1688. There is hardly an aspect of the nation's life in which the transition is not to be discerned.

The first master of classical English prose was John Dryden (1631–1700). Dryden was twenty-six years younger than Sir Thomas Browne, twenty-three years younger than John Milton; but his thought and his writing move in a different world from theirs. His aim is not splendour but practical lucidity. We no longer hear the organ-music of the older writers; we are introduced to a prose which is simple, flexible, above all spare and athletic. Sir George Clark speaks of Dryden's prose as 'clear, free, and informal, expressing his copious flow of thought without visible effort'.[1] This was the line that was to be followed by Addison, Steele, and Swift. In reading their works, we are hardly conscious of the span of years by which we are separated from them.

In music Henry Purcell (1659–95) is the first English

1. *The Later Stuarts 1660–1714* (1934), p. 346.

master of the modern idiom. His genius was so great that it could compass various styles, and could submit to various influences – French, Italian, and others – without losing its own proper character. But his work, at its best, is marked by a freshness, a gaiety, a depth of feeling, that for a brief moment restored England to its primacy as the first musical nation of the world, and still speaks direct to the modern ear without any impression of strangeness or distance.

Among influences on thought none is more significant than the *Royal Society of London for Improving Natural Knowledge*. The Society was founded in 1660, and therefore had roots in the period before 1688. But it was in the last years of the seventeenth century and the early years of the eighteenth that it attained to its greatest glory and its most far-reaching influence. The great thinkers of earlier years had aimed at producing systems of the universe; and these could not but be, in the nature of things, very largely the products of human intellection. Has not Descartes told us how he shut himself away from all outside influence, in order to attempt to penetrate to an understanding of himself and of the nature of his own intellectual powers? For all this the men of the Royal Society would substitute *observation* – it might be the observation of humble mechanic minds; but, if it was true observation, it would contribute to the understanding of the world, and, without such understanding of observable phenomena, there could be no true philosophy. In a sense Plato was dethroned, and Aristotle had come into his own again. God and the soul were excluded from the purview of the Society; all else could come under its scrutiny.

So 'we find many noble rarities to be every day given in not only by the hands of learned and professed philosophers, but from the shops of mechanics; from the voyages of merchants; from the ploughs of husbandmen; from the sports, the fishponds, the parks, the gardens of gentlemen. . . . It suffices, if many of them be plain, diligent and laborious observers: such, who though they bring not much knowledge, yet bring their hand, and their eyes uncorrupted; such as have not their brains infected by false images, and can honestly assist in the examining and registering what the others represent to their view. It seems strange to me, that men should conspire to believe all things more perplexed, and difficult, than indeed they are.' So Thomas Sprat (1635–1713), later Bishop of Rochester and historian of the Royal Society.

The scientific temper is inevitably in some ways a sceptical temper. It must ask for evidence, and must distinguish between many different degrees of certainty. The eminent Robert Boyle (1627–91) wrote a book called *The Sceptical Chymist*, in which he took leave to doubt many things commonly believed, on the simple ground that no adequate evidence for them had ever been presented: 'I must tell you, that in matters of philosophy, this seems to me a sufficient reason to doubt of a known and important proposition, that the truth of it is not yet by any competent proof made to appear. And congruously hereunto, if I shew that the grounds, upon which men are persuaded that there are elements, are unable to satisfy a considering man, I suppose my doubts will appear rational.' It is not surprising that some orthodox persons wondered whether the scientists would not carry their scepticism a good deal further than the harmless domain of the elements, and raised the cry 'Religion in danger'. But, in fact, for these pioneer scientists the problem of a war between religion and science simply did not exist. It is possible that some of them endured an uneasy dichotomy between their science and their faith; but all alike were convinced that all truth is from God, and that there can be no ultimate conflict between those truths that are learned by observation and those that are apprehended by faith. Did not Robert Boyle himself learn Hebrew in order more accurately to study the Scriptures? Did he not found by his will the Boyle lectures to defend the Christian religion against 'notorious infidels, viz. atheists, theists, pagans, Jews and Mahommedans'? Sir Isaac Newton did not believe in the doctrine of the Trinity, and would not therefore take holy orders in the Church of England; but he was ever a profoundly religious man, to whom it came natural to write in his *Opticks* that 'all material things seem to have been composed of the

hard and solid particles above-mentioned, variously associated in the first creation by the counsel of an intelligent
Agent. For it became him who created them to set them in
order. And if he did so, it's unphilosophical to seek for any
other Origin of the world, or to pretend that it might arise
out of a chaos by the mere laws of nature: though being once
formed, it may continue by those laws for many ages.'
The greatest work of the greatest of English botanists, John
Ray (1627–1705), was entitled *The Wisdom of God Manifested
in the Works of the Creation* (1691).[1]

Yet perhaps those who feared for the future of religion
were not altogether without grounds for their anxiety. The
two works which more than any others may be regarded as
typical of this new period are John Locke, *On the Human
Understanding* (1690), and Sir Isaac Newton, *Principia
Mathematica* (1687). Locke's work was 'the first extensive
attempt to estimate critically the certainty and adequacy of
human knowledge when confronted with God and the
universe . . . he sought to make a faithful report, based on
an introspective study of consciousness, as to how far a
human understanding of the universe can reach'.[2] Newton,
by bringing an enormous number of apparently complex
phenomena under a small number of easily apprehensible
principles, demonstrated that the universe is a world of law
and regularity, and not either of mere chaos or of the unintelligible designs of a wholly inscrutable providence. These
books, and others that followed them, spread all over
Europe, and made the new England more than at any other
time before or since the teacher of the new Europe.[3] Seventeenth-century man had been 'a stranger and afraid in a
world I never made'. Eighteenth-century man lived in a
universe in which it was possible to be at home, even comfortable. As the secrets of nature were disclosed, he became

---

1. It is interesting that the only complete biography of John Ray is by
Canon C. E. Raven (1942), the most brilliant example in our day of the
combination of the man of science with the theologian.

2. *Encyclopaedia Britannica* (14th ed.), s.v. 'John Locke'.

3. All this is set out with wonderful brilliance in Paul Hazard, *La
crise de la conscience européenne* (1935, E.Tr. *The European Mind*, 1953).

ever more profoundly impressed by the beauty of its order, by the wonderful adaptation of its parts to one another and to the whole. A little blind to the contrary evidences of disorder, conflict, and imperfection, if he continued to believe, as most men still did, in one supreme invisible Spirit, the Author of nature, he was inclined to think that the word which best described that Spirit, and the character which would most commend men to fellowship with Him, was benevolence. It was through the logical deductions that might be drawn from such lines of thought as these that the Church found itself thrust into the most violent and dangerous of the eighteenth-century controversies concerning religion.

The settlement of religion on the arrival of William III could not but be severely Protestant in character.[1] The Declaration of Rights laid down that for the future England could not be governed by a sovereign who was himself a papist, or married to a papist; the ruler of England must be in communion with the national Church. The form of oath which William III took at his coronation included, after the vow to maintain the laws of God and the true profession of the Gospel, the words 'the Protestant Reformed Religion established by Law'. Some objection had been raised to the inclusion of the term 'Protestant' in such a document; but the objection was overruled on the ground that a crypto-Romanist sovereign might put a 'false and subdolous' construction on the oath unless this word were included; and so it stood.[2]

The first and saddest consequence of the Revolution for the English Church was the separation from it of those who

1. The word is used here in its seventeenth-century sense. Until the nineteenth century, the opposite of 'Protestant' was held to be, not 'Catholic', but 'Papist'.

2. See the lengthy discussion of the question in D. Ogg: *England in the Reigns of James II and William III* (1955). Mr Ogg concludes: 'Its dual phrase . . . has served to enunciate one of the essential characteristics of Anglo-Saxon civilization. The alternative was Bourbon-Stuart civilization, a totalitarian system, having as its agents the priest, the dragoon, and the hangman, a system to which many Englishmen were determined not to submit.' Op. cit., p. 239.

came to be called the Non-jurors. Many Anglicans held a high view of the dignity and authority of the King, even when that King was a Roman Catholic and an enemy of the Church which they served. In 1689 nine bishops and about four hundred priests found that, having once taken the oath of allegiance to James II, they could not, according to their consciences, during his lifetime take the oath of allegiance to any other sovereign. Among those who took this view were the Primate Sancroft, and Thomas Ken (1637–1711) of Bath and Wells, the most saintly of the bishops of the reign of Charles II.[1] The Government did its best to make things easy for these men. The oath to the new sovereigns was proffered in the form 'I A.B. do sincerely promise and swear to bear true allegiance to Their Majesties, King William and Queen Mary'; the sovereigns were not described as 'lawful and rightful' monarchs, and no mention was made of their successors. But even to this these good men found that they could not swear; and so in due course it came about that they were deprived, and successors appointed to them in their diverse cures.

This need not, of itself, have led to a schism. The Non-jurors might have retired into lay communion with a Church in which they could no longer officiate; and this was, in practice, not far from the attitude adopted by Ken in his quiet retreat at Longleat. But others took a different and harsher view, most vocal among them George Hickes (1642–1715), who had been Dean of Worcester. These men held that the Church of England had lost all claim to be considered a true Church, and that the Non-jurors alone could be regarded as the true, catholic, and apostolical succession in England. Unlike their predecessors, the Laudians of the Commonwealth days, the Non-jurors decided to continue among themselves the episcopal succession, though they had no organized Church to which bishops could be nominated. In 1694 George Hickes and Thomas Wagstaffe were nominated by the exiled James II to the episcopate, and consecrated Bishops Suffragan of

1. He was consecrated only twelve days before the death of Charles II.

Thetford and Ipswich respectively, titles which the Non-jurors had no real authority to bestow.[1] From that time on the Non-jurors were unmistakably in schism, and no longer had any claim to be regarded as a part of the Church of England.

A good opportunity to end the schism on the death of James II in 1701 was unfortunately lost. Another opportunity was lost when the death of Wagstaffe in 1712 left Hickes as the only living Non-juring bishop. But Hickes was not the man for any kind of conciliation. In 1713, with the help of two Scottish bishops, he consecrated three new bishops, 'all the Catholic bishops of the English Church having died, except the Bishop of Thetford'; this time no attempt was made to confer any title – these men were bishops at large, the episcopate being for the first time identified with the man rather than with the Church within which he was to exercise his office. From this time on, the existence of the Non-jurors became ever more shadowy, and their shadow-episcopate died out with the death of Charles Booth in 1805.

A most interesting record can be compiled of the services of the Non-jurors to learning,[2] of their liturgical experiments, of the bitter controversies that arose among them over liturgical uses between the 'Usagers' and 'Non-usagers', and of their attempts to make contact with and perhaps to secure recognition from the ancient Orthodox Churches of the East. It was a grievous loss to the national Church that the great services of which these men were capable were rendered outside the Church, and had so little effect upon it. The one man among them, and one who was not in the strict sense of the term a Non-juror, to exercise a wide influence outside the narrow circles of the schism was William Law (1686–1761).

1. The Non-juring Archbishop Sancroft and Bishop Lloyd had nominated these men as their suffragans, and the semblance of legality had therefore been preserved.
2. The fiery George Hickes, for example, devoted a part of his enforced leisure to the preparation of a grammar and dictionary of the ancient Scandinavian languages (1703–05).

Law had hardly begun his ministry when he lost his fellowship at Emmanuel, and thereby the opportunity of exercising that ministry, through his unwillingness to take the oath of allegiance to George I. For the rest of his long life he lived in retirement of one sort and another, being at one time tutor to no less a person than Edward Gibbon the historian; and this long leisure he employed in the use of a ready and singularly effective pen. In early life he engaged in the controversies of the time, with some of which we shall later be concerned. Later he came under the influence of the German mystic Jakob Boehme, and produced mystical works, of which the best known is *The Way to Divine Knowledge*. The basis of his theology was that love is all; it is only through knowing that God is love that we can come to know Him at all, and through love all the problems of divinity can be resolved. But the book on which Law's fame has chiefly rested, and deservedly, is his *Serious Call to a Devout and Holy Life* (1729), a book so simple that anyone can read it, so full of vivid delineations of character that everyone can find himself somewhere in it, and so Christian that anyone who lived according to it would not fall far short of being a perfect Christian. Dr Johnson recorded that this was the book that first directed him to the life of religion. Gibbon remarked that 'if Law finds a spark of piety in his reader's mind he will soon kindle it into flame'. Perhaps it is not fanciful to suppose that, in his picture of the holy priest Ouranios, Law is setting forth what he wished and hoped himself to be: 'He now thinks the poorest creature in his parish good enough, and great enough, to deserve the humblest attendance, the kindest friendships, the tenderest offices, he can possibly shew them. He is so far now from wanting agreeable company, that he thinks there is no better conversation in the world, than to be talking with *poor* and *mean* people about the Kingdom of heaven.'

With the withdrawal of the Non-jurors, William III was faced by the necessity of filling many vacant sees. It is not surprising that appointments were made almost exclusively from the ranks of the Latitudinarians; the first to be elevated was Gilbert Burnet, who became Bishop of Salisbury. In a sense these were political appointments, but none of them was discreditable. Of the new bishops, Mr Ogg has remarked a little tartly that 'they were, on the whole, more tolerant than their predecessors, because better educated: none of them very eminent, they possessed a quality which had been denied to many of their predecessors, that of ordinary Christian charity'.[1]

Holding the views that they did, it was natural that the bishops should attempt to promote a return of the more reasonable of the Nonconformists to the bosom of the national Church. With the Toleration Act of 1689 they had good success. This did not go far, but it at least secured those Nonconformists of whose political loyalty there was no doubt against the worst excesses of the Clarendon Code, and assured them of some legal recognition of their rights of separate worship in that their chapels or meeting-houses were now to be certified to the bishop or the archdeacon of the diocese. This was in fact a highly revolutionary step. The Revolution carried further the process, which had begun at the Restoration, of the recognition of the possibility of the peaceful coexistence of several Churches within a single State. Up till this time almost everyone in England had believed that there should be one Church and one only; and, in the century and a half since the Reformation, Papists, Anglicans, Presbyterians, and Independents had each striven, by bitter persecution of all the others as opportunity offered, to establish their own monopoly. Now men had recognized that that could not be. Certainly the ideal is that all Christians should be able to live and worship together in a single Church. If we cannot have the ideal, it is far better that Christian bodies should learn to live peacefully together, albeit in somewhat uneasy juxtaposition, than that they should strive endlessly in mutual intolerance and persecution.

With a Comprehension Bill the bishops had very much less success. Some proposals were drawn up which, it appears, would have attempted to draw Nonconformists back to the Church by a measure of laxity in regard to the three principal stumbling-blocks – the sign of the Cross in baptism,

1. Op. cit., p. 235.

kneeling to receive the Holy Communion, and the wearing of
the surplice. Some talked of these things as of things indifferent in themselves; others darkly of the thin end of the
wedge, exactly as they do today. The proposals were not
presented to Parliament but remitted to the Convocations
of the Clergy. For nearly a generation Convocation was the
storm centre of the life of the Church.

Historically, the Convocations had met at the same time
as Parliament, in order that the clergy might vote their own
taxation separately from the High Court of Parliament; they
might also transact other business by express permission of
the King. In 1665, by private agreement between Archbishop Sheldon and the Crown, the right of separate taxation was abandoned. The question naturally arose as to
whether Convocation need ever meet again. There was a
further problem within Convocation itself. The general
view of the bishops was that a Convocation was a single
assembly, over which the Archbishop presided, although the
two houses, of bishops and of other clergy, met separately.
Many in the Lower House held the view that their House
could claim such an independence of the Upper House of
Bishops as was maintained by the House of Commons in
relation to the House of Lords. Here were materials enough
and to spare for controversy – and in 1689, whereas the
bishops were in the main Latitudinarian, in the Lower
House there was a 'high church' majority.

This is the first moment at which it is correct to use this
term (though we have used it earlier of the Laudians),
since it was about this time that it came into common par-
lance.[1] The ecclesiastical division corresponded to a deep
division of political conviction. The Latitudinarian was
whole-heartedly committed to the Revolution and to the
Protestant succession to the throne; many high churchmen
were at best no more than half-hearted in their allegiance

1. All this has been dealt with, admirably and with great erudition, by
Bro. George Every in *The High Church Party 1688–1714* (1956). To this we
can now add the temperate and erudite account in N. Sykes, *William Wake,
Archbishop of Canterbury, 1657–1737* (1957) Vol. 1, pp. 80–156, of highly
complex controversies, which I have here perhaps unduly simplified.

to the new sovereigns, and some of them cast more than a glance towards James II, and then to 'James III', the king over the water. In practical affairs the immediate rock of division was the problem of the Nonconformists. The Latitudinarian would gladly have made some concessions to bring them back; the high churchman, raising the cry of 'the Church in danger', stood on the letter of Anglican principle, and would not have the dissenters back on any terms other than those of total submission. In addition, the high churchman had coined for his opponent the contemptuous term 'low churchman', and for the purposes of propaganda insinuated that the unscrupulous low churchman would stop at nothing to betray even the most sacred foundations of the faith that he had been commissioned to defend.

In Convocation the principal high church firebrand was Francis Atterbury (1662–1732), who was eventually made Bishop of Rochester by the 'high church' Queen Anne. But the man who attained to the greatest notoriety was Dr Henry Sacheverell, a foolish and shallow cleric, whose only title to fame is his folly. Of this gentleman, Sarah, Duchess of Marlborough, who for good reason did not like him, wrote that 'he had not one good quality that any man of sense ever valued him for. . . . He had a haughty insolent air, which his friends found occasion often to complain of; but it made his presence more graceful in public. His person was framed well for the purpose, and he dressed well. A good assurance, clean gloves, white handerkchief well managed, with other suitable accomplishments, moved the hearts of many at his appearance. . . . The weaker part of the ladies were more like mad or bewitched than like persons in their senses.' On 5 November 1709 Sacheverell preached an intemperate sermon in St Paul's Cathedral, in which he averred among other things that the Church's 'Holy Communion has been rent and divided by factious and schismatical imposters, her pure doctrines corrupted and defiled; her primitive worship and discipline profaned and abused; her sacred orders denied and vilified; her priests and professors (like St Paul) calumniated, misrepresented and ridiculed, her altars and sacraments prostituted to hypocrites, deists, socinians and atheists'.

The Whig government, instead of taking no notice, were foolish enough to impeach Sacheverell, on the ground that he had preached and printed his sermon 'with a wicked, malicious and seditious intention to undermine and subvert her Majesty's government and the Protestant succession as by law established . . . to create jealousies and divisions among her Majesty's subjects, and to incite them to sedition and rebellion'. After a lengthy trial Sacheverell was found guilty by a majority of seventeen, but sentenced to nothing worse than suspension for three years from the office of preaching. Sir Robert Walpole remarked that 'I think they had as good as acquitted him'. The Dean of Christ Church added that 'in attempting to roast a parson the Whigs had scorched themselves'. For the moment it might seem that the Tories and the high churchmen had had it all their own way. But the Whigs had only to bide their time. With the accession of George I they came back to power, and held it for a very long time. After 1717 the Convocations did not meet again until 1741; then, owing to the revival of the old disputes, they were again prorogued, and this time their silence lasted for 114 years – till 1855.

In one other matter the high churchmen took advantage of their temporary power under Queen Anne to score a victory over the Nonconformists. It had long been the custom of godly dissenters occasionally to receive the Holy Communion in the Church of England, as a mark of Christian fellowship with a body of which they did not wholly disapprove, though they could not bring themselves to enter into full communion with it.[1] But occasional conformity was also used for a very different purpose. Under the Test Act, no one could hold office, either civil or military, without having first received the Holy Communion according to the use of the Church of England – 'an insidious degradation to which the Anglican Church in its alarm submitted, and

1. On this see N. Sykes in *A History of the Ecumenical Movement 1517–1948* (1954). The practice seems not to have entirely died out until the nineteenth century.

from which it was not reluctantly delivered until the nineteenth century was well on its course.'[1] A number of Nonconformists had accepted the necessity, in order to qualify
themselves for office, of this occasional conformity without
ceasing to be regular worshippers at their own conventicles –
a blameworthy policy, no doubt, but much less blameworthy
than that of the Church which allowed its most sacred ordinance to be prostituted to political ends. In 1710 the high
churchmen decided to put an end to this undesirable state
of affairs, not as they should have done by pressing for the
repeal of the Test Act, but by imposing penalties on those
who had so far managed to make the best of both worlds.
The 'Occasional Conformity Act' of 1711, more properly
called 'An Act for Preserving the Protestant Religion by
Better Securing the Church of England as by Law Established', laid it down that any office-holder who, after
receiving the Sacrament in the Church of England, should
'knowingly or willingly resort to or be present at any conventicle, assembly or meeting within England, Wales,
Berwick-on-Tweed or the isles aforesaid [Jersey and
Guernsey] for the exercise of religion in other manner than
according to the liturgy and practice of the Church of
England' should be fined forty pounds, and disqualified
from holding office. The Act was repealed in 1719; the
wounds that it had caused were not so quickly healed.

It has been necessary to give some space to these rather
squalid controversies, not only because of the storms and
passions that attended them, but because they draw attention to permanent factors in the religious situation in England. And, as so often happens, Christians were wearing one
another out and squandering the spiritual resources of the
Church in controversy at a moment at which it was urgently
necessary that all good men should stand together. The real
question of the times was not as to which group or party
should have predominant influence in the Church; it was,
whether there should within a few years be any Church for
anyone to belong to at all. The deist controversy of the
eighteenth century introduced one of the gravest periods of

1. G. N. Clark: *The Later Stuarts, 1660–1714* (1956), p. 76.

peril through which the Church, in all its long history, has been called to pass.

This did not become immediately apparent. Some of the deist propaganda was crude and superficial, as in the hands of such men as Toland (*Christianity not Mysterious*, 1696). But some of it was temperate, rational, and well expressed, and doubly dangerous because it could make its own so much of the contemporary language of orthodox divines. Thus the ablest of the deists, Matthew Tindal (1657–1733), in his *Christianity as Old as Creation* (1730), could quote extensively from such men as Tillotson in favour of the use and authority of reason, though with a view to establishing conclusions that Tillotson would have eschewed. It was all too easy to overlook the profound difference between the Platonic 'reason', with its large admixture of imagination and adoration, and the cool, secular, logical reason of the 'Age of Reason', which could lead in the end only to the scepticism of Voltaire and Hume.

'What is the purpose of religion?' asked the deists, and their answer was that the aim of religion is to make men virtuous. And how are men to be made virtuous? By the acceptance of certain sound general principles men could become good; natural religion would accept, as generally valid principles, belief in God, in human immortality, and in some system of rewards and punishments after death (though Shaftesbury in his *Inquiry Concerning Virtue* (1699) was prepared to abandon the third of these as otiose). How are these principles to be learned? The answer is that they are self-evident, as can be seen from their general acceptance in all the religions and all the countries of the earth. Locke had already remarked that the moral teachings of all the religions were much the same, and that Christianity had only the advantage of presenting them in a more orderly and coordinated form. But, if such principles are self-evident, there is no need for a special revelation; and the traditional Christian arguments from miracle and from prophecy are irrelevances and inconveniences rather than aids to true religion.

At this distance of time, we can see that the defect of all

forms of deism was that they treated religion as being a system of ideas and a code of moral precepts. The heart of the Christian faith is personal communion with a living God, redemption from sin, and the redemption of history through the personal interposition of God Himself in it through the incarnation. But part of the difficulty experienced by the Church in answering the deists was that most of the champions of the Church had themselves come to think in the same categories as the deists, and were therefore at a great disadvantage in meeting an adversary whose use of the available weapons was perhaps more skilful than their own.

One possible method of meeting the situation was to change the battle-ground by affirming with Pascal that the heart has its reasons, of which the head knows nothing. This is not necessarily an evasion of the issue. Logical mathematical reasoning cannot deal with all the worlds in which man is capable of experience. This may be illustrated by a consideration of the general eighteenth-century attitude towards poetry. What is there that can be better expressed in poetry than in prose? Inevitably the honest eighteenth-century answer was, 'Nothing'. Poetry was the natural form of expression in less cultured ages; but, now that thought has been developed and expression defined, poetry can have only an ornamental value in rendering agreeable that which can in reality be better expressed in other forms. The true understanding of poetry is that it is the vehicle for the expression of those immensities of the human spirit for which symbolic and evocative language alone is appropriate. True religion is far more akin to those immensities than to the neat Palladian architecture and the formal gardens in which eighteenth-century man delighted. Any insistence on these other dimensions is likely to be condemned, by the devotees of reason, as a 'flight from reason'; it may, for all that, be in truth a return to reality.

But, first, it was desirable that the deists should be met on their own battle-ground and defeated with their own weapons. Their logic was more plausible than sound. The universe is not such a tidy place as they imagined, and their assumptions often rested on nothing more solid than unproved assertion. Many things that had been paraded as certainties could be shown to be unsupported by anything that could rightly be called demonstration. Such rigorous argument is rarely inspiring, but it is often the indispensable preliminary to a recovery of faith. In a word, Joseph Butler (1692–1752) had to come before John Wesley (1703–91). A study of the work of these two great men will open the way to an understanding of almost the whole of eighteenth-century religion.

Butler, the greatest of all the thinkers of the English Church, was the son of a dissenting tradesman and had been educated at a dissenting academy. At the age of twenty-six he received the orders of the Church of England, and, after the tenure of a number of quiet cures, in which he had abundance of leisure for quiet thought and study, became bishop of the sees, in succession, of Bristol (1738) and of Durham (1750). He was a lonely thinker, ever wrestling with the fundamental problem of the Being of God and the nature of man, patient, scrupulous and exact, fair-minded towards the position of those whom he wished to confute and always inclined rather to understate than to overstate his case.

His most famous work, *The Analogy of Religion*, was published in 1736. This great book is more often praised than read; and it is unfortunate that it is so often robbed of the courtesy of being cited under its full title, *The Analogy of Religion, Natural and Revealed, to the Constitution and Course of Nature*. The title makes exactly clear the object which Butler had set before himself. Natural and revealed religion are not contrasted with one another; each is considered analogically to the world around us, and to the knowledge which we have of that world.

Butler starts from the current unbelief: 'It is come, I know not how, to be taken for granted, by many persons, that Christianity is not so much as a subject of inquiry; but that it is now at length discovered to be fictitious. And accordingly they treat it as if, in the present age, this were an agreed point among all people of discernment; and nothing remained, but to set it up as a principal subject of mirth and ridicule, as it were by way of reprisals, for its having so long interrupted the pleasures of the world.' But, asks Butler, has Christianity really been disproved? And has anything else been proved in its place?

The idea that 'natural religion' is demonstrably true is baseless. We may overlook the mysterious elements in the world, but by doing so we do not prove that the world is no longer mysterious. All the arguments which the deist brings against Christianity can be applied with equally deadly effect to his own moral and religious beliefs. There is a close analogy between what we know of nature and what can be affirmed of the Christian faith: 'It is then but an exceeding little way, and in but a very few respects, that we can trace up the natural course of things before us to general laws. And it is only from analogy that we conclude the whole of it to be capable of being reduced into them: only from our seeing that part is so. It is from our finding that the course of nature, in some respects and so far, goes on by general laws that we conclude this of the rest.' No scientist of today would wish to write otherwise. Now comes the application to Christianity. 'Upon the whole, then, the appearance of deficiencies and irregularities in nature is owing to its being a scheme but in part made known, and of such a certain particular kind in other respects. Now we see no more reason why the frame and course of nature should be such a scheme than why Christianity should. And that the former is such a scheme, renders it credible, that the latter upon supposition of its truth, may be so too.' All our arguments together will not give us demonstrative certainty – for any kind of religion, natural or revealed; but that is because we are in a state of probation, and in such a state the best that can be given us is reasonable probability. If it can be shown that such a probability exists, the obligation rests on men to give it the most careful consideration. Action on the basis of such probability is the very nature of faith. But this action rests on a certain moral disposition. 'If this be a just account of things, and yet men can go on to vilify or disregard Christianity, which is to talk and act, as if they had a demonstration of its falsehood; there is no reason to think they would alter their behaviour to any purpose, though there were a demonstration of its truth.' These are the concluding words of the book.

The first thing that must strike any reader of Butler is the almost painful integrity of his mind and of his style: 'At whatever cost of clumsiness and awkwardness, Butler always says that thing that he means to say – that thing and not another. "Everything is what it is, and not another thing", is an odd, characteristic axiom of his, which in various forms he is very fond of. "Things and actions are what they are, and the consequences of them will be what they will be; why then should we desire to be deceived?" . . . This honesty and reality of writing is the outward vesture and form of the mind within, severely trained to the strictest truth and honesty of thinking. Open-eyed, cautious, watchful against the tricks and idols of his own thoughts, the things that come before him he tries to see as they are – to see them, not as they are talked about, or appraised by temporary or accidental opinion, but in their solid, plain simplicity.'[1]

It would be a mistake, however, to think of Butler only as a remote and austere thinker, a kind of theological Newton. His sermons reveal the depth and earnestness, indeed the controlled passion, of his inner faith and of his devotional life. Nothing could be less emotional than his two sermons on the love of God; but has any mystic ever better expressed what it will mean to good men to reach the end of their pilgrimage, and to dwell in the presence of God for ever? 'Would not infinite perfect goodness be their very end, the last end and object of their affections, beyond which they could neither have nor desire, beyond which they could not form a wish or thought? . . . What then will

---

1. R. W. Church: *Pascal and Other Sermons* (1895), pp. 29–30. All theological students should be compelled to read Butler, not necessarily in order to think the same thoughts as Butler, but in order to learn how to think theologically. As a theological teacher in South India I used to make my students translate selections from Butler's Sermons into Tamil, a task which I think we all found difficult but profitable.

be the joy of heart, which his presence and the light of his countenance, who is the life of the universe, will inspire good men with, when they shall have a sensation that he is the Sustainer of their being, that they exist in him; when they shall feel his influence to cheer and enliven and support their frame, in a manner of which we now have no conception? He will be in a literal sense their strength and their portion for ever.'

At some time in 1739 Bishop Butler accorded at Bristol an interview to the Reverend John Wesley, Fellow of Lincoln College, Oxford, in the course of which he remarked (according to Wesley's account of the discussion): 'I once thought you and Mr Whitefield well-meaning men; but I cannot think so now. . . . Sir, the pretending to extraordinary revelations and gifts of the Holy Ghost is a horrid thing, a very horrid thing.'[1] It was hardly to be expected that the two greatest Anglicans of the eighteenth century should understand one another. What is significant is that they were both Anglicans; and it might be debated to the end of time which of the two has rendered greater services to the cause of Christ not only in England, but throughout the world.

The Methodist movement set on foot by John Wesley and his colleagues is often treated as an isolated phenomenon; but this is to misunderstand its true character. Although the movement would not have been what it was without the impress of the character and of the experience of Wesley, he himself can best be studied as the meeting-place of a whole range of trends and traditions.[2] The serious concern of the English Puritans for salvation was one of the living

1. This famous conversation is so much misquoted that it should be looked up and read in full in *Wesley's Works*, Vol. xiii (1831), pp. 464–6. Wesley at once replied that he was in no way responsible for Mr Whitefield, and that 'I pretend to no extraordinary revelations or gifts of the Holy Ghost; none but what every Christian may receive, and ought to expect and pray for'.

2. In the literature known to me this has been most fully carried out by the German biographer of Wesley, Martin Schmidt, in *John Wesley*, Vol. I (1953). An English translation of this valuable work appeared in 1963. Vol. II appeared in German in 1966.

forces in religion in the seventeenth century. It crossed the sea to influence the German, Philipp Jakob Spener, who through the publication of his *Pia Desideria* in 1675 gave the impulse to the great movement known as pietism. Wesley was influenced by both forms of pietism, – its more churchly form with its great centre at Halle under the leadership of the Franckes, and the more sentimental Moravian form which it took on at Herrnhut under the leadership of the pious Count Zinzendorf. Nor must we forget the tradition of the 'societies', small groups formed for the reformation of manners or the propagation of pure Christian zeal, each with its own rules, which were popular at the end of the seventeenth century, and with which the name of Anthony Horneck (1641–97), a German graduate of Oxford living in London, is specially associated. And, for some reason, 'revival' was in the air; several movements entirely independent of Methodism sprang into existence in the period with which we are now dealing.

John Wesley (1703-91) experienced what he regarded as conversion in a meeting of Moravians held in Aldersgate Street on 24 May 1738. Here, after a varied and troubled pilgrimage, he found peace. What had happened was that he had discovered the apostolic doctrine of justification by faith; on that he rested for the remainder of his long life. But, being untroubled by most of the controversies that had confused the period of the Reformation, he went on to see and to state, with a clarity that has rarely been equalled, that any genuine spiritual experience must forthwith find its outward expression in practical holiness of life. The aim of his life was to promote Scriptural holiness throughout the land.

Wesley early became convinced that his special mission could be accomplished only by an itinerant ministry, in which the parochial organization of the Church was not so much denied as disregarded. As early as the interview with Bishop Butler already referred to, he had maintained that 'being ordained as Fellow of a College, I was not limited to any particular cure, but have an indeterminate commission to preach the word of God, in any part of the Church of

England'.[1] In fifty years of travelling he visited almost every part of the country, with powerful effects. But nothing could be less like the reality of the work of Wesley than the popular picture of a wandering revivalist. The three main emphases in his work were theology, education, and discipline. The theological content of his sermons is such that 'the Sermons of Mr Wesley' are almost as much a theological standard for Methodists as Calvin's *Institutes* for Presbyterians. The educational level of eighteenth-century England was deplorably low. Serious Christians, according to Wesley, must learn to read, and must be provided with suitably edifying reading material. Believers must be organized in societies, provided with leaders qualified to nurture their spiritual growth, and stimulated, through the discipline of the weekly class-meeting, to greater heights of Christian achievement.

In all this there was nothing schismatic or anti-Anglican; and to the day of his death John Wesley regarded himself as a faithful minister of the Church in which he had been ordained. But perhaps he himself underestimated the separatist tendencies which from the beginning had been latent in the movement that he had created. As early as 1746 he had reached, like Luther before him, the conclusion that in the apostolic age there was no distinction between the episcopate and the presbyterate, and that there is no ecclesiastical act commonly attaching to the episcopal office which the presbyter, in case of necessity, cannot equally well perform. It was only at the end of his life that he ventured to ordain men to the presbyterate in Scotland and in England (for the latter no more than three). On 2 September 1784, in a private room at Bristol, Wesley set apart Thomas Coke, a Welshman, as Superintendent for America. Immediately on arrival in America Coke ordained Francis Asbury as deacon, priest, and superintendent. In

1. It is interesting to note the way in which Wesley tried to reconcile his unconventional methods with the principles and practice of the Church of England. On 21 October 1738 he went to see Edmund Gibson, Bishop of London, in order to ask whether 'religious societies', such as he had a mind to form, could be regarded as 'conventicles'.

1787 he persuaded the American Conference to change
the title to 'bishop'. Wesley strongly objected to the adoption of this title; but in 1784 Coke had already been
twelve years in holy orders in the Church of England, and
the ceremony of setting him apart as 'superintendent'
would have been meaningless, unless Wesley had believed
himself to be conveying to Coke some authority other than
that which he had possessed through his earlier ordination.[1]
All this was, of course, completely irreconcilable with the
most tolerant interpretation of Anglican principles.

It is possible to imagine that, with a little more vigour on
the part of the Church of England, and a little more flexibility on the part of Wesley, it might have been found practicable to retain the gifts and graces of Methodism within the
Church of England. Almost all Anglicans deeply regret that
the separation between Anglican and Methodist took place.
But it is important not to overlook the power and significance
of that part of the Evangelical Revival which did remain
within the Church of England. It has become almost
traditional to regard the Anglican Evangelicals as a comparatively small group of Methodists who remained within
the Church of England while the others moved out. This is
an almost complete misconception. Anglican Evangelicalism
was a distinct movement, with its own marked characteristics, which have continued to be the characteristics of
the Evangelical wing of the Church of England till the
present day. Some of the Evangelicals had come under the
influence of Wesley, others had not; and on the whole they
spent a great deal more of their time in criticizing the great
man than in agreeing with him.

Evangelicals in the Church of England have never been a
party.[2] They have always been obstinate individualists –
this is their strength, and in part also their weakness. Like-
minded men have found one another out and coalesced into

1. This is the origin of the episcopal succession of the great Methodist
Church of America. For Coke and his subsequent attempts to re-unite
the Methodists with the Episcopal bodies, see D. N. B. s.v. 'Thomas Coke'.
    2. In spite of the title of G. R. Balleine's well-known book *A History of
the Evangelical Party in the Church of England* (1908).

groups; but the vigorous assertion of their own independence has always, and fortunately, made impossible the formation of a unified party organization. A surprising number of the early Evangelicals came to their peculiar tenets through some purely individual experience, related to the Word of God alone, and not to the word of any living man. Typical in this way was William Grimshaw (1708–63) of Haworth (later the home of the Brontës), who after his ordination and before his conversion, was so far respectable that 'when he got drunk, [he] would take care to sleep it out before he came home'. After years of darkness, illumination came to him through careful study of the Bible one night between a Saturday and Sunday in 1742. On the testimony of Wesley himself, 'All this time he was an entire stranger to the people called Methodists. . . . He was an entire stranger also to their writings.'

Somewhat similar was the experience of Charles Simeon (1759-1836) thirty-seven years later, as an undergraduate at King's College, Cambridge: 'In Passion week [1779] the thought came into my mind, What, may I transfer all my guilt to another? Has God provided an Offering for me that I may lay my Sins on His head? Then, God willing, I will not bear them on my own soul one moment longer. Accordingly I sought to lay my sins upon the sacred head of Jesus . . . on the Sunday morning, Easter Day, April 4, I awoke early with these words upon my heart and lips, "Jesus Christ is risen to-day! Hallelujah! Hallelujah!" From that hour peace flowed in rich abundance into my soul.'

In several points the Evangelicals differed from the Methodists.

In the first place, they on the whole rejected John Wesley's idea of itineration, and regarded the parish as the place where the work of the Lord was primarily to be carried on. Here we may once again cite the work of Grimshaw at Haworth, typical in its effectiveness, though exceptional in the vigour with which it was prosecuted. In the inimitable reminiscences of John Newton, the converted slaver (1725-1807), we read the following note on a Sunday at Haworth: 'A friend of mine passing a public-house on a Lord's Day saw several persons jumping out of the windows and over the wall. He feared the house was on fire, but upon inquiring what was the cause of the commotion he was told they saw the parson coming. They were more afraid of the parson than of a Justice of the Peace. His reproof was so authoritative and yet so mild and friendly that the stoutest sinner could not stand before him.' With typical Evangelical regard for the Holy Communion, Grimshaw introduced monthly Communion, a rarity in Hanoverian England. When questioned by the Archbishop of York in 1749 about his relations with the Methodists, he reported that, on coming to Haworth, he had found twelve communicants, but now 'in winter, between three and four hundred, according to the weather. In summer sometimes nearer twelve hundred.' The Archbishop concluded that there was no need to interfere with the work of such a man.

The Anglican Evangelicals were Calvinists, and the Methodists were Arminians. It has always been the weakness of Evangelicals to quarrel over trifles, and to vilify one another for alleged adherence to tenets which have never been either precisely defined or adequately understood. As always, the Calvinists supposed that the Arminians were in some way derogating from the honour due to God alone, and arrogating to man a greater share in procuring his own salvation than man can rightly claim. But, in point of fact, their own Calvinism in most cases amounted to no more than the doctrine, once again in the words of John Newton, 'which ascribes to God the whole of a sinner's salvation, from the first dawn of light, the first notion of spiritual life in the heart, to its full accomplishment in the victory over the last enemy', a doctrine to which John Wesley would whole-heartedly have subscribed. That the differences were minimal is made plain in the famous interview between Wesley and Charles Simeon on 30 October 1787, which was brought to a conclusion by Simeon saying, 'Then, sir, with your leave, I will put up my dagger again: for this is all my Calvinism ... and therefore, if you please, instead of searching out terms and phrases to be the ground of contention between us, we will cordially unite in those things wherein we agree.' The men of lesser rank were of a lower temper; they delighted to revile one another in the rival pages of the *Gospel Magazine* and the *Arminian Magazine*; and the cause of true religion was weakened and discredited.[1]

What, in point of fact, were the special doctrines held by the Evangelicals of the Church of England? The answer is that, then as now, they had no special doctrines; they were simply men who took seriously what they read in the Bible and Prayer Book. Sir James Stephen summarized their character by defining 'an orthodox clergyman as one who held in dull and barren formality the very same doctrines which the Evangelical clergyman held in cordial and prolific vitality; or by saying that they differed from each other as solemn triflers differ from the profoundly serious'.[2] Evangelicals were profoundly convinced of the sinfulness of man, of his inability to save himself, and of his need for a Saviour. At the end of the century, one of the greatest of the Evangelicals, William Wilberforce, in his *Practical View of the Prevailing Religious System* (1797), pointed out that 'the bulk of Christians are used to speak of man as being naturally pure, and inclined to all virtue'. 'Complaining that the nominal Christians had lost all knowledge of real Christianity, Wilberforce with sure instinct picked out the decay of the sense of sin as the basic cause.'[3] The Evangelicals presented God not as an Idea, but as the God who cares, the God who has acted in history, the God who in Jesus Christ has met man with provision for his every need. Man's response must be through obedience and holiness. It is impossible to understand these men at all unless it is realized that what they were primarily interested in was holiness. Often narrow in their theology and illiberal in their ideas, they kept always before their own eyes and the eyes of their hearers the ideal of the man of God, unspotted from

1. Rather surprisingly, one of the bitterest controversialists on the Calvinistic side was Augustus Toplady, the writer of the hymn 'Rock of Ages, cleft for me'.
2. *Essays in Ecclesiastical Biography*, Vol. II, p. 155.
3. R. N. Stromberg: *Religious Liberalism in Eighteenth-Century England* (1954), pp. 119–20.

the world, and in all his thoughts and actions serviceable to God and man. Any reference to feelings and emotions was likely to be deprecated and discouraged. John Newton is reported once to have said, 'Don't tell me of your feelings. A traveller would be glad of fine weather, but, if he be a man of business, he will go on.' Religion was to be not feeling but action, directed to the saving of souls and to practical help to men in their various needs.

It is important not to exaggerate the effects of the Evangelical Revival of the eighteenth century, whether in its Methodist or in its Anglican form. In 1790 it was reckoned that there were about 70,000 regular members of the Methodist movement, a comparatively small result when we consider the immense energy that had been put into the evangelistic campaigns of the Wesleys and their colleagues. Evangelical clergy were never more than a small minority in the ranks of the Established Church. But they were a leaven. So good a judge as Professor E. W. Watson reckons that by 1770 they already formed the dominant group in the Church, a position which they were destined to hold for almost exactly a century.

But, then as now, the great majority of English churchmen, whether clerical or lay, adhered to no group or party, and cannot be classified under any of the familiar labels. They were just 'Church of England', and nothing else. In any age the effectiveness of a Church must be judged in terms of the ideals and the practice of its ordinary members rather than as seen reflected in the achievements of its great men and its saints. But, just because the ordinary man is generally inarticulate, it is always extraordinarily difficult to make a fair assessment. The eighteenth-century Church has suffered from the accident of falling between the intensities of the seventeenth century and the widespread and varied revivals of the century that followed it. Less than justice has been done even to the modest merits to which it can lay claim.

There was certainly much in the Hanoverian Church which may rightly be criticized. Appointment to the episcopate was in many cases more a reward for political services

than a recognition of spiritual merit, and the scrambling and
jostling of those who regarded themselves as suitable candidates was far from edifying. The bishops had to spend far
too much time in London attending on their political duties.
Yet few of them were really bad men. They visited their
dioceses. They administered Confirmation to large numbers
of children, though often without the solemnity which we
should today regard as desirable.[1] They held ordinations,
and took steps to ensure at least a reasonable standard of
education in the candidates.

Then as now, every kind of clergyman was found in the
parishes, from the saint to the reprobate, with the majority
perhaps falling somewhere near the middle line between
these two extremes. Yet the services of the Church were
regularly read, and sermons preached, not perhaps always
to the edification of the hearers. In 1746 it was noted that in
London and Westminster the practice of saying Mattins
and Evensong daily was maintained in fifty-eight churches.
In most parishes, Holy Communion was celebrated only once
a quarter, though there were notable exceptions, eleven
churches in the London and Westminster areas observing
the Communion weekly in 1728. But, if the Sacrament was
celebrated only rarely, the number of communicants was
surprisingly large, and, proportionately to the population,
far larger than at the present day. Dr Norman Sykes collected statistics from many areas of the country, showing that
numbers at the regular celebrations were considerable, and
at the Easter season outstandingly large. His most remarkable
evidence is from the diocese of Bangor, commonly reckoned
to have been at that time one of the deadest and dullest in
the whole country. In the parish of Llanfechell in Anglesey,
with no more than ninety families, there were 230 communicants at Easter in 1734, and 217 in 1760.[2]

1. Norman Sykes (*Church and State in the Eighteenth Century*, 1934
pp. 429–36) gives detailed and astonishing figures. For instance, in July
1786 Bishop Ross of Exeter held thirteen confirmations in Cornwall, at
which 9,133 persons were confirmed. In the years 1953, 1954 and 1955,
confirmations in the diocese of Truro numbered respectively 1,516,
1,355, and 1,589.

2. N. Sykes, op. cit., pp. 251 ff.

In certain directions the eighteenth century proved a pioneer, though it did not always succeed in living up to its own best insights.

As we have seen, the truth of which educated man in this age was most firmly convinced was the benevolence of God, as manifested in the wonderful adaptation of His works in the creation to their ends, and in particular to the needs of men.[1] No better way of responding to the benevolence of God was open to men than that of himself becoming benevolent. The favourite form that this benevolence took was the foundation of Charity Schools. 'It is the most striking of the many social experiments of the age. To it was applied the new method of associated philanthropy, and the new device of joint-stock finance.'[2] So great was the enthusiasm evoked by these manifestations of philanthropy that Richard Steele wrote of them in 1712 that they were 'the greatest instances of public spirit the age had produced', and Joseph Addison a year later, 'I have always looked upon the institution of Charity Schools, which of late has so universally prevailed throughout the whole nation, as the glory of the age we live in'. But the age had not the grace of persistence in well-doing; and it has to be recognized with regret that as early as 1723 the first popular enthusiasm had been spent; and that, though there is a line of connexion between the Charity Schools and the great work of Hannah More and the Evangelicals at the end of the century, it was a rather thin and tenuous line.

Reference has already been made to the growth of the tendency, at the end of the seventeenth century, to form voluntary societies for religious and social purposes. Most of

1. It is interesting to observe the catastrophic effect on this conviction of the Lisbon earthquake of 1 November 1755, in which not less than 10,000 people lost their lives. Those who are old enough to remember the sinking of the *Titanic* will have no difficulty in realizing how hard the men of that age found it to reconcile such a tragic event with the idea of God's providential ordering of the world. See T. D. Kendrick: *The Lisbon Earthquake* (1956).

2. M. G. Jones: *The Charity School Movement in the XVIII Century* (1938), p. 3. It is to this excellent study that we owe most of our knowledge of this subject.

these were small and of ephemeral character. But it must not be forgotten that the period now under review saw the foundation of two great voluntary societies, which have survived to our own day, and are among the glories of the Anglican Communion.

The great man by whom the impulse was given was Thomas Bray (1656–1730),[1] who had for a time been the Bishop of London's Commissary in North America. Seriously impressed by the harm accruing to the Christian faith through the deep ignorance in which the majority of professing Christians were sunk, he drew up in 1696 *A General Plan of the Constitution of a Protestant Congregation or Society for Propagating Christian Knowledge.* Bray's original plans were too ambitious; but on 8 March 1699 *The Society for Promoting Christian Knowledge* was born: '*Whereas* the growth of vice and immorality is greatly owing to 'gross ignorance of the Christian religion, we whose names are underwritten do agree to meet together, as often as we can conveniently, to consult, (under the conduct of the Divine Providence and assistance) how we may be able by due and lawful methods to promote Christian knowledge.' The Society set itself to provide missionaries 'for the Plantations', a task soon handed over to the S.P.G., to publish and circulate books and Bibles, and to found and direct schools. Within a few years the work had grown to considerable proportions, and had been extended to the neglected areas of Wales, Ireland, and the highlands of Scotland.[2]

Not content with one great achievement Bray followed it up with another. In April 1701 he directed a petition to King William III, requesting the formation of a Body Politic and Corporate for the Propagation of the Gospel in Foreign Parts. The King was agreeable, and on 16 June 1701 a Charter was issued to Dr Bray and his associates. The S.P.C.K. was a voluntary society of individuals; the S.P.G. a regularly incorporated and chartered society, with the

1. On whom see the recent life by H. P. Thompson: *Thomas Bray*, (1954), esp. pp. 36–42 for the founding of the S.P.C.K.
2. The concern of the S.P.C.K. with missionary work in Asia will come before us in the following chapter.

support and authorization of both Church and State. The primary aim of the Society was to provide for an orthodox clergy to live in the British colonies overseas, and to care for the spiritual needs of the colonists; but the needs of 'the heathen' were not forgotten. In the first Annual Sermon of the Society, preached by the Dean of Lincoln, Dr Willis, these two aims were set forth in clear and unmistakable terms: 'The design is, in the first place, to settle the state of Religion as well as may be among our *own People* there, which by all accounts we have, very much wants their pious care; and then to proceed in the best methods they can towards the *Conversion* of the *Natives*. . . . This may be a great Charity to the souls of many of those *poor Natives* who may by this be converted from that state of *Barbarism* and *Idolatry* in which they now live, and be brought into the Sheepfold of our blessed Saviour.' Two hundred and seventy years later the S.P.G. is still faithful to the two great principles on which it was founded.

By the end of the eighteenth century the Church of England had become isolated from the rest of the Christian world to an extent which is in marked contrast with its earlier history[1]; but until the middle of the century it was still consciously a part of the world-wide *Una Sancta*, maintained its contacts in many and varied directions, and was engaged in some remarkable attempts to restore the unity of the divided Church.

One of the ecumenical leaders of the time was Henry Compton of London, 'the Protestant bishop'. Then as now, England was a haven of peace for persecuted refugees from the Continent. Compton's efforts on behalf of the French refugees were such as to earn for him an almost extravagant eulogy from a minister of their Church: 'You gather our wandering and frightened sheep; and you employ yourself with an indefatigable zeal to procure for them spiritual and bodily pasture; so that the faithful in foreign lands are equally the object of your paternal kindliness along with

---

1. I believe that an interesting study could be made of the extent to which the disappearance of Latin as the common language of all educated men contributed to this disintegration of the Christian world.

those of your own nation; thus we can say with truth that you are (in effect) bishop of the world.'[1] For some years Compton was carrying on a somewhat ineffectual correspondence, through one J. C. Werndley, Chaplain to the English Envoy to the Swiss Cantons, with leaders in the Swiss Churches. But this led to nothing, since the Swiss wanted a clear answer to the question whether this was an official approach, or merely a friendly personal discussion; and also, then as now, to the question: 'Why are such Priests as are ordained by the Roman Catholic Bishops admitted into the Sacred Offices in the Church of England, when the French and other foreign ministers are not admitted without a Re-ordination tho' they were ordained by a Protestant Synod?'[2]

The eighteenth century knew, however, a far greater ecumenical figure than Compton – William Wake, Bishop of Lincoln (1705–15) and Archbishop of Canterbury (1716–37). It is a sad illustration of time's revenges that, though Wake was perhaps the greatest churchman of his time, there is no account of him either in the *Encyclopaedia Britannica*[3] or in the German *Realenzyklopädie*.[4] Even to enumerate the directions in which Wake's interests reached out would take more space than can be allotted in this book.

He corresponded ardently with friends in Paris who maintained that Gallican tradition of loyalty to Rome combined with independence of it, which in some respects was not dissimilar to the Anglican tradition. Holding to the principle that a distinction must be made between

1. Quoted in E. F. Carpenter: *The Protestant Bishop* (1956), p. 374.
2. ibid., pp. 351–2. We are given one tantalizing glance and no more at one Cyprian Appia, a Piedmontese (presumably Waldensian) who had been apparently admitted to episcopal orders in England by the Bishop of London, and hoped to introduce the English liturgy in Italy – though some members of that Church – then as now! – felt it to be 'very like Popery' (ibid., p. 356). An Italian translation of the Prayer Book had been published in 1685.
3. Fourteenth edition. But there *is* an article on Wake in the eleventh edition!
4. Full justice has at last been done to Wake in the massive biography by Professor Norman Sykes (2 vols., 1957), which appeared in the month in which this chapter was written.

fundamental articles of the faith and those that can be regarded as secondary, and that much liberty must be granted to individual Churches, Wake believed 'that a plan might be framed to bring the Gallican Church to such a state that we might each hold a true Catholic unity and communion with one another, and yet each continue in many things to differ as we see the Protestant Churches do'.

Wake was in close contact with the leaders of the Swiss Churches, and in particular with three remarkable and openhearted men, Turrettini in Geneva, Ostervald in Neuchâtel, and Werenfels in Basel. He encouraged foreign students to come to the English universities, and to receive episcopal ordination before returning to their own countries. But little progress was made beyond friendly personal relationships and a better understanding between the Churches.

Wake had also to pick up the languishing remains of an ambitious project for the unification of the Lutheran and Reformed Churches in Prussia on the basis of Episcopacy, and for a union of this united Church with the Church of England. The moving spirit in this plan was Daniel Ernst Jablonski, who had studied in Oxford between 1680 and 1683, and had in 1699 become a bishop of the *Unitas Fratrum*, the ancient reformed Church of Bohemia. Naturally, with the prospect of the Hanoverian succession, contact between Britain and Germany was close. But the project became involved in the endless complexities of politics; gradually the fair prospects of union faded away, and this has to be added to the long list of schemes for Church union which in the end came to nothing. The one lasting result of it was the publication in 1704 of a German translation of the Book of Common Prayer – a rather ponderous translation, which has yet served in a measure as a link between the German Churches and the Churches of the English-speaking world.[1]

If these many schemes led to nothing in the way of practical and visible results, they at least served to keep alive in England the idea of the *Una Sancta*, and to make other

1. On all this, see especially N. Sykes: *Daniel Ernst Jablonski and the Church of England* (1950).

Churches aware of that spirit of comprehensiveness which has been the lasting glory of the Church of England. Wake served his day and generation well; and he had every right to claim that 'when that day shall come, in which I stand before the judgement-seat of Jesus Christ, this will be not the least part of my confidence and hope: that though I may have been otherwise an unprofitable servant, nevertheless I have ever sought, counselled, and with all my zeal and effort pursued those things which belong to the peace of Jerusalem'.[1]

When the best has been said for it that can be said, the eighteenth century remains spiritually a depressing period. The Gospel of reason did not avail to bring to men victory over their sins, or to inspire them with any ideal of heroic saintliness. The Church still left large sections of the population without any regular or effective ministry. Profligacy abounded in all sections of the community. The Evangelical revival had touched only the fringe of human need. As the century closed, with the strains and terrors of the Napoleonic wars and the first evil effects of the Industrial Revolution, the prospects for the future of the Christian religion may well have seemed to be dark indeed.

But this must not be the last word on the century. If asked the question, 'Who was the most typical Englishman of the eighteenth century?', most people would almost certainly answer 'Dr Johnson', and they would be right. Johnson was typical in his sincerity, his dislike of cant and hypocrisy, his rough commonsense; and in nothing was he more typical than in his vigorous, wholly unsentimental, practical Christian faith.[2] He was regular in his attendance at divine worship, and prepared himself with almost painful earnestness for participation in the Holy Communion. The prayers that he composed breathe a spirit of deep penitence for sin, of manly resolution, and of a tender confidence in the mercy

1. Quoted by N. Sykes in *A History of the Ecumenical Movement 1517–1948* (1954), p. 167.
2. I cannot subscribe to the view put forward by C. E. Vulliamy in his book *Ursa Major* (1946) that Johnson's religion was a neurotic factor in his make-up, and that he would have been better without it.

of God. All his life he had lived a prey to strange panic fears; but, as death approached, he was delivered into a peace such as he had not previously known. A few days before his death, which took place on 13 December 1784, having asked his doctor for a plain answer as to the possibility of his recovery, and having been told the truth, he replied: 'Then I will take no more physic, not even my opiates; for I have prayed that I may render up my soul to God unclouded.' On the testimony of the same physician, 'For some time before his death, all his fears were calmed and absorbed by the prevalence of his faith and his trust in the merits and *propitiation* of Jesus Christ. He talked to me often about the necessity of faith in the *sacrifice* of Jesus, as necessary beyond all good works whatever for the salvation of mankind.'

It is inconceivable that Samuel Johnson should have been anything but an Englishman; and the life of this great and good man shows that, even in her worst days, the Church of England was still, what she has ever been, *Ecclesia Anglicana mater sanctorum.*

CHAPTER 8

Anglicans Abroad

DURING the reign of Queen Elizabeth Englishmen developed that explosive energy which has carried them into every corner of the earth, habitable and uninhabitable, and with them the English language and the English Church.

When Sir Francis Drake set out on the famous voyage that was to end only with the circumnavigation of the globe, he was accompanied by his chaplain, Francis Fletcher. There are few scenes in English Church history more macabre than that of Drake, in the remote Patagonian port of St Julian, receiving the Holy Communion, and then dining with Thomas Doughty, whom he had just sentenced to be hanged for mutiny. On 17 June 1579, at a later stage of the same voyage, Drake made a landfall in a convenient harbour on the coast of California, in latitude 38° 30′, and stayed in that neighbourhood for a month. There he himself read from his Prayer Book, and led in the singing of the Psalms appointed for the day, thus holding what is believed to have been the first Anglican service ever held on the continent of North America.

Early Anglican expansion can be conveniently summarized under the not unattractive association of 'Gain and the Gospel'. English interests were propagated by companies of merchant adventurers; where these went, they took their chaplains with them, and in most countries secured the right to maintain worship according to the order of their own Church, though often in the face of considerable opposition and under rigid limitations.

One of the oldest Anglican congregations abroad is that of the ancient Hanseatic city of Hamburg. This was established in 1612, by permission of the Senate, in the teeth of the strenuous opposition of the Lutheran clergy of the city, who had no wish that their city of 'pure doctrine' should be polluted by the presence within it of this impure worship. Participation in Anglican worship was a privilege strictly limited to members of the Worshipful Company of the Merchant Adventurers of England and their families; but not unnaturally other British residents of Hamburg broke the law in order to attend the services with which they were familiar. Except for some interruptions in time of war, the English church in Hamburg has maintained the continuity of its life and worship, and flourishes today as one of the most remarkable examples in the world of cooperation between the Church of England and the Church of Scotland.

Probably the merchants of the Levant Company were rather prosaic and uninteresting people; but it is hard not to think romantically of them and of the chaplains who served them in the fabulous lands of the Middle East. We have already noted that a number of the Laudian exiles under the Commonwealth served as chaplains in this region. There seems never to have been a lack of chaplains, and a surprising proportion of them were men of outstanding distinction.[1]

The most famous of them all was Edward Pococke (1604–91), almost the first of the great succession of English orientalists. As chaplain at Aleppo Pococke devoted himself to the study of Arabic, of which he became a perfect master in both its classical and colloquial forms, and began that work of the collection of manuscripts in many oriental tongues in which he has hardly ever had a peer. In 1636 Laud, one of whose merits was his interest in oriental studies, founded a lectureship in Arabic at Oxford, and invited Pococke to become the lecturer. A further visit to the East (1637–40) was mainly spent in Constantinople, where Pococke enjoyed the friendship of the eminent and unfortunate patriarch Cyril Lucar.[2] The rest of his long life was passed in the promotion of oriental studies in England.

1. Biographies of the various chaplains referred to in this section will be found in the *Dictionary of National Biography*.
2. Best known in England for his gift to James I of the famous Alexandrine manuscript of the New Testament now in the British Museum.

One of his many achievements was the preparation of an Arabic version of the Book of Common Prayer, which was published in 1672.

Even more remarkable than Pococke, from the point of view of the dissemination of Anglican ideals in the East, was Isaac Basire (1607–76), rightly described by John Evelyn as 'that great traveller, or rather French apostle, who has been planting the Church of England in divers parts of the Levant and Asia'.[1] Basire was in the East for nearly fifteen years (1647–61), and throughout this period his main interest was the communication to Eastern churchmen and hierarchs of correct information about the Church of England, its worship and its order. 'It hath been my constant design,' he wrote, 'to dispose and incline the Greek Church to a communion with the Church of England, together with a canonical reformation of some grosser errors' – it is to be supposed that he did not unduly stress the second part of this purpose in his intercourse with his Orthodox friends. In 1652 he 'passed over the Euphrates and went into Mesopotamia, Abraham's country, whither I am intending to send our catechism in Turkish to some of their bishops'. For seven years he was Professor of Divinity in the University of Alba Julia (Weissenburg) in Transylvania, surely the oddest appointment ever held by an Anglican. Everywhere he showed himself friendly, inquisitive, and above all indomitably devoted to that Church of which he held 'what David said of Goliath's sword, "There is none like it" both for primitive doctrine, worship, discipline, and government'.

A briefer notice of three other chaplains must suffice. Robert Frampton (1622–1708) was chaplain at Aleppo from 1655 to 1667, where he learned Arabic, and lived on friendly terms with the leading Muslims as well as with the European community. After his return to England, having heard that the plague had broken out at Aleppo, he at once retraced his steps and spent a further three years ministering to the sufferers at considerable danger to his own life. In 1681

1. It seems probable, though it is not certain, that Basire really came from Jersey.

Frampton became Bishop of Gloucester, but was deprived in 1690 for his refusal to take the oath of allegiance to William and Mary.[1] John Covell (1635–1722) was chaplain in Constantinople between 1671 and 1676, and travelled widely before his return to England in 1679. His book, based on materials gathered during this period, *Some Account of the present Greek Church, with Reflections on their present Doctrine and Discipline, particularly on the Eucharist*, did not appear till 1722, and by that time English interest in the Eastern Churches was beginning to die down. Thomas Smith (1638–1710), a man of immense learning, was in Constantinople as chaplain to the ambassador for only three years, 1668–71; yet he remains one of our best sources of information with regard to the Churches of the Levant in the seventeenth century. His two books, *Remarks upon the Manners, Religion and Government of the Turks* (1678; originally in Latin, 1672) and *An Account of the Greek Church under Cyrillus Lucaris* (1676), were widely read, and directed the attention of many English churchmen to these ancient Churches which, like the English, claimed to be Catholic, but were free from the errors of popery.

One curious result of this renewed interest in the ancient Orthodox Churches was the project, under the episcopate of Henry Compton, to build a church in London 'for the nation of the Greeks'. The church was built in Crown Street, Soho, and dedicated in 1677. Bishop Compton intended that this church should remain under his own jurisdiction; and, in order to bring it to some extent into line with Anglican usages, laid down the following conditions for its use: that there should be no pictures or ikons in the church; that every priest officiating there should repudiate the doctrine of transubstantiation; and that there should be no prayers to the saints. When the news of all this reached Constantinople, the Patriarch naturally had his own stipulations to make, and asked whether the Greek Church in London could be placed under the general jurisdiction of the Oecumenical Patriarch. To this the

1. It is interesting to note that Queen Anne, at her accession, wished to 'translate' him to the See of Hereford.

British ambassador replied that this demand was 'extravagant and unreasonable, as they could not be exempt from his Majesty's jurisdiction nor would the Bishop of London be deprived of his rights'. After this, it is not surprising that patriarchal interest in the scheme died away.[1]

This, and similar abortive projects, showed that the English Church, though still intensely English, was very far from being insular; but it also reveals how very far indeed the separated Churches were from really understanding one another.

On 31 December 1600 Queen Elizabeth conferred a charter on 'The Governor and Company of Merchants of London, trading into the East Indies'. Nothing was laid down in the charter about the provision of religious ministrations, nor were any chaplains appointed until 1607. But from the beginning the godly merchants in London took account of religious concerns, and in their commission to their 'General' for each of the first seven voyages laid it down that 'for that religious government doth best bind men to perform their duties, it is principally to be cared for that prayers be said every morning and evening in every ship, and the whole company called thereunto with diligent eyes, that none be wanting'.

The first Anglican clergyman to reside in India appears to have been the Reverend John Hall, Fellow of Corpus Christi College, Oxford, and chaplain to that good and ecumenically-minded man, Sir Thomas Roe (1581–1644), the first British ambassador at the court of the Great Mogul. Hall survived only one year; and, on his death in 1616, the pious ambassador wrote to the Company's factors at Surat, 'Here I cannot live the life of an atheist. Let me desire you to endeavour me supply, for I cannot abide in this place destitute of the comfort of God's Word and heavenly Sacraments.' Edward Terry, also a fellow of Corpus Christi College, was already on his way to India, and on his arrival was sent up to Ajmere to join the ambassador. Terry was an excellent man, who lived to return to England, and to become a member of the Westminster Assembly of

1. Details in E. F. Carpenter: *The Protestant Bishop* (1956), pp. 357–64.

divines. More relevant to our purpose, he left behind one of the best of the early accounts of English residence in India.[1]

In 1614 the East India Company's factory was founded at Surat, then the leading port of western India. A factory in those days was a little like an Oxford or Cambridge college. No servant of the Company was allowed to take his wife with him to India. The men lived a communal life, with a common table; attendance at morning and evening prayers was insisted on, and penalties were imposed for swearing, brawling, drunkenness, or not returning to the factory at night before the gates were shut. But, like undergraduates, the younger members of the Company soon found that there were other ways into the factory than through the gates.

It may be imagined that it was difficult, in that distant land, to uphold English standards of decency and order; yet the records show that there was a constant effort on the part of the authorities to maintain Christian traditions of worship and conduct in all their establishments. In 1679 the new factory at Hughli was visited by the President of Madras, the redoubtable Streynsham Master, and strict rules were laid down for the conduct of all who resided in the Company's House. Absence from prayers, whether on Sundays or on weekdays, was to be punished by a fine of twelve pence for each offence; and 'if any by these penalties will not be reclaimed from their vices or any shall be found guilty of adultery, fornication, uncleanness or any such crimes or shall disturb the peace of the Factory by quarrelling or fighting, and will not be reclaimed, then they shall be sent to Fort St George, there to receive condigne punishment'.

The number of chaplains was never adequate to the need, and not all of those who were sent out were able to resist the temptations to idleness and dissipation by which they were perpetually surrounded. Their remuneration was very small; and, though they were forbidden to trade, comparatively few refrained from using the opportunity to enrich themselves by doing what everyone around them was

1. *A Voyage to East India* (London, 1655; reprinted 1777).

engaged in doing. Perhaps it is not unduly malicious to think that there is some connexion between a letter of 1692 concerning the Reverend John Evans – 'Mr Evans having betaken himself so entirely to merchandising, we are not willing to continue any furthur salary or allowance to him after the arrival of our two Ministers' – and the fact that in 1701 Evans was appointed to the bishopric of Bangor in his native Wales, which he exchanged in 1716 for the more profitable see of Meath in Ireland.

Divine service was not always conducted with seemly reverence. A note from Calcutta, early in the eighteenth century, reads as follows: 'Ministers of the Gospel being subject to mortality, very often young merchants are obliged to officiate, and have a salary of £50 a year, added to what the Company allows them, for their pains in reading Prayers and Sermons on Sundays.' It is to be hoped that those who chose the sermons to be read took as much care in the matter as Sir Roger de Coverley. There were, for a long period, no proper places for worship. The factories had no chapels; some room used for another purpose during the week was hurriedly put into some kind of order for the Sunday services. And there were no churches. It was a great event in the life of the British community in India when, by the efforts of the same Streynsham Master, St Mary's, the Old Church in the Fort in Madras, was completed, and solemnly consecrated on the Feast of St Simon and St Jude, 1680.[1]

So far we have spoken only of the care of the East India Company for its European employees; if nothing has been said of attempts to evangelize the Indians, that is because there is nothing to say. There was some interest in this subject in England, and one of those who took up the matter was Robert Boyle, whom we have already encountered as a pillar of the Royal Society. Bishop Fell of Oxford, writing to Archbishop Sancroft on 21 June 1681, reports a conversation that he had held with Boyle: 'It so happened that he

1. St. John's Church, Calcutta, was consecrated on 24 June 1687, much of the money required for the building having been raised by a lottery.

fell into the discourse of the East India Company, and I enlarged upon the shame that lay upon us, who had so great opportunities by our commerce in the East, that we had attempted nothing towards the conversion of the natives, when not only the Papists, but even the Hollanders had laboured herein.' Dr Prideaux, Dean of Norwich, warmly took up the matter, and when the Charter of the Company was next renewed, in 1698, the result of these efforts was seen. It was now laid down that 'All such Ministers shall be obliged to learn within one year after their arrival the Portuguese language, and shall apply themselves to learn the native language of the country where they shall reside, the better to enable them to instruct the Gentoos that shall be the servants or the slaves of the Company, or of their Agents, in the Protestant Religion.'[1] At the same time a beautiful prayer for the East India Company was authorized by the Archbishop of Canterbury and the Bishop of London (our old friend Henry Compton), including the words, 'That we adorning the Gospel of our Lord and Saviour in all things, these Indian nations among whom we dwell, beholding our good works, may be won over thereby to love our most holy religion, and to glorify Thee our Father, which art in heaven.'[2]

It is easy to make rules, harder to ensure that they are observed. There is no evidence that any chaplain ever set himself to evangelize the Gentoos. In the whole of the seventeenth century, we have record of only one baptism of an Indian according to the rites of the Church of England. In 1614 the Reverend Patrick Copeland brought back to England a Bengali boy, who was very 'apt to learn'. The Company therefore decided 'to have him kept here to school to be taught and instructed in religion, that hereafter being well grounded he might upon occasion be sent unto his country where God may be so pleased to make him an

1. Gentoos, apparently from the Portuguese *gentio*, the term generally used at that time for 'Hindu'.

2. At the request of the Company, a version of the Prayer Book had been published in 1695 in Portuguese, the *lingua franca* of the people of mixed Indian and European descent in the Indian dominions.

instrument in converting some of his nation'. Two years later the Archbishop of Canterbury was consulted as to whether the boy might suitably be baptized, and gave it as his considered opinion that he might. The event aroused intense interest. King James himself chose for the boy the baptismal name of Peter. On 22 December 1616 the Lord Mayor and Aldermen of London, with a very large crowd of less distinguished people, attended the church of St Dionis Backchurch, Fenchurch Street, where the baptism took place.[1]

If the English Church did not directly take part in missionary work in India, it was not long before it found means indirectly to set forward the great cause. The foundation of the S.P.G. aroused considerable interest on the continent of Europe, and the news of it stimulated King Frederick IV of Denmark to take in hand the evangelization of his Indian dominion of Tranquebar. As no Danes were found ready or equipped to go, he turned to the Franckes at the great German pietist centre of Halle, which for the next century was to supply the majority of the non-Roman missionaries in South India. The landing of the first pair, Bartholomew Ziegenbalg and Henry Plütschau, at Tranquebar on 6 July, 1706 is one of the great events in the history of the universal Church. The reports of the missionaries were translated into English and aroused wide interest. Gifts, including a printing-press, were sent to them, first through the S.P.G., and after 1710 through the S.P.C.K., which, as a private society, was less fettered in its operations than the officially chartered S.P.G.

A new phase in the work opened when the S.P.C.K. went beyond the supply of financial aid to a foreign mission and began to appoint missionaries in South India as its own agents. The King of Denmark felt himself responsible only for the souls of his subjects in the tiny Danish territory of Tranquebar. Inevitably the missionaries began to look to

1. Details in F. Penny: *The Church in Madras* (1904), pp. 14–16. Three letters of a later date from Peter have been printed. From these it appears that he had added to his Christian name the incongrous surname 'Pope'!

wider fields of work, and, as they did so, new arrangements
had to be made. The first missionary to come directly under
the S.P.C.K. was Benjamin Schultze, the (rather unsatisfactory) translator of large parts of the Old Testament into
Tamil. In 1726 Schultze removed himself from Tranquebar
to Madras, which from the start had been a British settlement, and wrote to the S.P.C.K., asking that permission to
reside there be obtained for him from the Directors of the
East India Company. Permission was granted, and in 1728
the S.P.C.K. appointed Schultze as its own agent, taking full
financial responsibility for his support. In 1730 the Directors
wrote to Madras for information as to the work and progress
of the Danish missions; and, on receiving the reply that the
Danish missionaries were quiet and modest, and had two
hundred converts, they noted, on 11 February 1732, that
'the behaviour of the Danish Missionaries being so agreeable
to their profession is pleasing to us; and we hope all in your
several Stations will give due countenance to their laudable
undertaking'.[1]

Christian Friedrich Schwartz (or Swartz, as he signed
himself in later years), the most distinguished of all Protestant missionaries in India, was for the greater part of his
career in the service of the S.P.C.K. He arrived at Tranquebar in 1750, and served in India without a break till his
death in 1798. The first twelve years of his service were
passed quietly in Tranquebar; but in 1762 he took up residence at Trichinopoly, the centre of one of the most
important British garrisons in South India. Here he was
instrumental in the building of Christ Church, which was
dedicated on 18 May 1766, and was persuaded to remain
there, combining the work of a missionary with that of a
chaplain to European troops and other residents. This
arrangement had the full approval of the authorities in

1. The term 'Danish missionaries' has led to endless confusion and
endless inaccuracies in the books of missionary history. The mission was
the Royal Danish Mission, and most of the missionaries were ordained by
Danish bishops. But with eight exceptions the missionaries were Germans,
and an increasing proportion of the financial support came from England. Gradually that part of the work which was wholly dependent on the
S.P.C.K. came to be known as 'The English Mission'.

Madras: 'There being at present at Trichinopoly a large body of Europeans for whom we have no Chaplain, it is agreed to request of Mr Schwartz, one of the Danish Missionaries, who has long resided in that part of the country, speaks English perfectly well, and bears a most unexceptionable character, to officiate at that garrison . . . and to allow him £100 per annum to be paid monthly by the Commissary-General.'

In 1776 Schwartz moved to Tanjore, where he gradually acquired a dominating influence on all classes of society from the Rajah downwards. Although much of his time was occupied in secular and diplomatic concerns, he never forgot his character as a missionary, and devoted the most minute and faithful attention to the instruction of his converts and the spiritual care of his Christians. Such was the veneration he inspired that, on his death, the Company, which at the same time was venomously persecuting the Baptist missionaries in Bengal, ordered the erection of a memorial to him in the Fort church at Madras, in which they record that 'He during a period of fifty years "Went about doing good", manifesting in respect of himself the most entire abstraction from temporal views. . . . In him religion appeared not with a gloomy aspect or forbidding mien, but with a graceful form and placid dignity.'

Altogether, during the eighteenth and early nineteenth centuries twenty-eight missionaries served under the S.P.C.K. in India, all, except two Danes and one Swede, being Germans.[1] Seventeen of them served also as chaplains to the Europeans, troops and civilians, and were paid in part by the East India Company. All had Lutheran orders only. In many cases they used the Prayer Book of the Church of England; indeed, the first translation of the Book of Common Prayer into Tamil was made by these missionaries. They baptized children and celebrated the Holy

1. The S.P.C.K. did attempt from time to time to secure English missionaries, and in 1729 were very nearly successful in the attempt. See L. W. Cowie: *Henry Newman: An American in London, 1708–43* (1956), pp. 104–31, where much important information drawn from the S.P.C.K. archives is to be found.

Communion according to the Anglican rite. When necessary, they ordained both Indian 'country priests' and younger missionaries. One of the Indians, Satthianadhan, had a distinguished career as the second founder of the Tinnevelly Church, and became known in England through the publication of the sermon that he had preached at his ordination on 26 December 1790. In 1787 Schwartz ordained to the ministry his foster-son, John Caspar Kohlhoff, who had been born in Tranquebar in 1762. Kohlhoff lived in Tanjore till 1844, and was in charge of the work there both under the S.P.C.K. and under the S.P.G., to which the mission was transferred in 1829. Although the first Bishop of Calcutta, Middleton, arrived in India in 1814, the question of Kohlhoff's re-ordination according to the Anglican rite seems never to have been raised. Kohlhoff was survived for a few years by the last of the 'country priests' in Lutheran orders, and by the last of the non-English and non-Anglican missionaries in the service of the Church Missionary Society.[1]

What is to be thought of this strange episode, in which an Anglican society in the highest standing made use for considerably more than a century of the services of men, not one of whom, according to the strictest Anglican standards, was validly ordained? Naturally the question was raised by anxious Anglicans. The general policy of the S.P.C.K. seems to have been as far as possible to avoid controversy on the subject; they 'thought it their prudence and charity to avoid as much as they could putting it into the hands of the benefactors that the missionaries were Lutherans or ministers not episcopally ordained. . . . They considered that . . . it is rather to be connived at that the heathen should be Lutheran Christians rather than no Christians.'[2] When the

---

1. The last of the country priests, Nallathambi, died in 1857; the last of the missionaries, P. P. Schaffter, on 15 December 1861. J. C. Kohlhoff's father, J. B. Kohlhoff, spent more than half a century in the service of the Royal Danish Mission in Tranquebar (1737–90). His son, John Christian Kohlhoff, who was in Anglican orders, served in South India till his death in 1881.

2. Henry Newman to Mr Chamberlayne quoted in Cowie, op. cit., p. 127.

missionaries, in 1733, held their first ordination of a 'country priest', the Society declared its entire satisfaction with their proceedings, and refused to indulge in 'a comparison of the validity of episcopal with foreign ordinations, which could yield no service to the mission but might do abundance of mischief'. There was no bishop in India. Indian candidates could not be brought to England. 'The sending over candidates for orders to Europe would be attended with insuperable difficulties', wrote Newman. If there were to be any Indian ministers of the Gospel at all, no other procedure was possible than that which had been adopted by the missionaries.

It appears that the S.P.C.K. never made any formal statement on the validity or otherwise of Lutheran ordinations. The scholar who in recent years has gone most thoroughly into the question tells us that 'another difficulty in the way of establishing the Society's attitude on this question is that its South India Mission has not yet been subjected to a modern scientific investigation, based on the English as well as German and Danish sources. Any such thorough investigation is in present circumstances obviously impossible.'[1] But he continues: 'The most natural interpretation seems to be that by using these Lutherans the Society did recognize the validity of their Lutheran orders. It is difficult to contest this interpretation, since it turns out to be the conception of that time, as far as it is possible to find any reference to it.'[2] Recognition by a voluntary society would not, of course, in any case be the same as official recognition by a Church. It is clear that, throughout the century, and particularly towards its close, there was a strong feeling that the tradition of episcopal ordination for both missionaries and Indian ministers ought to be restored

1. H. Cnattingius: *Bishops and Societies* (1952), pp. 41–2. Dr Cnattingius is referring to the fact that the early archives of the Mission are at Halle in Eastern Germany. But these archives are being extensively studied with most interesting results by Dr Arno Lehmann, formerly a missionary in India, and sometime Dean of the Faculty of Theology of the University of Halle. See also N. Sykes, *Old Priest and New Presbyter* (1956), pp. 154–67, for further important information.

2. Ibid., p. 43.

and perpetuated. The Society's Report for 1791 contains words which have a singularly modern ring: 'How long it may be in the power of the Society to maintain Missionaries . . . is beyond our calculation . . . we ought to have suffragan Bishops in the country, who might ordain Deacons and Priests, and secure a regular succession of truly apostolical Pastors, even if all communications with their parent Church should be annihilated.'

When we turn from India to America, the scene is strangely different, and strangely complicated.

We are so accustomed to thinking of the United States as one single great country, that it is hard to remember that until the separation from England in 1776 the colonies which had gradually grown up along the Atlantic seaboard were thirteen separate countries, distinct in origin, in constitution, in outlook, and in manner of life. Separated as they were by the great distances and by the difficulties of communication, it was by no means certain that they would ever coalesce into unity. Their diversity in religion was only a reflection of their diversity in every other respect. Many of the early colonists had crossed the Atlantic to escape from oppression and persecution in England, but the last thing they had in mind was the establishment of religious toleration. Massachusetts and Connecticut had established Churches, and the enjoyment of full civic rights depended on full membership in the Church – but the establishment was as far as could be conceived from the Anglican pattern. Maryland had been originally a Roman Catholic settlement, and its first ministers of religion were three Jesuit priests. Rhode Island won immortal glory by first establishing religious toleration as a principle of government. Pennsylvaia had been granted to William Penn the Quaker, but on the principle of liberty of opinion for all. In Virginia the Church of England was the established Church.

In Massachusetts, the home of the Independent stalwarts, the Anglican tradition naturally experienced great difficulty in finding a footing. In 1679 it was reported that there was not one episcopal clergyman in the whole of New England.

Although the connexion between Church and State was dissolved in 1691, the Anglican clergyman was not a welcome visitor, and the redoubtable Increase Mather (1639–1723; President of Harvard College) could still pontificate on 'the unlawfulness of the Common Prayer worship' – 'those broken responds and shreds of prayer which the priests and people toss between them like tennis balls'.

New York, the home of the Dutch, saw the foundation of its first Episcopal parish, Trinity, in 1697. Christ Church, Philadelphia, dates from 1695; Trinity Church, Newport, Rhode Island, from 1702; Burlington, the first church in New Jersey, from 1705.

North Carolina was isolated and backward. As late as 1730 it was reported that there was not a single Anglican minister in the colony, and in 1741 there were only two. Of Georgia, the scene of the early ministry of John and Charles Wesley, we read that the white colonists have 'very little more knowledge of a Saviour than the aboriginal natives'. In South Carolina things went better; but even there there were only three priests resident in 1706.

Naturally it was in Virginia that Anglican life was most vigorous and continuous. Jamestown was founded on 6 May 1607. On 20 June, the chaplain, Robert Hunt, 'an honest, religious and courageous divine,' celebrated the Holy Communion for the first time. 'We did hang an awning (which is an old sail) to three or four trees, till we cut planks our pulpit a bar of wood nailed to two neighbouring trees. . . . Yet we had Common Prayer morning and evening, every Sunday two sermons, and every three months the Holy Communion.' By 1671 the population had grown to forty thousand, and there were forty-eight parishes with organized vestries. But of the clergy it was said, 'Of all other commodities, so of this, the worst are sent us.' Later reports confirm this – the clergy were on the whole of poor quality, and tyrannized over by the laymen of their vestries.

The picture is not highly encouraging, but the difficulties under which the Church laboured must be fully borne in mind. Some of the priests had left England to escape from difficulties in which they had involved themselves at home. They lived under conditions of terrible isolation, in the midst of a population that had almost forgotten the existence of the Gospel under the brutalizing hardships of frontier life. Many of them had to struggle through a whole lifetime against desperate poverty. There was no ecclesiastical authority, other than the shadowy control of the Bishop of London – so shadowy that it was only in 1675 that even the clergy of Virginia were required to hold his licence.[1] Oftentimes the chaplains were at loggerheads with their parishioners, sometimes with unsympathetic Governors. Where the Church was established, the clergy lacked the spiritual independence necessary to their work; where the Church existed as a barely tolerated minority, they had difficulty in securing even so much as a place in which their congregations could meet.

Successive bishops of London regarded seriously their responsibility to care for the spiritual welfare of the colonies, and from time to time took the only step open to them – that of appointing commissaries to represent them with limited authority in those parts. The most famous of the commissaries was Thomas Bray, whom we have already met in connexion with the foundation of the S.P.C.K. and the S.P.G.; but Bray was in Maryland for little more than a year. Probably the most effective of the group was the Scot James Blair – 'that autocratic Scot' Mr Thompson calls him.[2] Blair, aged thirty, went to Virginia in 1685 at the suggestion of Bishop Henry Compton. Four years later he was appointed Bishop's Commissary, and held the office till his death in 1743 at the age of eighty-eight. The most spectacular of Blair's achievements was the foundation in 1693 of William and Mary College at Williamsburg, second in seniority to Harvard alone among American Colleges, and possessing as one of its chief glories the building

1. It was in 1685 that the East India Company first required that its chaplains should be licensed by the Bishop of London. The general responsibility of the Bishop of London for Anglicans abroad dates from the episcopate of William Laud.

2. H. P. Thompson: *Into All Lands* (1951), p. 58.

which was designed for it by Sir Christopher Wren. But Blair left the mark of his strong personality not only on the life of the Church, but also on the life of the colony as a whole. He fought successive Governors, brought clergymen out, largely from Scotland, defined the powers of vestries, and secured for his clergy a reasonable stipend from public funds.

If the history of the American Church in the eighteenth century is far brighter than in the seventeenth, that is due in the main to one cause and to one only – the foundation of the S.P.G., and the tender and unremitting care which the S.P.G. directed to all the needs of the colonies. In the course of the century the society sent 310 missionaries to America. There were some failures; but in general the calibre of the men was good; they laboured with great faithfulness in what were still very difficult conditions, and some of them set examples of truly apostolic zeal.

It is possible to select only one or two illustrations of the kind of work done by these good men, and of the many difficulties that they had to encounter.

One of the outstanding missionaries was Clement Hall of North Carolina, who, after a number of years of voluntary service as a layman, presented himself for ordination in 1743. In his report for 1750 he notes that 'In Easter-week I set off and journeyed about 427 miles through my south Mission, and in about 30 days preached 19 sermons, baptized about 425 white, and 47 black children, 3 white and 11 black adults, whom upon examination, I found worthy, and administered the holy Sacrament of the Lord's Supper to about 325 Communicants. The congregations were numerous, notwithstanding that many came very far, and some of the days there were continual rains. The people generally behaved very devout and orderly.'[1]

This quotation draws attention to one of the main concerns and problems of the S.P.G. in the eighteenth century – the welfare of the Negro slaves. In 1725 the Society sent out a strongly worded circular to its missionaries: 'It has been intimated to the Society that proper care hath not been taken to instruct in the Christian religion and baptize the

1. Quoted in H. P. Thompson, op. cit., p. 57.

Negroes in the plantations in America. . . . You will please, Sir, to take notice of this particular direction of the Society and also encourage and advise your parishioners who may have Negroes to let them be instructed and baptized.' Conditions varied from colony to colony, but everywhere there were two inescapable difficulties. Slaves could hardly be baptized without the consent of their owners. And how could Christian marriage of slaves be solemnized, when the masters had absolute right of sale, and at any moment and at their unchallenged will could sell husband and wife in different directions and break up the home? For all this, some progress was made, and Negroes in fair numbers became Christians. But the Episcopal Church has never had outstanding success with its Negro work, and has rarely appointed Negro bishops. The election of the Venerable John M. Burgess, in 1962, to be Suffragan Bishop of Massachusetts and of John T. Walker to be Suffragan Bishop of Washington, were promising steps in the right direction.

One offshoot of this work for Negroes deserves particular mention as the sole attempt by the Anglican Churches to carry on missionary work in Africa in the eighteenth century. In 1745 the Reverend Thomas Thompson, Fellow and Senior Dean of Christ's College, Cambridge, left his comfortable and estimable university life in order to become a missionary in New Jersey. In 1751 he wrote to the S.P.G., asking whether he could be appointed to 'a Mission to the Coast of Guiney, that I might go to make a Trial with the Natives, and see what hopes there would be of introducing among them the Christian religion'. The Society agreed, and on 13 May 1752 Thompson reached Cape Coast Castle. The work proved more difficult than he had expected, and with the Africans, whose language he did not know and could hardly find means to learn, he had scarcely any success. After less than four years on the Coast, he returned to England.[1]

Thompson seemed to have failed; but one remarkable

1. In 1937 the S.P.G. published a facsimile of Thompson's book *Two Missionary Voyages* (1758), one of the most fascinating of missionary records.

success remains to be recorded. In 1754 Thompson had sent three African boys to be educated in England. After eleven years one of them, Philip Quaque, was ordained, the first Anglican minister from among the peoples of Africa, and sent back to be a missionary to his own people. There were difficulties. In England Quaque had entirely forgotten his own language – an early warning of the dangers inherent in Westernization of the leaders in the younger Churches. He had to contend with indifference on the part of the Africans and opposition from the Europeans. But he held on faithfully for more than fifty years until his death in 1816, a solitary light in what was then a very dark land.

One of the most urgent needs of the Church west of the Atlantic was the provision of a regular supply of candidates for the ministry. Perhaps the first to see that this need could not be met by the occasional sending of missionaries from England, and that the care of the American Churches must rest upon America's own sons, was George Berkeley (1685–1753), the great philosopher, who later became Bishop of Cloyne in Ireland. He wrote in 1725 a tract entitled *Proposal for the Better Supplying of Churches in our Foreign Plantations and for Converting the Savage Americans to Christianity by a College to be created . . . in Bermuda.* Unlike most theorists, Berkeley was prepared to put his theories to the test by practical action, and wished to abandon his promising career in the Church of Ireland in order himself to undertake the direction of the college in Bermuda. As a first step he crossed to America in 1728, and took up residence in Newport, Rhode Island, where memories of him still remain, and some delightful traces of which are to be found in the *Alciphron*, the book with the composition of which he beguiled his enforced leisure. The promised money for the college was not forthcoming; delays were endless. Finally, in 1731, Berkeley returned to Ireland and philosophy, and no more was heard of the scheme. It is doubtful whether candidates would have been forthcoming, and whether Bermuda would have proved a satisfactory centre. But in this, as in so many other things, Berkeley was a man whose prophetic genius ranged far beyond the general concepts of his age.

The S.P.G. was not concerned only with the colonies on the mainland of America. One of its first acts was to make a grant to the single impoverished clergyman then living on the desolate coast of Newfoundland. It was engaged also in the supply of ministers to the increasingly prosperous West Indian colonies, of which Jamaica was the largest and most prosperous. In most of these colonies the Church of England was established, and adequate stipends were theoretically provided from government funds.[1] The difficulty here, as everywhere else, lay in the provision of a sufficient number of ministers, and of the spiritual supervision which was needed if their work was to be effective.

It was in this region that the S.P.G. was faced by one of the most remarkable and challenging opportunities that have ever come its way. In 1710 General Christopher Codrington, at one time Governor of the Leeward Isles, died, leaving to the Society his estates in Barbados, with a view to the formation of a college, with 'a convenient number of Professors and Scholars maintained there . . . who shall be obliged to study and practice Phisick and Chirurgery as well as Divinity, that by the apparent usefullness of the former to all mankind they may both endear themselves to the people and have the better opportunities of doing good to mens' souls whilst they are taking care of their bodys'. There were many delays in the fulfilment of Codrington's intentions. In 1745 the institution was opened as a school with about fifty students. In 1760 Richard Harris, one of the first students to be admitted, was ordained to the diaconate, the first of a great series of West Indian ministers, which continues to the present day.

The bitter paradox through two centuries was that an episcopal Church was endeavouring to develop great Churches west of the Atlantic without the introduction of episcopacy. This meant that Confirmation could not be

1. In Barbados the Church is still in a measure established; this is perhaps the only part of the world (except, oddly enough, Belgium) where the salaries of Anglican clergy, other than chaplains to troops and prisons, are provided by the State.

administered. No commissary, whatever his legal position, and this in many cases was doubtful or was questioned, could exercise the spiritual authority which attaches to the episcopate alone. Any American who wished to be ordained had to cross the Atlantic in both directions at a cost which it was very unlikely that he would be able to afford. In 1760 a convention of clergy in New York reported that 'not less than One out of five who have gone home for holy orders from the Northern Colonies have perished in the attempt, Ten having miscarried out of Fifty-one'.[1] The arguments in favour of bishops for America were overwhelming. And yet nothing was done.

It was not for lack of trying. As early as 1638 Archbishop Laud had taken up the question of a bishop for Virginia. After the Restoration in 1660, Dr Alexander Murray, who had been with Charles II in exile, was actually nominated Bishop of Virginia, but was never consecrated or sent out to the work. A generation later, it was rumoured that Jonathan Swift (1667–1745), cheated of his desire to be a bishop in Ireland, was to be consoled by becoming the first bishop in America. From the date of its foundation the S.P.G. had ardently taken up the question, and in 1712 went so far as to draft a Bill for 'the establishment of bishops in America'; the year before, they had actually bought a house for a bishop in Burlington, New Jersey, 'in the sweetest situation in the world'. In 1728 Bishop Edmund Gibson of London invited the clergy of Maryland to nominate one of their number whom he might consecrate as suffragan. This time the difficulty was from the American side; the Government of Maryland would not permit the chosen candidate to proceed to England, and so nothing could be done.

It is important to note where the real difficulty lay. If a bishop was to be appointed for America, the question must necessarily be answered as to what kind of a bishop he was to be. And that raised the question as to the significance of the office of a bishop in the Church of God. The question of an episcopate for America reveals plainly how far the

1. Quoted in H. P. Thompson, op. cit., p. 98.

Church of England was from having cast off the pernicious medieval heritage. The bishop was still an officer of state, a member of the House of Lords; he still had certain powers of coercive jurisdiction. He was not so much a Father in God as a judge, entrusted primarily with the task of maintaining discipline and repressing disorder. At this very time the great Bishop Thomas Wilson, Bishop of Sodor and Man from 1697 to 1755, was exercising almost dictatorial authority in the Isle of Man. Was this the kind of bishop that was to be provided for America?

Some of the difficulties at the English end were legal. Owing to the state connexion no consecration could be carried out without a mandate from the sovereign. In what terms was the mandate to be framed? There were no precedents. This comes clearly in the questions addressed in 1704 by the S.P.G. to the law officers of the Crown: '(2) Whether the Archbishops and Bishops of the Realm would be liable to any inconvenience or penalties from the Statutes or Ecclesiastical Laws should they consecrate Bishops for foreign parts endowed with no other jurisdiction but that of Commissary or the like? (3) Whether by the Act of Edward VI for the election of bishops, the Queen might not appoint suffragans for foreign parts within her dominions?' It was felt that the appointment of suffragans would avoid many difficulties, since it would not involve creating any new and independent jurisdiction.

Strong opposition to the creation of an American episcopate was to be expected from the Colonies themselves.

The ghost of Archbishop Laud still walked. The ancestors of many New Englanders had left England to find refuge from episcopal persecution of dissenters. They had no wish to find their dissenting Canaan invaded by episcopal Hittites or Assyrians. 'They feared that their Lordships would come endowed, as in England, with political as well as spiritual authority. New Englanders lost valuable sleep envisioning bishops' palaces arising on the sacred Yankee soil, Church courts to try Congregational "heretics" (thus reversing an old New England custom), a greedy clergy with control over marriages and wills, and even mitred bishops dominating the colonial assemblies.'[1] The *New York Examiner* for 1769 published a wonderful cartoon, depicting the frustration by loyal Americans of an attempt to land an English bishop on American shores; the bishop, arrayed in full robes, is seen in a little ship, clinging to the rigging, and remarking piously, 'Lord, now lettest thou thy servant depart in peace', while a mob hastens his departure with yells and clenched fists, and one among them flings after him a book appropriately entitled, 'The Works of Calvin'.[2]

But it was not only among dissenters that opposition was to be feared. It was not at all evident that the Anglican vestries in Virginia would welcome closer episcopal super-vision. They had become accustomed to managing their own affairs, even to the extent of excluding the rectors from membership of the vestry and from attendance at its meetings. They had carried on a running warfare with com-missaries, not altogether without success. It was not certain that they would welcome the necessity of engaging in conflict with a more formidable antagonist than a com-missary.

So matters rested until the American Revolution burst upon the Church. Many of the episcopalian clergy were loyalists, and were expelled from their churches for refusal to omit the prayers for the royal family. Thousands of lay people fled for refuge to Nova Scotia. The loss to the United States was Canada's gain, and it is really in this period that the history of the Anglican Church in Canada can be said to begin. But the American Church was gravely weakened, and its situation was precarious. It was at this point that many of the objections previously felt to episcopacy fell to the ground. Bishops in an independent America could not possibly have any political power; it became possible, as it was not yet possible in England, to recognize the office of bishop as a purely spiritual ministry within the Church.

1. D. H. Yoder, in *A History of the Ecumenical Movement 1517–1948* (1954), p. 231.
2. This can be seen as Plate xviii in Anson Phelps Stokes's monu-mental *Church and State in the United States*, Vol. 1 (1950).

The initiative came from the clergy of Connecticut, where the Church was less disorganized than in other regions. Early in 1783, ten priests met and chose as their candidate one of their own number, Samuel Seabury (1729–96), whom they despatched to London to seek consecration at the hands of the Archbishop of Canterbury. There was at that time no 'diocese' of Connecticut, and there could be no question of Seabury exercising any jurisdiction outside the State in which he had been called to the episcopate; even within that State he could exercise authority only over those ministers who voluntarily accepted that authority. The pattern of the American Church remained undefined, and all the details of its organization had still to be worked out.

In London the perplexity of the Archbishop of Canterbury was considerable. Seabury could not take the oath of allegiance to the House of Hanover, and the Archbishop could not legally consecrate anyone who could not take that oath. In any case, was the vote of ten priests in one of the American States sufficient ground on which he could proceed to a consecration? A way out of the difficulty was found by recourse to the persecuted Episcopal Church in Scotland.

Up till 1689 there had been only one Church in Scotland. At times it had been Presbyterian, at times Episcopal, at times a mixture of both. But in 1689 the Church of Scotland became, and has remained ever since, Presbyterian, and the Episcopalians became a dissenting minority. Most of the Episcopalians were loyal to the House of Stuart, and therefore failed to win the favour of William III. After the rebellion of 1745, because of its real or supposed complicity in the rebellion, the Episcopal Church was subjected to a period of bitter persecution; the penal laws were such as to imperil its very existence, and the organization of the Church in dioceses and parishes almost ceased to exist. But, as so often, persecution proved a blessing in disguise. The episcopal succession had been maintained, and, just because they were deprived of all political influence and wordly statue, the Scottish bishops had developed a far more apostolic understanding of the episcopate than was to be found in England, Wales, or Ireland. When the request of Seabury for consecration reached them, they agreed to act.

So it came about that, on 14 November 1784, in an upper room of Bishop Skinner's house in Aberdeen, three bishops of this persecuted Church, Robert Kilgour the Primus,[1] Arthur Petrie, and John Skinner, following apostolic order, raised to the episcopate the first Anglican bishop ever appointed to minister outside the British Isles.

Not long, after the Church of England took steps to enlarge its own episcopate and to regularize its operations overseas. In 1786 an Act was passed 'to empower the Archbishop of Canterbury, or the Archbishop of York, for the time being, to consecrate to the Office of a Bishop, Persons being Subjects or Citizens of countries out of His Majesty's Dominions'. This could still be done only by royal licence, but it could be done 'without requiring them to take the oaths of allegiance and supremacy, and the oath of due obedience to the Archbishop for the time being'. In the following year, on 4 February 1787, William White was consecrated as Bishop of Pennsylvania, and Samuel Provoost as Bishop of New York.[2] The special significance of the occasion was marked by the participation, most unusually, of the Archbishop of York as well as the Archbishop of Canterbury in the act of consecration.

The organization of the American Church still presented countless problems. Not all the clergy were prepared to accept the validity of the consecration of Seabury, which to their minds had been carried out in an irregular fashion. This was put right in 1789 by a vote of the newly formed General Convention of the Church. Furthermore, in 1790, James Madison was consecrated at Lambeth as Bishop of Virginia, and America had three bishops of the English, as well as one of the Scottish, succession. From this time on

1. This is the title which is still held by the chief bishop of the Episcopal Church in Scotland.

2. On 12 August 1787 Charles Inglis was consecrated Bishop of Nova Scotia; this was the first consecration of a bishop for the work of the Church of England abroad.

bishops for the American Church were consecrated in America without further dependence on England.[1]

This was the beginning of that world-wide expansion of the Anglican episcopate which was to be one of the most remarkable features of the Church history of the nineteenth century. If bishops of the Anglican Communion meet at Lambeth in 1978, less than a quarter of them will be bishops of dioceses in the British Isles.

1. The Anglican Churches, in this as in so many things far stricter than the Roman, have not once since the Reformation made an exception to the rule that a minimum of three bishops is required for the regular consecration of a bishop. In 1963 the Archbishop of West Africa was faced with the question as to whether he should proceed to a consecration with only two bishops present, but luckily the third bishop turned up in the nick of time, and the consecration could proceed in the normal fashion. It is interesting to note that William White had suggested to the Archbishop the possibility of presbyteral consecration of bishops as an interim measure, until regular consecration could be secured.

CHAPTER 9

The Nineteenth Century in England: 1

It is not easy for the historian to determine at what point the nineteenth century is to be regarded as having begun. A good case can be made out for taking as the division between the epochs the outbreak of the French Revolution in 1789; that was the beginning of a convulsion in which many ancient things, including the unquestioned recognition of privilege, were for ever destroyed, and many new things were born. But equally the year 1815 and the ending of the Napoleonic period marked a turning-point; in some ways we are nearer to 1815 than 1815 was to 1789. From the point of view of ecclesiastical history, however, the question is not one of any great importance. The period of the struggle with Napoleon was, in fact, a kind of parenthesis; so much of the energy of the British people had to be directed to the conduct of the war that it was hardly to be expected that great events would take place in other spheres; and, though within that period one or two happenings of considerable importance for the future of the Church can be recorded, what we mainly notice is the slow development of tendencies that were to find fuller and more open expression with the return of the days of peace.

The British emerged from the Napoleonic wars exhausted, but with a sense of mission and achievement. The war had been a war for national survival; but increasingly, as the years passed, it was seen to be a war between two opposing principles, principles which without too great exaggeration could be identified with right and wrong. 'The England of Pitt and Wellington,' writes Sir Arthur Bryant, 'rested on a juster and more enduring base than the France of Robespierre and Napoleon. . . . For in the light of our own apocalyptic experience, we can see that Britain's supreme asset was the inner respect of her people for moral law. . . . They waged war in conformity with the dictates of the individual conscience and individual commonsense. Theirs was that saving humility and wisdom called by the Hebrew seers "the fear of the Lord" and by the old Greeks "justice".[1]

The period from 1815 to 1914 constitutes a remarkable unity in national history. In that period Britain turned aside from Europe, and became isolated from the general currents of European thought and history, as never before or since. Only once, in the Crimea, did Britain become engaged in war on the Continent; and this was war in a remote area, by which the general life of the nation was not seriously affected. Britain became simultaneously insular and imperial – partly absorbed in the immense problems of her own interior development, partly occupied, somewhat inadvertently, in the acquisition of a world-wide empire. It was an age of apparently inexhaustible vigour and self-confidence. It was also an age of outstanding men and women in every department of national, intellectual, and religious life.

This was the great century of political experiment. At the beginning of the period England was governed by a close-knit oligarchy; by the end, it was experiencing the advantages and the perils of an advanced form of democracy. Overseas the miraculous discovery had been made that starved and dependent colonies could be transformed into independent and self-governing nations, within a world-wide commonwealth of nations bound together by mutual respect and loyalty; a discovery that was to have profound effects on the development of the Anglican Communion.

In the application of power to production, in invention and the application of science to industry, Britain led the world. The emperor Napoleon had not been able to travel any more rapidly than the emperor Augustus. Now speed was to abolish distance, and mechanical power to produce what seemed to be unlimited wealth. Unhappily the instruments of production fell at the start into the hands of hard and unscrupulous men, and the resulting suffering was terrible. But it is a mistake to imagine that the nation was insensitive to these evils. For one Victorian who was com-

1. Arthur Bryant: *The Years of Endurance 1793–1802* (1942), pp. 358–9.

placent, it is possible to point to three or four who were seriously disturbed by the horrors that resulted from the Industrial Revolution. The century is marked by an increasingly sensitive social conscience, and by a steady determination to see to it that, with the increase of the national wealth, the distribution of that wealth to all classes should become more equitable.

Science was directed, as never before, towards the easing of human suffering. The introduction and general use of anaesthetics must be regarded as one of the major revolutions of history. Sensible methods of nursing, and victory over an ever-increasing range of diseases, have increased the span of human life and immensely increased the sum-total of happiness; though the consequent problem of a rapidly growing population has taxed all the ingenuity of both State and Church to meet it.

The nineteenth century was an age of extraordinary intellectual activity. The Victorian Age in literature is second only to the Elizabethan, and a list of the books published in 1859 goes far to justify the claim that this was the most remarkable year in the history of English literature.[1] The ancient universities recovered their past glory as homes of profound and adventurous scholarship. In the study of the classics, of the oriental languages, of history, of law, England was second to no other country in the world.

A great deal of the abounding vigour of the English people went into their religion. The great revival, which had been steadily proceeding since the early years of the eighteenth century, had by the beginning of the nineteenth worked its way into wide areas of the nation's life. Where the Church of England had failed, the dissenters and the Methodists had been active. 'The nineteenth century was a very religious century, and if we are ever to attempt to understand nineteenth-century men and women and their doings, we must consider carefully not only their religious controversies, but their great preachers and their pious

1. The list includes such very different masterpieces as Charles Darwin's *Origin of Species*, George Eliot's *Adam Bede*, and George Meredith's *Ordeal of Richard Feverel*.

literature.'[1] As in other fields, so in the field of religion, this was a period extraordinarily rich in outstanding men and women.

We must sedulously avoid the temptation to simplify what was in reality a very complex scene. It will be convenient once again to work out a kind of spectrum of Anglican life and thought; and, as in the rainbow, we shall find it necessary to distinguish not less than seven colours. But, even when we have attempted to identify all these various strands in the Anglican complex, it will still remain true that the majority of Anglicans, then as now, cannot be identified wholly with any one of these tendencies. And such is the wealth of outstanding personalities that, in order to avoid overcrowding this chapter with names, it will be necessary to pass over without mention many men and women who exercised a really great influence in their own day and generation.

We may start our survey with the old-fashioned high church party (later somewhat unkindly called the 'high and dry'). As we have seen, in origin the terms 'high' and 'low' marked rather a political than a religious distinction. At the beginning of the nineteenth century there was hardly any difference between the high churchman and the low churchman as regards dogma or ritual. The characteristics of the high churchman were usually rigid orthodoxy in doctrine, strenuous assertion of the rights of the Church of England as the national Church, a strong dislike of dissenters, and an inflexible opposition to every kind of reform, whether in Church or State. In fiction the perfect high churchman is Trollope's Archdeacon Grantly: in life he was Henry Phillpotts (1778–1869), Bishop of Exeter for nearly forty years, to whom Queen Victoria once politely referred as 'that fiend, the Bishop of Exeter'.

To make the acquaintance of this pugnacious prelate, we cannot do better than look at Trollope's pen-portrait of him as the Archdeacon's second son, Henry: 'He was also sent on a tour into Devonshire. . . . Henry could box well and would never own himself beat; other boys would fight while

1. G. Kitson Clark: *The English Inheritance* (1950), p. 11.

they had a leg to stand on, but he would fight with no leg at all. . . . His relations could not but admire his pluck, but they sometimes were forced to regret that he was inclined to be a bully.'[1] Phillpotts was constantly engaged in lawsuits in the defence of what he regarded as his rights and the maintenance of the discipline of the Church. He nearly split the Church of England by the violence of his dealings with the Reverend Charles Gorham, whom he believed to hold unsound views on baptismal regeneration.[2] Yet even Phillpotts could speak kindly of Methodists (whom he regarded as separatists rather than as dissenters): 'Would to God that the narrow partition which divides them from us could be broken down! that . . . all who look for salvation *solely* to the Cross of our . . . Divine Redeemer would unite in one holy bond of fellowship. . . . If we are to be separated in worship, let us not be separated in feeling and in affection.'[3]

For the type of the old-fashioned low churchman, who would have been bitterly insulted if he had ever been mistaken for an Evangelical, we may turn to Sydney Smith (1771–1845), who was admitted Canon of St Paul's in 1831. Sydney Smith was well known as a wit, and had been one of the founders of the famous, or notorious, *Edinburgh Review*. But he was more than a wit and a pamphleteer. Canon Prestige has written of 'his intellectual honesty and passion for social justice, his independence of mind, his magnificently lucid prose, his exuberent imagination and antiseptic irony, his conquering though somewhat hard and imperious common sense, and his devastatingly respectful

1. *The Warden* (1855), Chapter VIII. The Archdeacon's two other sons, Charles James and Samuel, are less successful pictures of Bishop Blomfield (1786-1857) of Chester and of London, to whom we shall come later, and Bishop Wilberforce (1805–73) of Oxford and of Winchester.

2. The classic study of the issues involved is J. B. Mozley: *The Baptismal Controversy* (1862); a more recent account is J. C. S. Nias: *Gorham and the Bishop of Exeter* (1951). One result of this fracas was that H. E. Manning (1808–92), later Cardinal, decided to leave the Church of England.

3. Quoted in G. C. B. Davies: *Henry Phillpotts, Bishop of Exeter, 1778-1869* (1954), pp. 148-9.

disrespect of persons'.[1] Unlike Phillpotts, the vehemence of whose speech in the House of Lords against the Reform Bill of 1832 was a contemporary scandal, Sydney Smith the Whig was in favour of reform everywhere except in the Church. In 1826, he was championing the unpopular cause of Roman Catholic Emancipation; and gave to the British public wise advice, which they would have done well to heed on this and many another occasion: 'My firm belief is that England will be compelled to grant ignominiously what she now refuses haughtily. . . . *If you think the thing must be done at some time or another, do it when you are calm and powerful and when you need not do it.*'

The indelible blot on the fame of Sydney Smith is the cruel and cynical contempt which he poured out upon the noble Baptist missionaries, William Carey (1759–1831) and his colleagues, who were sweating and stewing in the steam-bath of Bengal while the Canon of St Paul's was enjoying his comfortable pluralities in England. 'If a tinker is a devout man he is infallibly sent off for the East, benefiting us much more by his absence than the Hindus by his advice', was one of the kinder of his remarks. Even now his words cannot be read without indignant fury. It was this failure to realize the significance of missionary work that was ever the sharp barrier between the low churchman and the Evangelical.

To the Evangelicals we will now turn as the third colour in our spectrum.

In 1815, as forty years earlier, the Evangelicals were the most powerful group in the Church of England, and their influence continued steadily to increase until about 1870.[2]

1. *St Paul's in its Glory 1831–1911* (1955), p. 1. Lord David Cecil refers to him as 'Sydney Smith, most humane of clergymen, crackling away like a genial bonfire of jokes and good sense and uproarious laughter', *Melbourne* (Reprint Society ed., 1955), p. 32.

2. It is possible to cite in support of this view no less an authority than Cardinal Newman, who in the Introduction to the French translation of his *Apologia pro Vita Sua* affirmed that in the middle 1860s the Evangelicals were still the strongest group in the Church of England. We may· quote also R. C. K. Ensor, *England 1870–1914* (1936), p. 140: 'By 1870 the religion which we have been describing had attained its maximum influence in England'; pp. 136–40 of this book give an excellent account of 'Evangelicalism' in the broad sense of the term.

But to say that they were powerful does not mean that they were popular. Evangelicals were still a disliked and derided sect, making their way everywhere in face of persecution. It was almost impossible for any clergyman known as an Evangelical to secure any kind of preferment in the Church. Henry Ryder did, indeed, become Bishop of Gloucester in 1815, dislike of his Evangelical proclivities being outweighed by the advantage that he happened to be the younger son of a peer.[1] It was not until Lord Palmerston, during his almost continuous tenure of the Premiership from 1855 till 1865, handed over the control of his ecclesiastical patronage to that doughty Evangelical layman Lord Shaftesbury that Evangelicals could hope for that measure of public recognition which was their due.[2] If Evangelicals everywhere made headway, this was the result primarily of devoted and self-sacrificing pastoral zeal.

The most famous of the Evangelical leaders was Charles Simeon (1759–1836). In all his long life Simeon never enjoyed any preferment other than his fellowship at King's College, and the small vicarage of Holy Trinity, Cambridge. But, as he thundered from his pulpit – 'Mama, what is the gentleman in a passion about?' inquired a small girl who had been taken to hear him preach – he was described as exercising a far greater influence than any primate or prelate in the land. He might have claimed as his own the promise 'Instead of thy fathers thou shalt have children, whom thou mayest make princes in all lands'. He was the spiritual father and guide of the 'pious chaplain', Henry Martyn (1781–1812), whose *Journal* is one of the lasting treasures of Anglican devotion, of Daniel Corrie (1777–1837), the first Bishop of Madras, and of countless others

1. Opposition to Ryder's appointment was very strong, but he wore it down by sheer goodness and pastoral zeal. It is recorded that William Wilberforce 'highly prized and loved Bishop Ryder as a prelate after his own heart, who united to the zeal of an apostle the most amiable and endearing qualities, and the polished manners of the best society'.
2. But Shaftesbury was careful to recommend men of distinction from many different groups in the Church, including the Broad Churchman, A. C. Tait, who after a highly distinguished episcopate in London became Archbishop of Canterbury in 1868.

who went out to spread Christian devotion in every part of
the British Isles and in lands beyond the seas. This was the
third great Cambridge movement in the history of the
Church of England.[1]

There had been a time in the eighteenth century when
Evangelical clergymen had followed lines of action doubt-
fully compatible with loyalty to the Church of their ordination. The influence of Charles Simeon swung the movement
the other way, and all the Evangelicals of the first half of the
nineteenth century were convinced and devoted churchmen.[2] Wherever their influence penetrated, the result was a
renewed devotion to the services of the Church, and a new
reverence for its sacramental ordinances.

The figures of Church attendance may seem wellnigh
incredible in this age of laxity and indifference. When
W. W. Champneys, later Dean of Lichfield, was appointed
to the poor parish of Whitechapel in 1837, he found a completely empty church. In 1851 it was recorded that three
well-attended services were held every Sunday, and that on
one Sunday, on which a careful count was taken, 1,547
worshippers were present in the morning, 827 in the afternoon, and 1,643 in the evening.

Attendance at an early morning celebration of the Holy
Communion was a favourite form of Evangelical devotion.
In most churches, celebrations were not very frequent, and
the ceremonial was bare almost to the point of inanition;
but the service was conducted in an atmosphere of still and
awed adoration. An exceedingly high standard of personal
preparation was expected; a much used book of devotions,

1. Evangelicalism has always been associated in the popular mind
with Cambridge; but Mr J. S. Reynolds, in *The Evangelicals at Oxford
1735–1871* (1953), a most valuable piece of research, has shown that the
Evangelical tradition at Oxford has been stronger and more continuous
than is generally supposed.

2. When Mr Gladstone remarked that, in the days of his Evangelical
upbringing, he had heard nothing of the doctrine of the Church, he
meant of course that high mystical doctrine of the Church which he had
later learned from his friends in the Tractarian movement. The timid,
quasi-nonconformist low-churchism, which sometimes passes itself off
as Evangelicalism, came into the Church only after 1860, when low
churchmen and Evangelicals had joined hands.

prepared by Ashton Oxenden, later Bishop of Montreal, provided forms of preparation for a whole week before the reception of the Holy Communion, and thanksgivings for a whole week after it. A great many Evangelicals of that period could have echoed Bishop F. J. Chavasse's recollections of the 'Communion Sundays' of his boyhood: 'My father would be extra quiet all day, and shut himself up in his room both before and after the service. I have seen him come down from the rail with tears in his eyes.'[1]

One of the greatest contributions of the Evangelical bishops was the restoration of Confirmation to its rightful dignity among the ordinances of the Church. At the beginning of the nineteenth century, according to Dean Burgon, 'hurriedly to lay hands on row after row of children before the Communion-rails, and, at each relay of candidates, to pronounce the words of blessing once for all was regarded as the sum of the Bishop's function.' Bishop Ryder was the first to change all that: 'He sent a printed letter to the parents and god-parents of every candidate before the event, and . . . his chaplain gave a second letter to each of the candidates as they rose from under the imposition of hands. Then followed a short pastoral charge from the pulpit, which the bishop repeated from place to place with marvellous power.'[2]

Evangelicals of the Church of England have, happily, never been a party – their inveterate habit of biting and devouring one another over microscopic differences of opinion makes it unlikely that they ever will be. They have never had a central organization; they have never had one single organ of opinion. But it is a mistake to imagine that Evangelicalism is a form of a religious individualism. The Evangelicals certainly insisted on the vital significance of

---

1. J. B. Lancelot: *Francis James Chavasse, Bishop of Liverpool* (1929), p. 6. Evening Communion, though highly popular among Evangelicals, was not originally instituted by them, but by the clergy of Leeds, under the influence of the high churchman, Dr W. F. Hook, in 1851.

2. See S. L. Ollard in *Confirmation or the Laying on of Hands* (S.P.C.K. 1926), Vol. 1, pp. 219 ff. This is a first-class piece of research. Samuel Wilberforce was the great popularizer of the new style of Confirmation; but he was putting into practice what he had learned from the Evangelical friends of his father.

the relation of the individual to God, and each maintained
his own rugged independence of all others. But true Evangelical religion always tends to intense and intimate fellowship; and one of its characteristic manifestations is the
formation of societies.

There is hardly an end to the societies formed by the
Evangelicals of the Church of England at the end of the
eighteenth century and in the first half of the nineteenth.
The first and greatest was the Church Missionary Society
(1799), to which we shall return in another chapter. This
was followed by a host of others – for the evangelization of
the Jews (1809), of the aborigines of South America (1844),
for Bettering the Condition of the Poor (1796), for Pastoral
Aid in the Parishes (1836), for strengthening the Church in
the colonies (1838). Anglicans cooperated with dissenters
to found the Religious Tract Society (1799; now the United
Society for Christian Literature) and the British and Foreign
Bible Society (1804). If the Evangelicals had one single
centre, that was to be looked for rather than anywhere else
in the headquarters of the Church Missionary Society at
6 Salisbury Square; and, after the death of Simeon, their
most typical man was Henry Venn, Honorary Secretary
(and almost dictator) of the C.M.S. from 1841 to 1872.[1]

Even more characteristic than the Evangelical Society is
the Evangelical coterie, the group of like-minded friends
who hold together without organization and devote themselves together to pious works. The most famous of all these
groups was that fellowship of outstanding laymen which
came to be known as the Clapham sect.[2] By far the most
famous of the group was William Wilberforce (1759–1833),
the nightingale of the House of Commons, of whose voice it

1. The number of parishes supporting the C.M.S. is the one almost
infallible barometer of Evangelical influence in the Church of England.
By the end of the century the figure had reached 5,700 – nearly half the
parishes in the country.

2. The classic, and enchanting, account of these men is the Essay by
Sir James Stephen, who had known them all well in his boyhood,
reprinted in his *Essays in Ecclesiastical Biography*, a discouraging title for an
immensely entertaining work. A valuable recent account of the sect is
E. M. Howse: *Saints in Politics* (1952).

was said that it 'resembled an Eolian harp controlled by the touch of a St Cecilia'. But others were almost equally distinguished. Sir John Shore, Lord Teignmouth (1751–1834), had been Governor-General of India. Zachary Macaulay, 'the austere and silent father of the greatest talker the world has ever known',[1] had been Governor of Sierra Leone from 1794 to 1799. Henry Thornton (1760–1815) was a wealthy banker. These men lived for many years in a close association and amity to which there is no exact parallel in English Church history. The central interest of their lives was the practice of the Christian faith. Their watchwords were diligence, simplicity, and generosity. Henry Thornton habitually gave away two-thirds of his income, and in the midst of his busy practical life found time to spend three hours a day in prayer.[2] The Christian faith, as they understood it, was a simple, cheerful, vigorous, manly affair, entirely free from gloom or introspection. When the last of them was dead, the younger Stephen wrote, 'Oh, where are the people, who are at once really religious and really cultivated in heart and understanding – the people with whom we could associate as our fathers used to associate with each other? No "Clapham Sect" nowadays.'

It is already evident that what these men were more concerned about than anything else was holiness. But the nature of holiness as they understood it requires a little elucidation. They had passed far beyond that narrow concept of the Christian life which would identify it with the withdrawal of men from participation in the responsibilities of the world. As good Calvinists they held a theocratic view of the State, not in the sense that the Church should control the State, but that the State should be, or should be made, aware of its responsibilities as an instrument in the hands of God. It is the duty of the Church uncompromisingly

1. F. W. Cornish: *The English Church in the Nineteenth Century* (1910), Vol. 1, p. 13. The reference is, of course, to T. B. Macaulay the historian.
2. Here are notes of his spending over four years:
  1790 Charity £2,260 All other expenses £1,543
  1791 Charity £3,960 All other expenses £1,817
  1792 Charity £7,508 All other expenses £1,606
  1793 Charity £6,680 All other expenses £1,988.

to point out to the State and to society the requirements
of the law of God. These men believed in the reality of
'the conscience of a Christian nation'. They were convinced that, if the facts about contemporary evils could be
made known, the conscience of the nation would arise to the
destruction of those evils, and that then such legislative
action would become possible as would not make men good
by Act of Parliament, but would provide the conditions
under which every citizen could be assured of the possibility
of an honourable, Christian, and respectable manner of life.

To appeal to the conscience of a Christian nation – this
was both their aim and their strategy. Their first and, for
many years, their principal instrument was a quarterly
review, the *Christian Observer*, the first of all the English
quarterlies, first published in January 1802 at the price of
one shilling, a periodical which they hoped might contain
'a moderate degree of political and common intelligence'.
A large part of the paper was written by members of the
Clapham sect themselves.[1] Year after year, the British
public was served with well-informed, pungent comment
on every contemporary problem. In view of the intense and
timid conservatism later displayed by the Evangelicals, it is
interesting to read in the *Christian Observer*, at the time when
geology was first beginning to shake men's faith in Genesis,
that 'If sound science appears to contradict the Bible, we
may be sure that it is our interpretation of the Bible that is at
fault'. And, on the political responsibility of the Christian,
it expressed itself with the vigour of unquestioning certainty:
'We pity the heart and the head of any man, especially a
clergyman, who, when addressed upon the duty he owes to
God and his country in regard to such momentous topics as
Ecclesiastical Reform; the abolition of the anti-Christian
system of West-Indian slavery ... can affect to stigmatize
such considerations under the absurd name of "politics",
wrapping himself up in his own little selfish circle, perhaps
with a sneer at his friend's anxiety.'[2]

1. It is known that in fourteen years Henry Thornton contributed
eighty-three articles.
2. *Christian Observer*, 1832, pp. 437 ff.

It was in this spirit, and with these methods, that the Clapham saints advanced to the assault on the most formidable of all their dragons, West Indian slavery. The odds were tremendous. It was taken for granted by almost everyone that the abolition of slavery, even if it were possible, would mean the ruin of the country. John Wesley who had entered the lists himself in 1774 with his *Thoughts on Slavery*, a few days before his death in 1791 wrote to Wilberforce: 'Unless God has raised you up for this very thing, you will be worn out by the opposition of men and devils; but if God be for you, who can be against you?... Go on, in the name of God and in the power of His might, till even American slavery, the vilest that ever saw the sun, shall vanish away before it.' The saints had only two assets – the truth, and the fundamental decency that they believed to reside somewhere in the hearts of their fellow-countrymen. With courtesy, and with the utmost consideration for the views, the interests, and the feelings of those who differed from them, they persisted in their campaign of education, and in the end succeeded in creating a major revolution in the minds of men. It was a campaign of forty years, and victory was gradual. In 1806 the traffic in slaves was abolished. In 1833 the House of Commons carried what Professor Coupland calls[1] 'the noblest measure in its history', and abolished slavery throughout the British dominions. On 31 July 1834 every slave became free.[2] Wilberforce did not live to see the day; he had died on 29 July 1833.

Legend has it that the Evangelicals were so concerned about the souls of Hindus and the suffering of slaves as to be unaware of, or uninterested in, sufferings and evils nearer home, and particularly the immense evils for which the Industrial Revolution had been responsible.[3] Legend dies

1. *Wilberforce* (2nd ed. 1945), p. 429.
2. For scenes in the West Indies on that day, see J. McL. Campbell: *Christian History in the Making* (1946), p. 101.
3. I suspect that Charles Dickens's *Bleak House* is more responsible than anything else for the perpetuation of this myth; but it keeps cropping up in quite respectable works on the nineteenth century.

hard; but this one cannot stand against the most cursory study of the evidence. The Evangelicals were Whigs, and as such were in favour of reform in all directions, in politics and in the Church, in the law and in social affairs. Strenuous Protestants and Erastians as they were, they were unwearied in their advocacy of Roman Catholic Emancipation and of the relaxation of the disabilities under which the Nonconformists still suffered. The aristocratic Wilberforce spoke scathingly about 'our murderous laws', and condemned the game laws as 'so opposite to every principle of personal liberty, so contrary to all our notions of private right, so injurious and so arbitrary in their operation that the sense of the greater part of mankind is in determined hostility to them'.

The greatest champion of the workers between 1830 and 1850 was Antony Ashley Cooper, Earl of Shaftesbury (1801–85), who also stood in the Evangelical succession. But, though he was a great leader, he was not the pioneer. When Sir Robert Peel the elder brought forward in 1802 the first of the great series of Factory Acts – this proposed no more than to forbid the employment of apprentices in cotton or woollen mills before 6 a.m. or after 9 p.m., or for more than twelve hours a day – Wilberforce warmly supported it, and complained only that it did not go far enough. He also championed the cause of the boy chimney-sweeps, and the abolition of the press-gang for the Navy. When at last the Ten Hours Act was passed in 1847, its passage was rightly acclaimed as a triumph for Shaftesbury, though at the time he was not a member of Parliament; but he had been building on the great tradition of the Christian social conscience for which the earlier generation of Evangelicals had stood, and at almost every point could count on the support and backing of their heirs.

At two points the Evangelicals, even in the great generation, were preparing for the loss of influence which befell them in the later years of the century.

They failed to realize the vital importance of serious theological thought. There were men of considerable learning among them; but on the whole they occupied themselves only with the traditional learning, and not with the challenge of the new fields in philosophy, in natural science, and in the critical study of history that were being opened up in various countries of the Christian world. The result was that a certain eighteenth-century flavour attached to their piety even in the high Victorian age; and, whereas at an earlier date their message had had the vitality and excitement of novelty, by 1860, when grinding away at a few traditional themes had become all too much the rule in Evangelical pulpits, their style and manner seemed already fifty years out of date.

Secondly, they failed to challenge the false theories and philosophy that lay behind the evils of the industrial age. Here, too, they thought too much in the categories of the eighteenth century. They took it for granted that poverty would always exist, and that all that could be done was to mitigate the sufferings which accompanied it. They failed to see that the Industrial Revolution had for the first time made possible the abolition of human poverty. They fought like giants against evils within the system which they knew. It required a more prophetic mind and vision than theirs to recognize that the system itself must be changed, and could be changed, if the conscience of a Christian nation could be set to work in new and more revolutionary ways.

Yet, when all is said and done, the record of the Anglican Evangelicals is in every way memorable. To them more than to any other group or party in the Church was it due that, in the words of two secular historians, in the middle of the nineteenth century 'England became, perhaps, more nearly a Christian country than she had ever been before, perhaps more nearly than any comparable community before or since.'[1]

The three sections of the Church so far considered represent the continuance into the nineteenth century of movements and points of view that had become well established

1. G. M. Young and W. D. Handcock: *English Historical Documents*, Vol. XII (1956), p. 335.

at an earlier date. But the new age brought with it problems of which the Church had never previously had experience, and to meet which new adventures of the Spirit had to be made. For more than a century ecclesiastical politicians had been all too ready to raise the cry, 'The Church in danger'. Now the Church really was in danger – danger of the gravity of which even the most complacent Anglican was bound sooner or later to become aware.

Dr Kitson Clark is not exaggerating when he writes: 'Indeed in the second quarter of the nineteenth century a blast of hatred against the Church of England blows across English politics like a blast of hot scalding steam, when the cock of a boiler is opened.'[1] He quotes from the Methodist Stephens (1834): 'Ere long the very existence of the Established Church would be like a tale that is told, and remembered only for the moral evils which it had brought upon the country'; and from the Congregationalist Edward Miall (1841): 'The whole thing is a stupendous money scheme carried on under false pretences; a bundle of vested rights stamped for the greater security with the sacred name of Christianity.'

Various causes account for this bitter hostility of dissenters against the Church. There was, first, a new sense of their strength and their influence in the national life. In 1772 there had been about 380 Independent churches in England: by 1828 the number had risen to 799. In the same period Baptist congregations had increased from 392 to 532. There had been various threats even to the limited liberties which the dissenters enjoyed. In 1811 Lord Sidmouth had introduced into Parliament a Bill to explain and render more effectual sundry Acts relative to dissenting ministers, among whom, as he politely remarked, were to be found 'tailors, pig-drovers, and chimney-sweepers'. The dissenters not unnaturally formed in 1812 the 'Protestant Society for the Protection of Religious Liberty'. Even after the belated repeal of the Test Act in 1828, Nonconformists had ample grounds for resentment against the treatment accorded to them by members and ministers of the Estab-

1. G. Kitson Clark: *The English Inheritance* (1950), p. 125.

lished Church.[1] And now the dissenters were not without allies. On the one hand were the radicals, good men of the type of James Mill, who had rejected every form of revealed religion; on the other the Irish Catholics; an odd trio of confederates, but they were together strong enough to bring down a British government. Each group, for different reasons, would have been glad to take part in the overthrow and spoilation of the national Church.

What was to be done? There was so much in the life of the Church that was indefensible that complacency was impossible. Bold spirits saw that the only hope for the Church lay in penitence, rationalization, and reform. Those who followed this line may, in a very general sense of the term, be classed together as broad churchmen;[2] but they fall into three distinct groups, each of which has its heirs and assigns in the Anglican Communion today. It will tend to clarity if we devote some attention to each of the three groups separately.

First, there were those who, taking up again the old seventeenth-century idea of comprehension, believed that the Church could be saved only if its boundaries could be extended so as to include the majority of the orthodox dissenters.[3] The most intelligent and vocal of those who championed this view was Thomas Arnold, Head Master of Rugby School (1795–1842).[4]

In 1833 Arnold published a little book entitled *Principles of Church Reform*. Starting from the principle that the establishment of religion is a great blessing to a nation, but that in England this blessing has been largely lost, because

1. Particularly in such matters as the right of burial. For an Anglican to read B. L. Manning's *The Protestant Dissenting Deputies* (1952) is a soul-searing experience.

2. Though this term seems not to have been used until 1870. 'Broad Church' appears in 1853.

3. During the eighteenth century a strong tendency towards Unitarianism had developed in most of the dissenting bodies.

4. When Arnold remarked in 1832, in an oft-quoted phrase, that 'the Church as she now stands no human power can save', he was not referring to spiritual destitution, but to the impossibility of defending ancient privilege against a horde of angry enemies.

so large a part of the nation no longer adheres to the national Church, he asks whether it would not be possible 'to constitute a Church thoroughly national, thoroughly united, thoroughly Christian, which should allow great varieties of opinion, and of ceremonies, and forms of worship, according to the various knowledge, and habits, and tempers of its members, while it truly held one common faith, and trusted in one common Saviour and worshipped one common God'.[1] The first and 'most essential step towards effecting this and every other improvement in the Church, consists in giving to the laity a greater share in its ordinary government'.[2] To this end, Arnold recommended a reduction in the size of dioceses; the setting up of councils and assemblies in which the laity would take a full share; some control by the parishes over the election and appointment of ministers; and the admission to the ministry of the Church of men who had been too poor to receive a university education; thus, 'those who are now Dissenting ministers might at once become ministers of the Establishment, and as such would of course have their share in its government'.[3]

Arnold deplored 'the uninterrupted loneliness in which our Churches are so often left from one Sunday to another'. He believed that the remedy for this was to permit of a great variety of services, some typically Anglican, some following much more closely the Nonconformist tradition in worship. 'I believe that it would go a long way towards producing a kindly and united feeling among all the inhabitants of the parish, that the parish church should, if possible, be the only place of public worship, and that the different services required should rather be performed at different times in the same spot than at the same time in different places.'[4]

Arnold's ideals were not welcome in the Church of England, and met with harsh opposition from those very Nonconformists whom it was his purpose to win. But his ideas

1. T. Arnold: *Miscellaneous Works* (edn. of 1845), p. 279.
2. Ibid., p. 291.
3. Ibid., p. 298. Arnold seems not to have discussed the question whether this would involve re-ordination or not.
4. Ibid., p. 307.

lived on in his favourite pupil and biographer Arthur Penrhyn Stanley (1815–81), who on 9 January 1864 was installed as Dean of Westminster. It was always Dean Stanley's aim to make the Abbey a genuinely national home of religion, and to this end he invited churchmen of all shades of opinion, and some who were not churchmen (such as the great philologist Friedrich Max Müller, who lectured there in 1873), to speak within its walls. What brought upon Stanley stronger criticism than anything else was the invitation issued to all the members of the Company appointed to revise the English New Testament to attend the Holy Communion in the Abbey on 24 June 1870 before beginning their work; the Company included three Presbyterians, a Baptist, a Congregationalist, a Methodist, and one Unitarian. Naturally such an act of 'occasional conformity' aroused considerable commotion at the time.

As time passed, many of the ideals of Arnold came to be quietly adopted in the Church, though not yet in their fullness or in every detail; Arnold and Stanley would have felt themselves quite at home as members of the joint gathering of Anglicans and Free Churchmen which in 1950 produced the report *Church Relations in England*.

Reference may be made at this point to two men who, while not belonging to any group or party, maintained through the years of controversy that large, generous, catholic temper which is specially characteristic of the liberal tradition in the Church of England, and in their strength and their diversity serve to remind us of the astonishing wealth and variety of Anglican life in the nineteenth century.

Henry Hart Milman (1791–1868) had early won distinction as a poet and a scholar. As Dean of St Paul's from 1849 onwards he took his full share in renewing the life of that cathedral to meet the needs of the modern age. But his greatest services lay elsewhere. In his *History of the Jews* (1830), he was one of the first scholars in England to apply a reverent historical common-sense to the narratives of the Old Testament. By his *History of Latin Christianity* (1855) he permanently raised the standard of ecclesiastical history in

England. Dr Garnett rightly praises his qualities of 'liberal-
ity, candour, sympathy, and catholic appreciation of every
estimable quality in every person or party'.[1]

Connop Thirlwall (1797–1875) was one of the greatest
classical scholars of his age. In 1840 he was appointed by
Lord Melbourne to the remote See of St David's in Wales,
and there he remained until the end of his long life. Remote,
but not ineffectual. During his episcopate he issued to his
clergy eleven splendid Charges, in which every great issue
of those turbulent years was discussed with charity, calm,
and an astounding wealth of erudition. To this day there is
no better introduction to the history of the Church of England in the mid nineteenth century than the perusal of
Bishop Thirlwall's Charges.[2] The veneration in which 'St
David's' was held by his colleagues comes to light amusingly
in Archbishop Tait's account of the first Lambeth Conference (1867): 'New Zealand [G. A. Selwyn] committed
the great mistake of attacking the Bishop of St David's for
his charge reflecting on Capetown [Gray]. This produced a
storm, and let the Americans and Colonials understand that
St David's was looked on as a sacro-sanct. I first rose to the
rescue, and then the Bishop of Ely (Harold Browne) in great
emotion reproved New Zealand, declaring St David's to be
not only the most learned prelate in Europe, but probably
the most learned prelate who had ever presided over any
See.'[3]

Arnold had seen a great vision of an enlarged and comprehensive Church of England. That might be a bright hope
for the future. The immediate question was as to whether
the Church of England, as it then was, could be made to

1. *D.N.B.*, s.v. 'H. H. Milman'. Milman has recently come to life in
the writings of two contemporary scholars – C. H. E. Smyth: *Church and
Parish* (1955), Chapter 5, a wholly admirable chapter in one of the best
books on Church history ever written in English; and G. L. Prestige: *St
Paul's in its Glory* (1955), pp. 79–92. To Queen Victoria, Milman was
'the valuable, distinguished and excellent present Dean of St Paul's'
(letter of 18 September 1868).

2. *Remains of Bishop Thirlwall*, ed. J. J. S. Perowne, 3 vols., 1877.

3. R. T. Davidson and W. Benham: *Archibald Campbell Tait, Archbishop
of Canterbury* (ed. of 1891), Vol. 1, p. 379.

work. The answer was not obvious, but there were in the Church a certain number of resolute men who were determined that it could and should be made to work. The need for reform was very great. In the sixteenth century the Church had taken over the gigantic chaos of the medieval system, with its maze of property rights, its inequalities, its all too common divorce of clerical pay from the performance of clerical duties. Plurality had survived every attempt to suppress it. Bowyer Edward Sparke, Bishop of Ely from 1812 to 1836, had conferred on his eldest son endowments worth £2,634 a year, on his younger sons endowments worth £2,694 a year, and on his son-in-law another £2,213 a year; and in doing so had not supposed himself to be straying from the plain and estimable path of duty. Trollope has drawn an agonizing picture of the poverty of the Reverend Josiah Crawley as perpetual curate of Hogglestock on a stipend of £130 a year. But the investigation into clerical revenues carried out in the 1830s revealed that there were 3,500 parishes in England with an income of less than £150 a year.[1]

The great champion of financial and administrative reform was Charles James Blomfield, Bishop of London (1786–1857). Like his younger contemporary Thirlwall, Blomfield had been a brilliant classical scholar and Fellow of Trinity College, Cambridge, where even the great Porson called him 'a very pretty scholar'.[2] In 1828 Blomfield came to London as bishop, inspired, as Sydney Smith complained, by an 'ungovernable passion for business'. Later a brother bishop said of him, 'We all sit and mend our pens till the Bishop of London comes in.'

Reform of the Church was in the air. But how was it to be carried out, and who was to take the initiative? There could be only one answer. If anything was to be done, it could be

1. Many of the clergy, of course, supplemented their exiguous revenues by taking pupils and in other ways.
2. Professor A. E. Housman, in his Preface to Manilius, has lamented the successive strokes of doom which consigned Blomfield to the bishopric of Chester (Chester 1824–8; London 1828–57), and Badham to the antipodes. Charles Badam (1806–57) became professor of classics in the University of Sydney in 1867.

done only by Act of Parliament. The Convocations had not met for more than a hundred years, and, even if they had met, their powers were so limited that they could not have taken the urgently necessary action. In England, what Parliament has laid down by statute can be changed only by another statute – unless a revolution comes in which the whole of parliamentary government is swept away. Blomfield decided to recognize the facts, to cooperate with Parliament, and by this cooperation between Church and Parliament to deflect the bitter animosity of which the Church had so long been the object.

A Commission of Inquiry into the revenues of the Church had been appointed in 1831, and reported in 1835. In 1836 a permanent Ecclesiastical Commission was set up, to hold much of the property of the Church, and to see to 'the better distribution of ecclesiastical revenues and duties, the prevention or diminution of pluralities and non-residence, and the augmentation of poor benefices and endowment of new ones'. The Commission was much criticized at the time by conservative churchmen; but on the whole it did splendid work, and gave the Church of England a financial order under which it can live in the modern world. Sinecures were abolished; staffs of cathedrals greatly reduced; the incomes of bishops brought somewhat nearer to equality. (Friends of Charles James noted with pain that, though his successor would draw the mere pittance of £10,000 a year, Charles James would have till the day of his death an income of £20,000 a year – and that made him far richer than anyone can be in England today.) The bishoprics of Ripon (1836) and Manchester (1848) were founded.[1] In the first fifty years of its life the Commission endowed or added to the endowments of 5,300 parishes.

Another great concern of Bishop Blomfield was the building of new churches. The Church of England had become inflexible and immobile. In 1750 England had been in the

1. No new bishoprics had been founded, until this date, since the reign of Henry VIII. Of the forty-three bishoprics which now make up the provinces of Canterbury and York, twenty were founded in the hundred years following 1836. See G. F. A. Best: *Temporal Pillers* (1964).

main a country of villages, and its great centres of population were all south of the Trent. By 1850, with a greatly
increased population, it had become a country of teeming
towns, and the centre of gravity of the population was in the
north. But the Church was still trying to meet the problems
of a modern world with the creaking machinery of the
Middle Ages. In 1836 Blomfield launched his scheme for the
building of fifty churches in the London area, thus 'expatiat-
ing over the whole metropolis'. By 1846 forty-four churches
had been completed, ten were in process of construction,
and plans for a further nine had been laid. The work was
continued with undiminished ardour under Blomfield's
successor, Archibald Campbell Tait. The Evangelical
bishops, notably Baring of Durham and J. B. Sumner of
Chester, took up the problem with outstanding success. The
nineteenth century was the greatest period of church-building in England since the close of the fourteenth century.

Unhappily this effort came too late. In the areas of
greatest need, the agglomerations of population in the bleak
and cheerless new towns, the so-called working class had
been alienated from the national Church by more than a
generation of neglect. If these people were Christians at all,
they were probably adherents of the Methodist bodies, the
great proletarian Churches of the century. In a striking
letter to Bishop Samuel Wilberforce, written on 5 July
1843, Dr Hook of Leeds deplores the hatred felt by the men
of this class for the Church: 'They consider the Church to
belong to the Party of their oppressors; hence they hate it,
and consider a man of the working classes who is a Churchman to be a traitor to his Party or Order, – he is outlawed in
the society in which he moves. Paupers and persons in need
may go to church on the principle of living on the enemy;
but woe to the young man in health and strength who pro-
claims himself a Churchman.' The Church was faced by the
task of entering into this alienated world, and winning it
back to itself.

This problem forms the natural point of transition to the
thought of a man who, though he detested and criticized all

parties and systems, and equally those who would form a
'no-party' party to end all parties, stood in the true line of
Anglican liberalism, when that term is understood in its
proper sense of the demand that the Church should be in
every sense a living Church – Frederick Denison Maurice
(1805–1872). Maurice's thought is so wide and far-ranging
that it is hard to systematize it or to classify him as belonging
to any one school or tradition.[1] In his lifetime he was variously understood and misunderstood. A growing consensus
of opinion in the Anglican Churches regards him as the
greatest Anglican theologian of the nineteenth century.

Maurice had been brought up a Unitarian, and in early
years had had some connexion with almost every religious
body in England. His adherence to the national Church
was, therefore, not a matter of inheritance, but of sober,
serious, and intelligent conviction. He believed that the
nation as such is an ordinance of God, and that the State as
such is a servant of God; he had clear ideas as to the function
that a Church should perform within the life of State and
nation: 'A National Church should mean a Church which
exists to purify and elevate the mind of a nation; to give
those who make and administer and obey its laws a sense of
the grandeur of law and of the source whence it proceeds,
to tell the rulers of the nation, and all the members of the
nation that all false ways are ruinous ways, that truth is the
only stability of our time or of any time. . . . This should be
the meaning of a National Church; a nation wants a
Church for these purposes mainly; a Church is abusing its
trust if it aims at any other or lower purpose.'[2]

One of the points at which Maurice felt impelled to proclaim to the nation its duty was the crisis of the Industrial

1. Canon A. R. Vidler, in his excellent *The Theology of F. D. Maurice*
(1948), pp. 9–11, quotes a highly entertaining series of contemporary
judgements on him: e.g., J. B. Mozley (1853): 'Maurice has been petted
and told he is a philosopher, till he naturally thinks he is one. And he
has not a clear idea in his head.' Benjamin Jowett (1872): 'He was misty
and confused, and none of his writings appears to me worth reading.
But he was a great man and a disinterested nature.' Maurice himself
admitted that some people regarded him as a 'muddy mystic'.

2. *Lincoln's Inn Sermons*, II, 93 ff., quoted in Vidler, op. cit., pp. 197–8.

Revolution. The Evangelicals had devoted themselves to mitigating the evils caused by the system. Maurice looked deeper, and saw that a system which caused so much misery must itself be wrong. The political economy of Adam Smith, the *laissez faire* of the Manchester school, the philosophy of the Utilitarians, combined to soothe men's consciences with the belief that the unrestricted operation of economic laws would in the end work out to the greatest possible happiness of all, and that for the sufferings of the poor the poor were themselves in the main responsible. It was against such convictions and the appalling human wastage that resulted from them that Maurice felt impelled to protest in the name of humanity and of Christ. If the economists declared that competition was the sacred law of industry, the Church must stand up and declare that cooperation is a more sound and a more lasting principle.

The Christian Socialism, in which Maurice was associated, from 1848 onwards, with Charles Kingsley (1819–75) and other friends, was unpractical in many of its doings and limited in its immediate influence. It was of enormous importance in its creative influence on the whole future of the Church. It revealed to men a new world of Christian responsibility. Leaders like Bishop Westcott and F.J.A. Hort came under the influence of Maurice. They in their turn influenced Charles Gore and Henry Scott Holland. The influence of the thought of Maurice is to be traced in the Conference on Politics, Economics, and Citizenship held at Birmingham (1924); in the Industrial Christian Fellowship; in the Universal Christian Conference on Life and Work, Stockholm 1925; and wherever men protest against the view that there are areas of man's life, political, social, or economic, that can exist autonomously without becoming subject to the law of Christ.

We have described no less than six different groups and movements within the Church, of which four at least have left discernible traces in the life of the Anglican Churches today; and we have not yet said a word of that movement which was to attract more attention, and for a time to exercise greater influence, than any other – the Oxford, Tractarian, or Anglo-Catholic movement. For this method of treatment there is a reason; that great movement can be understood and fairly estimated only against the background of the abounding and varied spiritual life of the period in which it was to play so distinguished a part.

The Oxford Movement was, indirectly, a consequence of the great Reform Act of 1832. After 1832 Parliament was not what it had been up till that date. In the seventeenth century, members of Parliament had fought and killed one another; but they had still been able on occasion to attend St Margaret's Church, Westminster, as a corporate body, and to receive the Holy Communion together.[1] Now that Parliament had been opened to men of various religious confessions, this was no longer possible; Roman Catholics could not attend the Anglican Communion service; many Nonconformists would not attend, even if they had been regarded as eligible to receive such an invitation. But what kind of a body was this to be in control of the destinies of a national Church? Clearly the situation was intolerable. It still is intolerable. If the Church of England had not developed a capacity, unmatched in any other Christian communion in the world, for tolerating the intolerable, it would have been brought to an end long ago. In the past some Anglicans may have defended establishment on the ground that it was good for the Church. Since 1832, it is hard to see how any can have defended it, except on the ground that establishment is good for the State, and that endurance of the absurdities and limitations of establishment is a duty that the Church owes to the nation in order that from time to time the nation may have the privilege of manifesting itself before the world as a nation that still in some sense desires to be Christian.

Not all were prepared to see the extent of the revolution

1. For fascinating details of this, see C. H. E. Smyth, *Church and Parish* (1955), pp. 9–19. In 1833 the legislature was still nominally Christian, though not Anglican. It was not till 1888 that Jews and adherents of other non-Christian religions or of none were admitted to membership of Parliament.

that had taken place in 1832, or to draw drastic conclusions. Blomfield was so sure that reform must be carried out that he was prepared to cooperate with Parliament and the liberals, as they were coming to be called. A great many people thought that he was right. Not so the Tractarians, to give them at once the name which later became current.[1] In their eyes, that Parliament should touch two mites of the Church's property was sacrilege, and considerably worse than witchcraft.

The small spark which set off the explosion was the Irish Church Measure, put forward by the Government in 1833. The Anglican minority in Ireland was cared for by twenty-two archbishops and bishops, drawing revenues of £150,000 a year. 'The atrocious measure' proposed to reduce the number of bishops to twelve by combination of the smaller sees, and to use the revenues thus set free for the benefit of the poorer livings. This may not seem to us a very revolutionary step. But for certain Anglicans the question was not whether the measure was good or bad, wise or foolish. It had been put forward by Parliament and that was enough. John Keble (1792–1866) preached, on 14 July 1833, his Assize Sermon on 'National Apostasy', in which he accused Parliament of a 'direct disavowal of the sovereignty of God'. This day Newman considered and kept ever after as the start of the religious movement of 1833. A few months later, an 'Association of Friends of the Church' was formed. The first of the *Tracts of the Times* was published, and distributed widely among the clergy. The Anglo-Catholic movement had been born.

From the beginning the Tractarians, unlike the Evangelicals, formed a party. They had a centre, Oxford. They had a central organ of opinion, the *Tracts*. They had recognized and acknowledged leaders. Most eminent and unchallenged among the leaders was John Henry, later Cardinal, Newman.

1. After 1840. Before that date the supporters of the Oxford Movement were generally known, rather oddly, as Puseyites, after E. B. Pusey (1800–82), Regius Professor of Hebrew in the University, and friend and supporter of J. H. Newman. Even Modern Greek has the word *Pouzeismos*!

Newman (1801–1890) was certainly among the most remarkable and enigmatic figures in the history of the English Church. He had been brought up an Evangelical, and for a time had been secretary for the Church Missionary Society in Oxford. At the time of the publication of the early *Tracts*, he was Fellow of Oriel College, and Vicar of St Mary's Church. He was a strange combination of opposites – a penitent profoundly conscious of his own deep sinfulness, and a bustling party manager; a tender and affectionate friend, and a ruthless and unscrupulous controversialist; the most eloquent preacher and one of the best writers of his age; a sceptical intellect with an infinite capacity for devotion.[1] It was a formidable combination of gifts.

Newman's mind was subtle rather than strong, sensitive rather than profound. For some aspects of Christian truth he was all delicate attention; of others he was blandly unaware. A contemporary scholar has written, not unkindly, that Newman 'didn't have what it takes to be an Anglican'.[2] He lacked the willingness to suspend judgement, to recognize the role that doubt may play even in a confident and living faith, to be content with the kind of assurance that the Church of England offers, and to believe that there are many human questions to which God in His wisdom has given no clear and incontestable answer. Another scholar, deeply sympathetic to Newman, has commented that 'one gets the impression that his "fundamentalism" . . . was in fact symptomatic of a deep-seated craving for the support of an absolute external authority which, from the beginning and despite all his protests, he was dimly conscious of needing for his faith'.[3]

If the *Tracts* made Newman the leader of the party in the country, it was his sermons in St Mary's Church that

1. I think that F. J. A. Hort was right when he wrote of him, 'What Maurice said of him is profoundly true, that he was governed by an infinite scepticism counteracted by an infinite devoutness', *Life and Letters* (1896) Vol. II, p. 423.

2. C. H. E. Smyth: *Church and Parish* (1955), p. 164. The whole section, pp. 162–6, should be read with the most careful attention.

3. P. E. More in P. E. More and F. L. Cross: *Anglicanism* (1935) pp. xxx–xxxi.

enabled him to hold Oxford in the hollow of his hand. Few, if any, Anglican preachers have exercised so great an influence for good on so many young men. As Newman preached, men saw visions and dreamed dreams; their hearts were kindled, as they saw Christ risen and dwelling in the Church which is His body. Even a cursory acquaintance with the literature of the time makes it clear that what the Tractarians were more concerned about than anything else was holiness, a new kind of holiness, ascetic, austere, demanding, pledged to devotion to Christ to the uttermost whatever the cost. If the Evangelicals seemed still to be living in the eighteenth century, this was piety clothed in the radiance of the Romantic Movement and the Gothic Revival. Whatever criticisms churchmen felt impelled to make of other aspects of the Tractarians, here criticism was stilled. Even Bishop Bagot of Oxford, at whose request the series of *Tracts* was brought to an end, said of Tractarian principles: 'There they are . . . and they are forming at this moment the most remarkable movement, which, for three centuries at least, has taken place among us. . . . The system in question, instead of being an easy comfortable form of religion, adapting itself to modern habits and luxurious tastes, is uncompromisingly stern and severe – laying the greatest stress upon self-discipline and self-denial – encouraging fasting, and alms-deeds and prayer, to an extent of which the present generation, at least, knows nothing, – and inculcating a deference to authority which is wholly opposed to the spirit of the age.'[1]

The *Tracts*, published anonymously, were a strangely miscellaneous collection. Some of them consisted of no more than reprints of selections from the works of those seventeenth-century divines with whom the Tractarians claimed affinity. Those contributed by Dr Pusey, and signed with his initials, were complex theological documents. But all were alike in recalling the clergy to a sense of the Church as the Bride of Christ, and to their own vocation as a gift from God, independent of any connexion with, and certainly not

1. *Charge of May 1840 to the Clergy of the Oxford Diocese on Tractarianism and Tract 90.*

subordinate to, the will of any State: 'Christ has not left His Church without claims of its own upon the attention of men. Surely not. Hard Master He cannot be, to bid us oppose the world, yet give us no credentials for so doing. There are some who rest their divine mission on their own unsupported assertion; others who rest it upon their popularity; others on their success; and others who rest it upon their temporal distinctions. This last case has, perhaps, been too much our own; I fear we have neglected the real ground on which our authority is built, – OUR APOSTOLICAL DESCENT.'[1]

Here is the point at which the Oxford Movement has made its great and permanent contribution to the life of the English Church. All else – greater frequency of the celebration of the Holy Communion, liturgical experiment, more ornate ceremonial, the practice of 'sacramental' confession, the foundation of the theological colleges, the renewal of the monastic life, a stronger sense of continuity with the past – all these things, important as they may be, are secondary to the central issue. Newman and his friends asked the question, and made other men ask the question, 'What is the Church?' At a time when all too many in the Church of England were prepared to regard it as a branch of the Civil Service, or the religious aspect of a nation's existence, the Oxford men brought back to life the forgotten doctrine that the Church is the body of Christ, that the life it lives is His life, and that, outside of the divine authority that He has given it, it should neither desire nor seek for any other authority.

If the Oxford Movement had gone no further than this, it might have encountered opposition, but it would not have let loose the storms that distracted the life of the Anglican Churches for half a century. Did Newman and his party really stand in the succession of the great divines of the seventeenth century? Where did they stand in relation to the Reformation? These were questions that men asked, and had a right to ask. Those earlier divines, convinced as

1. *Tract I: Thoughts on the Ministerial Commission, Respectfully addressed to the Clergy*, 9 September 1833.

they were of the Catholic character of the Church of England, had never hesitated to call themselves Protestants,
and, regarding doctrine as more important than order, had
found their natural allies in those Churches of the Continent
which also accepted the supremacy of Scripture. But now it
seemed that the principles were being reversed. The Reformation was decried. The episcopate and the apostolic
succession were treated as such essential constituents of the
Church that only those Churches which had retained them
were regarded as living branches in the Catholic Church of
Christ. Where could this end, except in an assimilation of
the Church of England to what the vast majority of Anglicans at that time regarded as the errors of Rome? Very few in
the Church of England had put forward such views at any
time since the Reformation. This is not to say that the views
were necessarily erroneous; but those were right who maintained that the Oxford doctrines were innovations, and
represented a position but rarely held in the Church of
England since its separation from the Church of Rome.

'I am every day becoming a less and less loyal son of the
Reformation. It seems plain to me that in all matters that
seem to us indifferent or even doubtful, we should conform
our practices to those of the Church which has preserved
its traditionary practices unbroken.' So men could read in
the *Remains* of Newman's friend Richard Hurrell Froude
(1803–36). The way was being prepared for the catastrophe
of Tract 90.[1]

Everyone knows that the language of the Anglican Articles is at points vague and difficult of intepretation; reverently
vague, perhaps, when dealing with such mysteries as Predestination; studiously vague, perhaps, in order that as
many men as possible might be able to sign them without
doing violence to their consciences. The rigidly Protestant
interpretation had become almost traditional in the Church;
Newman set himself in Tract 90 to see how far the Articles

1. In a letter to Newman, written as early as 28 December 1835,
Froude had said: 'Really I hate the Reformation and the Reformers
more and more, and have about made up my mind that the rationalist
spirit they set afloat is the ψευδοπροφήτης of the Revelation.'

could be read in such a sense as not to conflict with the 'catholic' views which he held in 1842. This would, perhaps, have been legitimate as an academic exercise. But Newman was the leader of a great party in the Church. Nothing in this whole history is stranger, nothing shows more clearly how incredibly remote was the Oxford in which he lived from the world of ordinary men, than Newman's apparently total unawareness of the storm that the publication of his Tract must necessarily cause. 'The Protestant Confession was drawn up with the purpose of including Catholics; and Catholics now will not be excluded. What was an economy in the reformers, is a protection to us. What would have been a perplexity to us then, is a perplexity to Protestants now. We could not then have found fault with their words; they cannot now repudiate our meaning.'

They could and did repudiate his meaning. Men were inclined not so much to admire the ingenuity of Newman's reasoning as to charge him with disingenuousness and downright dishonesty. The pugnacious Henry Phillpotts expressed himself with characteristic forthrightness: 'The tone of the Tract as regards our own Church is offensive and indecent; as regards the Reformation and the Reformers absurd, as well as incongruous and unjust. Its principles of interpreting our articles I cannot but deem most unsound; the reasoning with which it supports its principles sophistical; the averments on which it founds its reasoning at variance with recorded facts.' But even the gentle Bishop Bagot felt constrained to tell his clergy: 'I cannot persuade myself, that any but the plain obvious meaning is the meaning which as members of the Church we are bound to receive; and I cannot reconcile myself to a system of interpretation which is so subtle, that by it the Articles may be made to mean anything or nothing.'

Neither friends nor opponents understood what Tract 90 really was. It was Newman's desperate attempt to assure himself that, holding the views he then held, he could continue to be a minister of the Church of which the Thirty-nine Articles were the charter. Newman loved Oxford and the Church of England; he knew that to turn his back on them would almost certainly mean to exclude himself for ever from happiness. It was a forlorn hope. The tremendous uproar occasioned by the Tract assured him that the hope had failed. As he himself wrote long after in his *Apologia pro Vita Sua* (1864), he was already on his deathbed as far as the Church of England was concerned. It took him a very long time to die; but clearly the illness was mortal; on 9 October 1845 he was received into the Roman Catholic Church.

The shock to Anglicans everywhere was immense. Such was the eminence that Newman had created for himself that many people doubted whether the Church of England could survive. But that Church is an anvil that has worn out a good many hammers. The less stable elements in the Oxford Movement followed Newman and exchanged the Anglican for the papal allegiance. All the great leaders stood firm. 'Oh, Pusey! we have leant on our Bishops, and they have broken down under us,' cried Newman in his distress. Dr Pusey had not leant on any bishops; he had leant on the Word of God and on the Church. He, and John Keble, and younger men like R. W. Church (1815–90), later Dean of St Paul's and historian of the movement, stood firm and rallied the shaken forces. Anglo-Catholicism had been saved for the Church of England.

The Church of England had been shaken, but it had not fallen. Indeed, there were reasons, at the mid-century, for thinking that its position in the life of the nation was stronger than it had been at any time since the Reformation. Among the Parliamentary Papers for 1852–3 is a unique document – the Census Report of 1851 on Religious Worship. This gives us more precise information as to the provision for religious worship, and as to the attendance of Christians at church, than is available for any earlier or later period. The figures show that, taking the population as a whole, and places of Christian worship throughout the country as a whole, accommodation was almost sufficient for as many people as were likely to wish to attend worship at any one particular time; but, since distribution was uneven, some centres like the City of London having far more places than could possibly be used, it had to be reckoned that, in spite of the building of 2,029 churches between 1831 and 1851, some areas were still seriously ill-provided with facilities for worship.[1] Even more striking were the statistics of attendance at church. On Sunday, 30 March 1851, there were just under eleven million attendances at church. Since a considerable number of people were known to have attended two services, those responsible for the census reckoned that 7,261,032 persons, out of a population of just under eighteen million, had been in church at least once on that Sunday. This is a larger number, and of course a far higher percentage of the population, than could be expected to appear in church on the corresponding Sunday a century later.

England had gone far to pull itself together after the lethargy of the eighteenth century, and to qualify itself to take rank as a Christian nation. But what, in point of fact, was the position of the Church of England in the nation, and what part had it played in this epic of Christian recovery? Probably 70 per cent of the population was nominally Anglican. But, as Nonconformists tartly pointed out, of the available sittings only 53 per cent were to be found in Anglican churches, and of those attending service on the day on which the census was made, only 3,773,474, or just under 52 per cent, were Anglican. If the Church of England was the national Church, clearly it was only in a qualified sense that it could be taken to be the Church of the English people.

In the century that followed, the claim of Britain to be a Christian nation, and the claim of the Church of England to be the Church of the English people, were to be more sharply challenged than at any time in its earlier history.

1. Of the available 10,212,563 sittings, the Church of England provided 5,217,915, the other Christian bodies 4,894,648. Under the system of 'pew-rents', many Anglicans paid money for the right to sit in a particular place in church. This was one of the things that made it impossible for members of the poorer classes ever to feel at home in an Anglican place of worship.

CHAPTER 10

The Nineteenth Century in England: 11

IT is generally supposed that the trouble began with the publication, in 1859, of Darwin's *Origin of Species*; but in reality the conflict between Christian and non-Christian views of the nature of man and the universe had begun a good deal earlier, certainly not later than the appearance in 1838 of Charles Lyall's *Elements of Geology*. It is also generally supposed that the Church flew into an unnecessary panic over the attempts of innocent and disinterested scientists to understand the secrets of the world. Nothing could be further from the truth. In the closing decades of the nineteenth century the Christian faith was the object of the unremitting, skilful, and malevolent attacks of enemies who wished for nothing more ardently than the total disappearance of that faith from the earth. Of such attacks Darwin himself must be declared wholly innocent; but what is true of Darwin is very far from being true of all his disciples.

In 1867 John Morley became editor of the *Fortnightly Review*. For fifteen years this remained the citadel of the anti-Christian forces. The Victorians were a violent and uninhibited race; they threw themselves into their controversies with the same whole-hearted passion that they manifested in other fields, and all later religious controversy looks pale and thin in comparison. Most vigorous and skilful of all the debaters was W. K. Clifford (1845–79; another of these Trinity men), one of the most regular of Morley's contributors in the *Fortnightly*; to him, Christian theology was 'the slender remnant of a system which has made its red mark in history, and still lives to threaten mankind', 'the seed of that awful plague which has destroyed two civilizations, and but barely failed to slay such promise of good as is now struggling to live among men.'[1] These men,

1. W. K. Clifford: *Lectures and Essays* (1879) Vol. 1, p. 253.

with their mechanistic view of the universe, would have left us without a doctrine of creation, without human freedom, without hope, and without God in the world.[1]

Disraeli, after all, was right. In his famous speech in Oxford on 24 November 1864, he said: 'I am not prepared to say that the lecture-room is more scientific than the Church. What is the question now placed before society with a glib assurance the most astounding? The question is this – Is man an ape or an angel? My Lord, I am on the side of the angels.[2] . . . It is between these two contending interpretations of the nature of man, and their consequences, that society will have to decide. Their rivalry is at the bottom of all human affairs.'[3] Man either is a free spirit, created with a view to eternal fellowship with a personal God, or he is not. Disraeli, who was an extremely clever man, had not merely coined a piquant and memorable phrase; he had summed up in a sentence what, in our own day, is still the major issue of faith or unbelief.

It was not only from the side of physical or biological science that Christian belief appeared to be threatened. Rationalistic criticism was laying hands on the Bible and shattering all traditional ideas of the life and mission of Jesus. As early as 1846 George Eliot had cooperated in the translation of Strauss's *Life of Jesus*; but that was a big book and technical, and made its impression only in limited circles. In 1863 Renan produced his *Life of Jesus*. Renan's great erudition, combined with the wonderful charm of his style, ensured for the book immediate popularity, and it passed through edition after edition in all the main languages of Europe. Renan's first principle was that the supernatural must be discounted; Jesus was the Son of Man, but He could not be in any transcendent sense the Son of God.

In 1875 Max Müller began his publication of the *Sacred Books of the East*. Men became clearly aware, as earlier they

1. A. W. Benn, in *English Rationalism in the Nineteenth Century* (2 vols, 1906), gives a clear account of this anti-Christian assault, and reckons that it reached its climax in 1877.

2. Darwin's *Descent of Man* was not published till 1871; Disraeli was not directly answering him.

3. *The Life of Benjamin Disraeli* Vol. IV (1916), p. 374.

had been dimly aware, of vast systems of religion and philosophy more ancient and venerable than the Christian Gospel. Tyler's *Primitive Culture* appeared in 1871. The parallels between certain parts of the Old Testament and the customs of other primitive peoples were too plain to be denied. The uniqueness of the Bible and of the Christian faith could no longer be regarded as a self-evident truth.

Many good men found the acceptance of the Christian Gospel no longer possible, and this was the great age of the virtuous agnostics. Three men most typical of the time were Matthew Arnold (1822–88), Leslie Stephen (1832–1904), and Henry Sidgwick (1838–1900). Stephen, son of the great Evangelical James Stephen, had gone so far as to take deacon's orders in the Church of England, but then sincerely and conscientiously withdrew. He wrote the great agnostic charter, *Essays on Free Thinking and Plain Speaking* (1873). Henry Sidgwick, Professor of Moral Philosophy at Cambridge from 1883 onwards (his best known book *Methods of Ethics*, 1874), went through agonies of indecision, but in the end found that he could not subscribe to the Thirty-Nine Articles, at least as they were understood in his day.[1] But Arnold, who to the end of his life held on to a shadowy Christian faith, expressed more poignantly than any other what the men of that generation suffered through the disappearance of that robust faith by which their fathers had been upheld. In *Dover Beach* he wrote:

> The Sea of Faith
> Was once, too, at the full, and round earth's shore
> Lay like the folds of a bright girdle furl'd.
> But now I only hear
> Its melancholy, long, withdrawing roar,
> Retreating, to the breath
> Of the night-wind, down the vast edges drear
> And naked shingles of the world.

1. For the cruelly superficial remarks on this of John Maynard Keynes, see the *Life* by R. F. Harrod (1951), pp. 116–17: 'He never did anything but wonder whether Christianity was true and prove that it wasn't and hope that it was. . . . He didn't seem to have anything to be intimate about except his religious doubts. And he really ought to have got over that a little sooner; because he knew that the thing wasn't true perfectly well from the beginning.'

This was the epitaph of a tragic generation,
> Swept with confused alarms of struggle and flight,
> Where ignorant armies clash by night.

What had the Church to offer in the way of enlightenment and comfort to all this suffering and disillusionment?
For the most part the answer has to be, Nothing at all.

From the Evangelicals nothing was to be expected. Their
trouble was that they had become timid and that they had
become Tory.[1] Their failing was that they tended to pontificate on subjects, such as Hebrew roots, which they had
never really taken the trouble to master.[2]

Nothing was to be expected from the Anglo-Catholics.
They had done good service in antiquarian research and in
the revival of patristic study. But they were innocent of any
acquaintance with scientific theology. Dr Pusey in early
years had gone to Germany to acquaint himself with the
modern approach to the Old Testament; but a slight tendency, on his return, to depart from the strictest canons of
verbal inspiration had so alarmed his friends that Pusey
abandoned the temptation to apply his massive intellect to
fundamental problems.[3]

The efforts of churchmen less bound by the past to speak
to the needs of the age did not meet with any signal success.
In 1860 a volume entitled *Essays and Reviews* appeared. It
was not a very good book; and it is not true, as is often
stated, that its alleged heresies became the commonplaces
of the thought of a later age. But three of the contributors to
it, Frederick Temple (1821–1902), Mark Pattison (1813–
84), and Benjamin Jowett (1817–93), were among the
greatest intellects of the century; and the book was a sincere
and honest attempt to begin to think theologically in face of
the new and burning problems which the Church was bound

1. *The Record*, founded in 1828, was the, sometimes vitriolic, expression
of Tory Evangelical views.

2. That admirable barometer the Church Missionary Society reveals
the decline of Evangelical influence during these years in the decrease in
the number of suitable candidates for missionary work overseas.

3. Nevertheless there is still a great deal to be learned from Pusey.
When I was a boy, 'Pusey on the Minor Prophets' was to be found in the
library of every self-respecting Evangelical clergyman.

to face. It did not deserve the indiscriminate abuse and condemnation that were heaped upon it. It was condemned by Convocation; eleven thousand clergymen signed a declaration of faithfulness to the strictest doctrine of the verbal inerrancy of Scripture, and to the literal accuracy of the account of creation in Genesis.[1] In 1865 Sir John Seeley published anonymously his *Ecce Homo*. This, also, was not a very good book; but it was a reverent and sincere attempt to deal with certain aspects of the humanity of Jesus. This did not save it from the most violent vituperation and obloquy.[2]

At such a time of danger it was essential that the Church should be united; unhappily, it was more sharply divided than at any earlier period in its history. The first generation of the Tractarians had not been interested in clothes or frills. Mr Newman, as an Anglican, had never celebrated the Holy Communion in any other manner than standing at the north side of the Holy Table, and wearing a black scarf. Dr Pusey adopted the 'eastward position' in Christ Church Cathedral for the first time in 1871, nearly forty years after the beginning of the movement, Dr Church only after becoming Dean of St Paul's, also in 1871. The attitude of the men of the second generation was very different. They introduced a great many changes in ceremonial and in the manner of conducting the services; and, as liturgical science was at that time almost non-existent in the Church of England, much of what they introduced was direct imitation of the liturgically least defensible medieval practices of the Roman Catholic Church. Now irregularity in doctrine may pass almost unnoticed by a congregation; any change in the accustomed manner of conducting the services, even

1. For *Essays and Reviews*, see the quite excellent account and discussion in V. F. Storr: *The Development of English Theology in the Nineteenth Century* (1913), pp. 429–54.

2. It must not be supposed that the entire Church was on this level of ignorance and obscurantism. The Bampton Lectures of Frederick Temple, then (1884) Bishop of Exeter, on the relations between Religion and Science, are well informed and surprisingly modern in tone; and the work of such men at T. G. Bonney, the geologist (1833–1923), should not be forgotten.

if it be no more than the introduction of a new tune for a familiar hymn, at once attracts attention, and is likely to arouse strenuous opposition; especially if that ancient rallying cry of the British people, 'No Popery', can be raised. The doings of the Anglo-Catholics were accompanied by unseemly and disgraceful riots in various places. It was urgently necessary that steps should be taken to restore order.

Many Anglo-Catholics were convinced that they had made no unlawful changes in the order of service, and that some of the unfamiliar practices they had introduced, such as the wearing of Eucharistic vestments, were in fact required by the Ornaments Rubric in the Prayer Book, though they had long fallen into desuetude. They were quite willing that the question at issue should be tried out in the ecclesiastical courts. But, as case after case was decided against them, they declared themselves unable to comply with the decisions of the courts, which they could not accept as genuinely spiritual courts. For this they had not unreasonable grounds; the final court of appeal in ecclesiastical cases in England was the Judicial Committee of the Privy Council, a body which had no spiritual authority, and which could hardly be regarded as well qualified to determine cases in which theological matters were at issue.[1]

But if clergymen declare themselves unwilling to comply with the decisions of the highest courts in the land, what is to be done with them? In self-governing Anglican Churches no difficulty arises. In England the problem is extremely complex. The beneficed clergyman enjoys what is called the 'parson's freehold'. Once he has been duly instituted and inducted, he is in legal possession of this benefice, and it is extraordinarily difficult to get him out. He can be deprived of his living, by legal process, for heresy or for grave misconduct, but this is a costly, difficult, and uncertain process.[2] Some things can be said in favour of the parson's

1. It was only in 1833 that the Judicial Committee became the final court of appeal in ecclesiastical cases. The problem of Church courts for the Church of England has at last (1963) been in a measure solved.

2. Recent legislation has somewhat reduced the difficulties in the way of removing an undesirable parish priest.

freehold, and some against. It has certainly been of value in making it possible for unpopular opinions and movements to hold their own in the Church of England. It protects the parish priest against undue pressure from, or oppression by, his bishop. It makes almost impossible the formation of such 'one-colour' dioceses as are to be found in many other parts of the Anglican world. On the other hand, it makes possible depths of idleness and incompetence in the parishes such as would not be tolerated in any other Church in the world. And it made things very difficult for the bishops ninety years ago. The virtuous Anglo-Catholics were not guilty of heresy or of immorality; and yet they refused to obey the decisions of the courts. What was to be done with them?

An attempt was made to solve the problem in the Public Worship Regulation Act of 1874, under which in certain circumstances clergymen could be sent to prison for irregularities in the conduct of public worship. It was the old story of the Puritans over again in reverse. In the seventeenth century the Puritans were prepared to suffer imprisonment and worse, because they would not wear as much in church as the bishops thought it suitable that they should wear. In the nineteenth century the Anglo-Catholics were prepared to go to jail because they regarded it as essential to wear more in church than the bishops were prepared to allow them to wear.[1] In each of these periods good men made a principle out of something indifferent in itself, and were prepared to die rather than surrender what they had come to regard as a principle. In each case the martyrs triumphed, and those who made martyrs of them were made to appear as cruel tyrants. A number of clergymen, whose virtues were perhaps in excess of their wisdom, were

---

1. At this time the dress of the bishops was very simple. Purple cassocks and pectoral crosses were still unknown. Bishops wore mitres on their notepaper and on their carriages, but not yet on their heads. It was in Central Africa that things began to change. The ultra-Protestant martyr Bishop Hannington of Uganda was pleased to find Bishop Smythies of the Universities' Mission to Central Africa 'keen on heart conversion, in spite of mitre and cope'. E. C. Dawson: *The Life of Bishop Hannington* (1887), p. 318.

put in prison. This public opinion would not tolerate; and before long prosecutions under the Act came to an end.

The Anglo-Catholics had triumphed; but we must be under no illusions as to the price paid by the Church for their triumph. In many parishes the people, confronted by services they did not want and could not understand, simply ceased to come to church and have never returned. Conversely, in parishes where a devoted Anglo-Catholic following had been built up, the whirligig of private patronage might bring in a successor of very different opinions, and the faithful might be dispersed by the brutal rejection of the traditions to which they had grown accustomed. The bishops were reduced to that dignified impotence which is their lot in England, though not in other anglican provinces. Worst of all, this was the period in which extreme Protestantism became characteristic of the Church of England. The excessive rigidity of the Act of Uniformity, and the apparent impossibility of altering it, led to widespread liturgical chaos. The difficulty of observing all the rubrics has led many clergymen to suppose that there are no rubrics that need be observed. Private judgement has taken the place of order; and any worshipper in England attending a church with which he is not familiar will have little idea in advance of what relationship, if any, the service in which he is about to take part will bear to the Book of Common Prayer. Although there has been great improvement in recent years, the Church of England still presents a scene of unparalleled liturgical disorder, and no group within the Church can put in a legitimate claim to be the true party of obedience and good order.

In this sad period, the strongest religious influence in England was exercised by three remarkable men, all Fellows of Trinity College, Cambridge, and intimate friends – Joseph Barber Lightfoot (1828–89; Bishop of Durham 1879–89), Brooke Foss Westcott (1825–1901; Bishop of Durham 1889–1901), and Fenton John Anthony Hort (1828–92). Of the three, the intellect of Lightfoot was the most vigorous, that of Westcott the most delicate, that of

Hort the most profound. Hort had performed the legendary feat of taking first-class honours at Cambridge in the schools of classics, mathematics, philosophy, and natural science (botany), before taking up his life's work in theology. Nothing could be less like these men than the traditional picture of the remote and ineffectual scholar. Westcott was for many years of his life a master at Harrow, Hort a country clergyman; and though they were at home in the most abtruse technicalities of scholarship, they could descend when necessary into the forum, Lightfoot with *Essays in Supernatural Religion* (1874–7; as a book, 1889), in which he pulverized the second-rate but influential rationalistic work published anonymously under the title *Supernatural Religion* (1874); Westcott by the publication of such series of sermons as *The Gospel of the Resurrection* (1886). Hort never had in the same degree the common touch; but his Hulsean Lectures for 1871, *The Way, the Truth, and the Life* (published 1893), are among the few golden books of Anglican theology which can be read and re-read and never lose their freshness. Lightfoot's appointment as Professor at Cambridge in 1861 marked the beginning of a new epoch in English theology.

It was a time of grave uncertainty, in which it seemed hard to find any resting-place between an obscurantism which refused to ask any questions, and a rationalism which was hardly willing to admit the possibility of any answers. These three men chose and found the middle way. The task they set themselves was no less than to survey the whole of primitive Christian antiquity, and to discover there the historical bases on which faith can firmly rest. Westcott and Hort worked together for twenty-eight years to establish the principles of the textual criticism of the New Testament. Lightfoot devoted many years to the study of the Apostolic Fathers, – a great scholar of a slightly later date, Dr C. H. Turner, has described the eagerness with which he awaited the appearance, in 1885, of Lightfoot's edition of the Letters of Ignatius.[1] But the greatest aim of all was the

---

1. *Catholic and Apostolic* (1931), p. 35. 'I don't think I ever looked forward to the appearance of any book with quite the same feverish expectation as to Lightfoot's *Ignatius*.'

production of a commentary on all the books of the New Testament. This was never completed; but the fragments are memorable; Westcott on Hebrews and St John, Lightfoot on Philippians and Colossians are among the greatest biblical commentaries in English or indeed in any other language.

It is almost impossible to exaggerate the influence exercised by these three men on every aspect of English religion between 1860 and the end of the century. Themselves adhering to no party, they extended their influence to all, and were the admired friends and counsellors of men of every school and of none. They showed the men of their time that it is possible to face every assault on the faith without anxiety and without resentment. They scorned every weapon other than the weapons of the truth, and believed that the truth has only to be made manifest to triumph. All their own work was based on a confident, reasoned, humble faith in Jesus Christ; the most minute consideration of the exact meaning of the smallest Greek particle was an adoring homage that they rendered to the Holy Spirit, the Spirit of truth.[1]

About the year 1890 the ecclesiastical sky began to lighten, and two new movements have at this point to be recorded.

The scene shifts again from Cambridge to Oxford. In 1889 a group of younger Oxford men produced together a volume of essays entitled *Lux Mundi*. They were good essays; but what is most memorable in the volume is contained in about ten pages, written by Charles Gore (1853–1932; Bishop of Worcester 1902–05, of Birmingham 1905–11, and of Oxford 1911–19), in which he affirmed that it is not inconsistent with the Catholic faith to accept the reasonable results of scientific criticism of the Scriptures. Anglo-Catholicism had moved out of the obscurantist fundamen-

---

1. It is interesting to note that, when Lightfoot died, the outstanding German New Testament scholar Adolf von Harnack described his work as 'of imperishable value. . . . There was never an apologist who was less of an advocate than Lightfoot.'

talism in which Dr Pusey would gladly have imprisoned it. Liberal Catholicism had been born.

Lux Mundi was no sudden scintillation. The group of friends who had produced it, and others like-minded with them, held together over the next thirty years, and impressed a special character on what may be called the 'country-house period' of English Church history. All Oxford men, all about the same age, well connected, associated with one another by marriage, Gladstones, Lyttletons, Furses, Mauds, Talbots, Pagets, and others, lived in amity and concord. After Gore, the most distinguished in the group were, perhaps, Francis Paget (1851–1911; Bishop of Oxford 1901–11), one of the best preachers of the age, and Edward Talbot (1844–1934; Bishop of Rochester 1895–1905, of Southwark 1905–11, of Winchester 1911–24). Year after year members of the group met at Longworth, the home of J. R. Illingworth (1848–1915), to share their concerns and to plan for the welfare of the Church of England. Impatient of frills and flummery, of all extravagance and Romanizing tendencies, they stood for a Catholic Anglicanism, rooted in the Prayer Book, in Hooker and Butler, and in the great traditions of the undivided Church.[1] Above all, this group was deeply exercised by that social concern, that sense of responsibility for the well-being of all men in society, which Gore had learned as a boy at Harrow from Westcott, and Westcott in his turn from F. D. Maurice.[2]

The other movement was the rise of the Student Christian Movement in the universities. This movement was wider than the Church of England, and its ecumenical significance will come before us in another chapter. In early days, it was strictly, even narrowly, evangelical in character, and was not unconnected with the revivalistic work of the American evangelists Dwight L. Moody and Ira D. Sankey. Its first effect was to breathe new life into the rather bedraggled Evangelicalism of the late nineteenth century. Once again young men saw visions and dreamed dreams; once again

1. To the end of his life Bishop Gore rarely celebrated the Holy Communion without reading the Ten Commandments in full.
2. The Christian Social Union was founded in 1889.

the call to heroic consecration to the service of Christ was to find a ready ear in the universities not only of Britain but of the world.

At first this new movement brought only limited renewal of strength to the Evangelical cause in England itself. The first emphasis of the Student Christian Movement was on missionary service overseas. The great American leader, John R. Mott, swept across the world with his watch-cry, 'The Evangelization of the World in this Generation'. An astonishing proportion of the ablest Evangelical leaders found their way into the service of the Church in the mission field. Men of the spiritual calibre and intellectual power of Douglas Thornton and Temple Gairdner of Cairo, of G. L. Pilkington of Uganda,[1] of W. E. S. Holland and Henry Holland of India, of E. H. M. Waller (Bishop of Tinnevelly 1914–23, of Madras 1923–42), of Alec Fraser of Uganda, Ceylon, and Ghana, were lost permanently, or for many years, to the Church of England in England. The world Church gained greatly by their many and varied services. Their scholarship was poured into translations of the Bible in many lands, such as the Luganda translation made by Pilkington; into the formation of an indigenous clergy overseas; and into the creation of a Christian literature for the growing younger Churches. More than ever, the headquarters of the Church Missionary Society was the centre of the Evangelical wing of the English Church. But the power and adventurousness of that wing in England were not equal to those of its extensions in the wider fields of the mission overseas.

It was characteristic, however, of the Church of England that its two most outstanding men at the turn of the century could not be fitted into any of the conventional categories of movements or parties.

Frederick Temple (1821–1902; Bishop of Exeter 1869–85;

1. It was of him that the Master of Pembroke College, Cambridge, wrote: 'I have never had any pupil who has gone out, in my opinion, so qualified spiritually, intellectually, and physically.' Filial piety may perhaps add to the list in a footnote the names of Dr Charles Neill (1868–1949) and Dr Margaret Penelope Neill (1870–1952).

of London 1885–96; Archbishop of Canterbury 1896–1902) had obtained at Oxford the high distinction of a double first, in both classics and mathematics. As a contributor to *Essays and Reviews* he had incurred dark suspicions of unorthodoxy, but all his writings give evidence of an intensely strong hold on all the central truths of the Christian faith. As Head Master of Rugby School he had exercised an extraordinary personal influence over both boys and masters, and was one of the first to modernize the curriculum by including in it natural science. He took a leading share in the development of popular and secondary education, and stood behind many of the liberal measures of the second half of the century. As he grew older, he gave increasingly the impression of rugged strength and inflexible determination; but this was balanced by extreme sensitiveness and tenderness of heart, and by an almost childlike faith in the Redeemer. His sermons to the Rugby boys, simple, manly, straightforward, delivered in a lapidary, almost monosyllabic style, are among the few nineteenth-century sermons which could be delivered from the pulpit today.

Mandell Creighton (1843–1901; Bishop of Peterborough 1891–97, of London 1897–1901) was much more of a high churchman than Temple; yet he too never allowed himself to be identified with any one school or party. He had won great distinction as Church historian, as parish priest for ten years at Embleton in Northumberland, as the first professor of Ecclesiastical History at Cambridge, before being called to the See of Peterborough. Never a popular preacher, in the sense of desiring to simplify Christian truth, he could at all times command the attention of the most thoughtful congregations in the country. He held strongly to the concept of the Church of England as the national Church. To him the Anglican *Via Media* was not a pale compromise, but a positive adherence to the Christian revelation, in which there could be wide tolerance for diversities of interpretation, provided that these were combined with loyalty to the great central principles of the Anglican way. His every utterance gives the impression of breadth of understanding and humble recognition of the many-sidedness of truth, with a deep inner piety, all the more impressive because the expression of it is so restrained. It was a tragedy that Creighton had to wear himself out in trying to deal with the eccentricities of the 'Ritualists', more chaotic and undisciplined in his diocese of London than anywhere else in England. There have been few men in the whole history of the Church whose early death has been more deeply and sincerely mourned. It was not without reason that Lord Rosebery described him as 'the most alert and universal intelligence in this land'.

In the period now under consideration, the first steps were taken to reform the government of the Church, and to provide the Church of England with a measure of that self-government which has long been enjoyed by the overseas Provinces of the Anglican Communion.

The first step was the revival of the Convocations of the Church. The Convocation of Canterbury met again, after an intermission of more than a century, in 1855, and that of York in 1861. Government by Convocation is a curiously clumsy business. The Convocations of Canterbury and York can meet together, but usually they meet separately. In each Convocation, the Upper House of bishops and the Lower House of other clergy generally meet separately; but Convocation is a single body, under the presidency of the Archbishop, and all attempts to secure for the Lower House such independent status as is enjoyed in Parliament by the House of Commons have been unsuccessful. Technically, the function of the Lower House is only advisory;[1] yet by custom the Lower House has acquired a veto on the acts of the Upper House, which are not valid without its consent.

As we have seen, at the time of the Reformation Cranmer had desired to see the revival of diocesan synods, in which he would have welcomed the participation of lay people. Just three centuries after his death, in 1857, Convocation

---

1. 'However Convocation may be organized, its authority resides in the Bishops alone . . . the lower clergy in a provincial synod are only a consultative body, whose function is to advise and, if necessary, to check the bishops, who alone can exercise the spiritual authority of the synod.' *Dictionary of English Church History* (1912), art. 'Convocation'.

discussed the question whether the 'counsel and co-opera-tion of the faithful laity' could be secured. No steps were immediately taken; but in 1885 Convocation agreed on a plan for the constitution of a House of Laymen, to be chosen by the lay members of the diocesan conferences, which in the second half of the nineteenth century had come into being in all the dioceses. The House of Laymen of the Province of Canterbury met for the first time in 1886, that of York in 1892. These Houses were not part of Convo-cation. They had no constitutional status – such status could be conferred only by Act of Parliament. They could not be more than advisory. This was only a very small beginning in the recognition of the place of the lay people in the life and ordering of the Church. But it was a step in the right direction; and it is against the background of these first beginnings that we shall have to consider the far wider developments that will come before us in another chapter.

In 1914, when the First World War broke out, there was still grave tension between the various parties in the Church. It was evident that large parts of the population had become, and remained, alienated from the Church. But the great movements of the previous century had left their mark on every part of the Church's life. The training and preparation of the clergy had been greatly improved by the foundation of theological colleges. The clergy in the parishes had never been more devoted or diligent. Church-going was still the fashion among educated people. The voice of the Church was making itself heard in all kinds of national and social concerns. Even though there were certain grounds for anxiety, the Church of England appeared to be almost the most stable part in the fabric of a nation, the most astonish-ing characteristic of which is unshaken stability in the midst of perpetual change.

CHAPTER II

Expansion in the English-Speaking World

THE Church of England was and is a national Church, established and in part controlled by the State. Before this body could serve as the nucleus of a world-wide fellowship, the question had to be asked and answered whether Anglicanism was merely a by-product of the English way of life and constitution; or whether it was a genuine form of Christian faith and practice, such as could maintain itself in total independence of the soil and of the State of England.

On the whole the State connexion has been a grievous hindrance to Anglican expansion. As long as the creation of new bishoprics required an Act of Parliament and bishops had to be consecrated in England under royal mandate, progress was slow, intermittent, and uncertain. The first consecration of a bishop for work in the British colonies overseas was that of Charles Inglis for Nova Scotia in 1787. In the fifty years that followed, only six other sees had been created;[1] and two more (Newfoundland and Toronto, 1839) in the two succeeding years. With the formation in 1841, largely under the impulsion of Bishop Blomfield, of the Colonial Bishoprics Fund, the pace began to improve, and in less than fifty years sixty-nine new bishoprics were brought into being, apart from those created by the Protestant Episcopal Church of America. But still the hand of the State rested on the greater part of what was not yet called the Anglican Communion; and it was only very gradually that the idea of independent national or regional Churches, within the one Anglican fellowship, began to take shape.

Yet the question of Anglican existence in independence of the State had already been asked and answered, not very far

1. Quebec 1793; Jamaica 1824; Barbados 1824; Madras 1835; Australia 1836; Bombay 1837.

away from England, in that small but venerable body the Episcopal Church in Scotland; and with that Church our survey of Anglican development outside England may appropriately begin.[1]

The Episcopal Church in Scotland

Twice after the Reformation, in 1610 and in 1662, episcopacy was reintroduced into the Church of Scotland, and there was hope that that Church might develop as an episcopal Church in fellowship with the Church of England. In spite, however, of the conciliatory policy of Robert Leighton, Archbishop of Glasgow (1611–84), and his colleagues, feeling against bishops was very strong, and it is uncertain whether their government could in any circumstances have endured. But in any case political changes changed also the situation in the Church. At the Revolution of 1688 the Scottish bishops felt themselves unable to take the oath of allegiance to William and Mary. In consequence, the Church of Scotland was established in the Presbyterian form that it has maintained ever since, and Episcopalians became a brutally persecuted minority.[2] In parts of Scotland, especially the north-east, this was a strong minority, but a dwindling minority, owing to the impossibility of training priests and maintaining regular ministrations. After the rebellion of 1745, matters were even worse; the penal laws were so severe that it was not possible for an episcopal priest to minister at one time to more than five persons without running the risk of transportation for life for the second offence. And yet, through all these bitter years, the

1. The survey follows roughly the chronological order of the development of the Churches as independent Provinces within the Anglican fellowship.
2. The venom of Presbyterian feeling is shown in a petition against a Bill for Toleration of Dissenters, introduced into the Scottish Parliament in 1703: 'being persuaded that in the present case and circumstances of this Church and nation, to enact a toleration for those of that way would be to establish iniquity by a law, and would bring upon the promoters thereof, and upon their families, the dreadful guilt of all those sins and pernicious effects both to Church and state that may ensue thereupon.'

Scottish Church kept itself in being, and maintained not merely the episcopal succession, but also its own liturgical tradition, and a strong consciousness of its own independent existence.

In 1788, on the death of Prince Charles Edward, the last legitimate Stuart claimant to the throne of England,[1] the Scottish bishops made their surrender to the house of Hanover, and the penal laws were gradually relaxed. But by that time the Episcopal Church had been reduced, in the words of Sir Walter Scott, to the shadow of a shade, with no more than four bishops and about forty priests. The diocesan and parochial systems had collapsed. Everything had to be begun afresh in the organization of a Church which had lived for many years without a clear confession of faith, with a confused order of worship, and with hardly any discipline.

The bishops took the initiative. In 1811 a synod was convened, attended by deans and other clergy, and declared to be the National Synod of the Episcopal Church in Scotland. A number of important Canons were passed, including a Canon in which the Thirty-nine Articles of the Church of England were accepted on behalf of the Scottish Church. This made possible a rapprochement with the Church of England. In particular it made possible unity in Scotland, through the gradual incorporation into the Scottish Church of the 'Qualified Chapels', in which during the days of the penal laws English and Irish priests had ministered, and which were exempt from the operation of the penal laws and from the jurisdiction of the Scottish bishops.[2]

The Scottish Church, now fully organized in seven dioceses, and with Canons of its own, revised and accepted by the Church in 1925, has certain characteristics which distinguish it from other Anglican Provinces.

---

1. Henry Cardinal Stuart survived till 1807; but, as a prelate of the Roman Catholic Church, he could not be regarded as a claimant to the throne of England.

2. The Thirty-nine Articles are not included in the Scottish Prayer Book, but every clergyman licensed in the Episcopal Church is required to assent to them.

Since the Revolution no archbishop has been elected in Scotland. In 1720 the bishops reverted to the ancient practice of electing a Primus, who holds office for life, but is not a metropolitan. Bishops take the oath of obedience not to the Primus, but to the synod of bishops.

The Scottish Church has not gone as far as other independent Provinces in recognizing the principle of the full participation of the lay folk in all the affairs of the Church. The Provincial Synod consists only of the bishops, the deans (who in Scotland perform the functions for which archdeacons are responsible elsewhere), and the representatives of the clergy. Laymen have a share in the election of a bishop. They have their place in the Consultative Council, which, as its name implies, is purely for consultation and has no executive function. They have membership also in the Representative Church Council; but when this was formed, it was clearly laid down that this Council is the organ of the Church in matters of finance only, and that it 'shall not deal with questions of doctrine or worship, nor with matters of discipline, save to give effect to Canonical Sentences of the Church'.

Since 1637 the Scottish Church has had its own liturgical tradition separate from that of England. As we have seen, the Prayer Book of 1637 was swept away almost before it began to be used; but it was not forgotten. Liturgical studies were a favourite pursuit of Scottish scholars, several of the bishops among them. In 1764 the bishops put out a new form of the Scottish liturgy, revised with the aim of bringing it 'to as exact a conformity with the ancient standards of eucharistic service as it would bear'. This was again revised in 1929. The English Prayer Book is fully authorized in Scotland; in a number of churches the two eucharistic liturgies are used side by side, but perhaps with a tendency for the Scottish liturgy to gain ground. Alone among the liturgies of the Anglican Communion, the Scottish liturgy follows the Rite of 1549 in keeping the great intercession as part of the Canon.

The greatest achievement of the Scottish Church was the consecration of Samuel Seabury in 1784. The Scottish bishops charged Seabury to introduce their form of the liturgy into America; and this was done, with the result that the American liturgy of today is nearer to that of 1637 than to that of 1552. There have always been feelings of close friendship between the Church in Scotland and the Church in America; in welcoming Seabury back to America, the clergy of Connecticut put on record words which well express the permanent feelings of the American Church: 'In the mysterious economy of His providence, He had preserved the remains of the old Episcopal Church of Scotland under all the malice and persecutions of its enemies. . . . And wherever the American Church shall be mentioned in the world, may this good deed which they have done for us be spoken of for a memorial of them.'[1]

Existing as a small minority Church (perhaps three per cent of the population) in an overwhelmingly Presbyterian country with a strong Roman Catholic minority, the Scottish Church exhibits both the merits and defects of minorities; on the one hand great devotion, on the other, a certain narrowness of outlook and a defensive mentality, on which participation in the wider life of the Anglican Communion, and of the ecumenical movement, is having a widening and mellowing effect.

The Protestant Episcopal Church in the United States

When the American Colonies decided to separate from England, thousands of loyalists, including a large proportion of the episcopal clergymen, moved north and east into Nova Scotia and Canada, in order to preserve both their political and their religious loyalty. The Church in the United States was left with a rudimentary episcopate, a group of faithful clergy, a small number of worshippers, and that was about all. As late as 1829 it was reckoned that the Episcopal Church had no more than 30,000 communicants.

The first task of the Church was to organize itself. An

1. The Bishop of Connecticut very suitably took part in the consecration of the Bishop of Aberdeen in 1956.

informal Convention was held in 1784; a more officially organized General Convention in 1785. It was at this Convention that the Church adopted the name 'The Protestant Episcopal Church', by which, in spite of many attempts to change it, it has ever since been known.[1]

The Church was for a long time the Church of the eastern seaboard. Its parochial organization and its rigid liturgical order rendered it less flexible, and less adapted, than some other Churches to deal with the problems of a rapidly advancing frontier, and the needs of a settler population that had almost completely lost such contact as it had ever had with the Christian faith. The real hero of those early days was the Methodist circuit-rider, moving freely with no more than he could carry in his own saddle-bags, and passing from settlement to settlement with a rudimentary and sometimes hair-raising Gospel. The Episcopal Church was usually far in the rear.

The whole land surface of the United States is now covered with dioceses and missionary districts of the Episcopal Church, and every Episcopalian is under the care of some bishop; but this is a process that has only very gradually been completed. The American Church will never forget the heroic labours of its pioneers – such men as Bishop Whipple of Minnesota, the apostle of the Indians; Bishop Tuttle, the first to contend with the Latter Day Saints (the Mormons) in their own chosen land of Utah; and Bishop Philander Chase, the crotchety and irascible frontiersman, who among other good works founded Kenyon College. But in many areas, when the Episcopalians arrived, other Churches were already firmly established in the field. In every part of the United States the Episcopal Church is a minority Church.

The Church was strong in Virginia, where it had once been the established Church, and fairly strong in other southern States. Inevitably, like other Churches, it felt fiercely the strain of the problem of slavery, and the bitter

1. The name 'Protestant Episcopal' appears to have occurred first in a petition from the clergy of Maryland to the legislature of that State in 1780.

divisions of the Civil War (1861–5). Yet, unlike the Methodists and the Presbyterians, the Episcopalians did not fall apart into two permanently divided sections. The eccentric career of the former army officer Bishop Leonidas Polk, who abandoned his episcopal office in order to serve as a general in the Confederate army, naturally aroused a great deal of criticism in the north; but gradually all the difficulties were overcome, and, not long after the return of peace, the unity of the Protestant Episcopal Church was secure.

This Church, like other independent Anglican Churches, has its own Prayer Book. The Communion office is a modification of that which Seabury brought back with him from Scotland. But in America, unlike Scotland, this is the only Communion office in use, and is accepted by all groups and parties in the Church. There is more adequate provision for family worship than in the English Book, and a wider range of special prayers and collects. The marriage service has been a good deal shortened. The form of absolution in the Office for the Visitation of the Sick has been omitted. The rubrics permit a good deal of variety in the shortening of services. But the Preface to the first American Prayer Book of 1789, which still stands unaltered, spoke the exact truth when it affirmed that it will 'appear that this Church is far from intending to depart from the Church of England, in any essential point of doctrine, discipline, or worship; or further than local circumstances require'. A worshipper from any other Anglican Church in the world feels himself immediately at home in the American Church.

Like all the other independent Anglican Churches, the American Church has certain arrangements which are peculiar to itself. It is governed by a General Convention, which meets once in three years, and consists of two houses. In the House of Bishops sit all bishops in active service, and all bishops retired on the ground of age or infirmity. Bishops, whether diocesan, coadjutor, or suffragan, have the right both to speak and vote.[1] The House of Deputies is

1. This system is markedly different from that prevailing in most provinces, where the House of Bishops consists of the diocesan bishops alone.

made up of the clerical and lay representatives of the dioceses and missionary districts.[1] In this most democratic of countries the Episcopal Church long resolutely refused to admit women as deputies. In 1952 three women were elected from the dioceses. The House resolved that they might sit but not vote; whereupon the ladies very properly walked out. In this respect the Church in the United States was less advanced than the Churches in Asia and Africa. On most matters voting is individual; but, where a vote by orders is required, the clerical and lay deputies vote separately, and thus practically constitute two separate Houses. The General Convention has taken over from the English Convocations the highly inconvenient parliamentary practice that the two Houses usually sit separately.

*The Episcopal Church is not alone in making a distinction between dioceses and missionary districts. A missionary district is a pioneer area which is not fully self-supporting and requires subvention from central funds. Some of these districts are in the United States, others in the overseas missions. At one point the position of the bishop of a missionary district differs markedly from that of his brother in a diocese. The American Church holds firmly to the ancient principle that a bishop is wedded to his see, and cannot be translated to any other. This has the advantage of putting an end to certain forms of episcopal ambition, but the drawback that, when one of the greatest sees falls vacant, the most obviously suitable candidates may not be available, as being already bishops of other dioceses. But the bishop of a missionary district, after five years' service in the mission field, is eligible for election to any diocesan bishopric in the Church.

The missions of the Church were organized in 1835. No distinction, in principle, was made between missions in the

1. The first informal Convention of the Church in 1784 had laid it down as a fundamental principle that 'to make Canons there be no other authority than that of a representative body of the clergy and laity conjointly'.

* Changes introduced in 1971 made the greater part of this paragraph obsolete.

undeveloped areas of the United States and missions overseas. But there was a tendency for the Evangelical groups to
look to China, Japan, and Latin America, and for Anglo-
Catholics to concentrate on the American fields. This distinction has long since ceased to have any reality; but traces of it
are yet to be found in the almost solid belt of strongly Anglo-
Catholic dioceses from north to south in the middle West.
One of the anomalies in the American system is that bishops
in the overseas missionary districts retain their full rights in
the American House of Bishops. Thus American bishops in
China and Japan, unlike the British and Canadian bishops,
had full rights of membership both in the Church which they
served, and in that from which they had been sent. But,
paradoxically, this privilege was not extended to the
Chinese and Japanese bishops who gradually took over the
dioceses from the Americans.

The chief bishop of the American Church is known as
the presiding bishop. He is not a metropolitan,[1] and has no
authority in any diocese which has a bishop of its own; but
he has great influence, as the presiding officer of the
General Convention, and as the chief executive of the
Church. For more than a century the senior bishop in order
of consecration automatically became presiding bishop. In
1925 the office was made elective; election is 'for life', but
the presiding bishop must retire after the General Convention which follows his sixty-eighth birthday. In 1943 the
presiding bishop in office at that time informed General Convention that he could not combine the offices of diocesan
bishop and presiding bishop. It was immediately arranged
that he should resign from his diocese, but continue in office
as presiding bishop; and it is now required that any bishop,
on election to the office of presiding bishop, shall cease to be
a diocesan bishop. This is an arrangement without precedent, and must be judged to be a very bad one. Under the
present system, the presiding bishop of the Church is without jurisdiction and without pastoral responsibility. He is
almost completely excluded from exercising any part of the

1. He is therefore, properly, 'The Right Reverend', and not 'The Most
Reverend'.

episcopal office except the consecration of other bishops. It is to be hoped that the American Church will recognize that this anomalous situation is undesirable, and will arrange that the presiding bishop should have, in at least a limited form, genuinely episcopal responsibilities.

The American Church is second to none in its sense of fellowship with the Church of England and the See of Canterbury. It does not always realize as clearly as it might the desirability of cooperation and consultation in areas where both are at work. Both the Church of England and the Protestant Episcopal Church have maintained chaplaincies for English-speaking residents in Europe and the Latin American countries. It is only in very recent years that any attempt has been made to coordinate these activities, or to bring them under a common jurisdiction.

The Protestant Episcopal Church ranks only seventh among the Churches of the United States. But its influence is not to be reckoned only in terms of its numerical strength. In part this is due to social considerations. The Episcopal Church has always been to some extent the Church of an aristocracy. George Washington himself was an Episcopalian, as was John Jay, the first Chief Justice of the United States, and in more recent times Franklin D. Roosevelt. But this is not the whole story. The leadership of the Episcopal Church in recent years has been intelligent and courageous. Its provision for the training of its clergy is adequate, and on the whole better than that available in England. As America moves out of the pioneer period and the revivalism naturally associated with it, as it comes to demand more thoughtful preaching and a more sober, ordered, type of worship, it seems likely that the American Church will be called to play an increasingly important part in the variegated spectrum of American religious life.

The Church of the Province of New Zealand

New Zealand is the most distant from Britain of all the British dominions; yet, in spite of its mixed population, it continues to be in some ways the most British of them all.

The Gospel seems first to have been preached in New Zealand on Christmas Day 1814, when the apostolic Samuel Marsden held service at Ruatura: 'A very solemn silence prevailed; the sight was truly impressive. I rose and began with singing the Old Hundredth Psalm, and felt my very soul melt within me when I viewed my congregation and considered the state they were in.' Missionaries of the Church Missionary Society followed, and the evangelization of the Maori people went ahead. It was discovered that New Zealand was suitable for European settlement, and under the impulse of that strange, restless man Edward Gibbon Wakefield (1796–1862) a large European population came into existence. Through wars and strifes and tensions New Zealand went forward to become, what it is today, one of the very few happy examples in the world of a multi-racial society, based on mutual toleration and respect.

The most important event in the ecclesiastical history of the Islands was the arrival in 1842 of George Augustus Selwyn (1809–78), the first Bishop of New Zealand. A Cambridge man, young, aristocratic, able, determined, tireless, devout, Selwyn was a prince among men. He had come under the influence of the Tractarians, but retained his independence of judgement,[1] and throughout his episcopate found it possible to cooperate heartily with the missionaries of the C.M.S. But from the day of his arrival it was his purpose to develop the Church committed to his care on what he believed to be the true, ancient, and apostolical model.

One of his first acts was the summoning in 1844 of an informal synod of his clergy. This was attended by three archdeacons, four other presbyters, and two deacons, and was intended 'to frame rules for the better management of the Mission and the general government of the Church'. This attracted widespread attention as the first attempt to

1. 'Three mighty men . . . went forth to draw water for us from the well of primitive antiquity; but one was taken captive by the foreign armies which had usurped the well.' *Primary Charge* (1847), quoted in *Life*, Vol. 1, p. 240.

restore synodical government anywhere in the Churches dependent on the Church of England.[1] It was followed by a more formal synod in 1847.[2]

But these were only preliminaries to a far larger and more significant adventure. In 1852 New Zealand had received a certain measure of political self-government, the first step towards what at a later stage came to be called dominion status. The *Church* of New Zealand was legally simply a part of the Church of England. But how could such a status be consistent with the independent authority now coming to be exercised by the colonial legislatures? In 1853 a Bill was brought forward in Parliament to enable bishops, clergy, and laity in the colonies to meet together and make whatever ecclesiastical regulations they might regard as necessary, provided that the standards of faith and worship and the supremacy of the Crown were duly maintained. This met with strong opposition, from, among others, Henry Venn, the great secretary of the C.M.S., who maintained that all resolutions of colonial synods should require confirmation by colonial legislatures – so hard does Erastianism die in the Church of England. The Bill failed to become law; it was becoming increasingly difficult to get any ecclesiastical legislation through Parliament. Was there any way in which the Churches in the colonies could obtain their urgently needed freedom?

Light came suddenly from an unexpected quarter. Mr Labouchère (later Lord Taunton), in a dispatch on the situation in the Canadian Church, used the words: 'I am aware of the advantages which might belong to a scheme under which the binding force of such regulations should be simply voluntary.' This chimed with advice given a little earlier by Mr Gladstone to the Churches in the colonies that they should 'organize themselves on that basis of voluntary

1. Not of course, as is often stated, in the Anglican Communion. It is astonishing how invariably Scotland and America get forgotten.

2. This Synod was attended only by Europeans. Bishop Selwyn was very cautious in the matter of ordaining Maoris; the first Maori deacon was ordained in 1852; and the first priest, by Bishop William Williams of Waiapu, in 1859.

consensual compact which was the basis on which the Church of Christ rested from the first'. It is extraordinarily difficult for those who have always lived in a State Church, in which the coercive power of the State lies behind the discipline of the Church, to believe that the Church can really exist at all if the support of the State is withdrawn, or that the Church can act on its own without prior consultation with the State. Once it is seen that this is an illusion, men simply cannot understand how it has come about that they have lived so long in the world of illusion. 'All at once, but upon some sooner than others, the light of their actual freedom dawned, and they saw that their bondage had been mainly self-inflicted; like some bedridden hypochondriacs, they suddenly believed that they could rise and walk, and they did so.'[1]

In May 1857 Selwyn summoned at Auckland a conference, consisting of two bishops,[2] eight priests, and seven laymen, to draw up a Constitution for the Church of New Zealand. On 13 June 1857 a document was duly signed, entitled 'CONSTITUTION for associating together, as a Branch of the United Church of England and Ireland, the members of the said Church in the Colony of New Zealand'. Selwyn had before him the example of the American Church. The Constitution was skilfully drafted, and has worked well with little need of revision. The General Synod is the supreme governing authority in the Church; but the title included in the Constitution makes it clear that there was no intention then, or at any later time, in any way to separate from the Church of England, and in fact the Constitution itself expressly disclaims any such intention: '1. This branch of the Church doth hold and maintain the doctrine and sacraments of Christ . . . as the United Church of England and Ireland hath explained the same in the Book of Common Prayer and the Thirty-nine Articles of Religion. And the general synod shall have no power to make any alteration in the authorized version of the Holy Scriptures, or in the above-named formularies of the

1. Dean Jacobs: *Church History of New Zealand* (1888), p. 167.
2. The diocese of Christ Church had been formed in 1856.

Church.' A later Act of the New Zealand Parliament (1928) has removed many restrictions on liberty. New Zealand is the most English of the dominions; the Church of New Zealand is in some ways the most English of all the Anglican Churches overseas.

One of Selwyn's chief concerns was the increase of the episcopate. Three new bishops arrived in time for the General Synod of the Church in 1859; and one of their first acts was the first episcopal consecration to take place in New Zealand, that of William Williams to be Bishop of Waiapu. But this episode serves as a further illustration of the grave hindrances placed by the State connexion in the way of the expansion of the Church; as the four bishops then in New Zealand had all been consecrated in England, they could not proceed to consecrate a colleague in New Zealand until permission had been received from the Crown and from the Archbishop of Canterbury. The necessary documents arrived at the last moment, while the General Synod was actually in session. An even more moving event was the consecration, in 1861, of John Coleridge Patteson (who was to die as a martyr in 1871) to be first Bishop of Melanesia. It was apparently through a clerical error in his Letters Patent (34° N. for 34° S.) that Selwyn was charged with the spiritual care of half the Pacific. It fitted well with the scope of his genius to take advantage of this error; and, as a result of his vision, the vast island worlds of Melanesia and Polynesia became ecclesiastically part of the Province of New Zealand (separate Province of Melanesia 1975).

Before Selwyn left New Zealand in 1868 to become Bishop of Lichfield, it was agreed that the bishops should choose one from among their own number to be primate, but that the primacy should not be tied to any one see. This arrangement still continues. It was only in 1922 that the title 'Archbishop' was attached to the primacy.

One special feature of New Zealand organization calls for comment. Although there is perfect equality, in Church as in State, between Maori and 'Pakeha', the Maoris are a minority people, in general on a rather lower level of culture than their European neighbours. There was a real danger, at one time, that they might fail to adapt themselves to new conditions and might die out. That peril is past; but there was still the danger that they might be overlooked in the general ministrations of the Church, and might fail to make their due contribution to its life. With a view to averting this loss, it was decided to appoint one Maori bishop, who would not have a diocese of his own, but would serve as a helper to all the bishops with their Maori work. In 1928 the Rev. F. A. Bennett became Bishop of Aotearoa ('the long low cloud', the Maori name of New Zealand).[1] The arrangement seems to have worked well; and after the death of Bishop Bennett, the second Maori bishop, Wiremu Netana Panapa, was appointed to the office in 1951.

Ireland

The Church has existed in Ireland in unbroken continuity since the days of St Patrick in the fifth century. With the breakdown of the old Roman civilization in the Dark Ages, culture and faith moved westwards to the farthest of the western isles, flourished in a marvellous tradition of art and learning, and flowed back in beneficent streams of missionary enterprise to the continent of Europe. There are some signs in the Middle Ages of a shadowy primacy of Canterbury over Armagh;[2] but the Church of Ireland steadfastly maintained its independence as a separate Church within the unity of Western Christendom.

In the reign of Henry VIII the Irish bishops accepted the changes made by him in the organization of the Church; and, after the episode of Mary's reign, all but two accepted the Anglican settlement introduced by Elizabeth. The Irish episcopal succession, through Hugh Curwen, Archbishop of Dublin (c. 1490–1568), is independent of the English, and

1. The Maori bishop is technically a suffragan of the Bishop of Waiapu.
2. Patrick, Bishop of Dublin (1074–84), was consecrated by Lanfranc, Archbishop of Canterbury, and in his 'profession' refers to Lanfranc as 'Primate of the Britains'. See *Journal of Theological Studies*, N.S. VIII. 1 (April 1957), p. 182.

free from the doubts and uncertainties that have surrounded the consecration of Matthew Parker. Unhappily, as a result of the bitter enmity between the Irish and the Anglo-Norman English-speaking population which prevailed throughout the reign of Queen Elizabeth, the Church of Ireland quickly became the Church of the English-speaking minority, and the true Irish returned to the ever-ready arms of the Church of Rome, being governed by titular bishops appointed by Rome to care for their spiritual welfare. All too little interest had been taken by the Elizabethan prelates in their Irish-speaking compatriots, and the opportunity, once lost, has never returned. The translation of the New Testament into Irish was undertaken in 1573 by Nicholas Walsh, later Bishop of Ossory (d. 1585); but it was not completed and published until 1603, fifty years too late. The Old Testament tarried till 1685, the complete Bible till 1690.[1] The Prayer Book appeared in 1608. By that time the religious division of the Irish people had hardened along lines that have changed little in three and a half centuries.

The Irish Church has always had its great men. Some of these, such as Jeremy Taylor (1613–67), were Englishmen transported to Ireland; others such as James Ussher (1581–1656) and George Berkeley (1685–1753) were Irish-born, though of the English-speaking minority. But all too often, especially in the eighteenth century, Irish bishoprics were used by the Government as pensions for its friends, most of whom found it more convenient to reside in England. Though rarely vicious, they were not very helpful to the Church; and there were so many of them that, if they had all resided in Ireland, there would have been hardly anything for them to do.

The Irish clergy have so often been the object of unjust and unfounded criticism that it may be well at this point to cite the considered judgement on them, on the eve of disestablishment, of one who was admirably placed to form an

1. Most of the translation of the Old Testament was carried through by the good William Bedel, Bishop of Kilmore (1571–1642), and revised by Narcissus Marsh, Archbishop of Dublin (1638–1713).

opinion. In 1867 the Roman Catholic Bishop of Kerry, Dr Moriarty, wrote in a letter to his clergy: 'In every relation of life the Protestant clergy are not only blameless, but estimable and edifying. They are peaceful with all, and to their neighbours they are kind when they can. . . . There is little intercourse between them and us; but they cannot escape our observation, and sometimes when we notice that quiet, and decorous, and moderate course of life, we feel ourselves giving expression to the wish "*talis cum sis utinam noster esses*".'[1]

We have seen in an earlier chapter how the parliamentary measure for the reduction in the number of Irish bishoprics was the spark that kindled the flame of the Oxford Movement. In 1865 the high churchman Mr Gladstone resolved upon a far more sweeping measure – the disestablishment of the Irish Church, and its complete separation from the State. This was part of his great campaign for the conciliation of the Irish people through the granting of Home Rule; and Gladstone, with his illimitable capacity for interpreting as a manifestation of eternal principles anything which he happened to believe to be right, came to regard the disestablishment of the Church as a sacred duty. 'In the removal of the establishment,' he said at Stockport in 1861, 'I see the discharge of a debt of civil justice, the disappearance of a national, almost a world-wide reproach, a condition indispensable to the success of every effort to secure the peace and contentment of that country; finally, relief to a devoted clergy from a false position, cramped and beset by hopeless prejudice, and the opening of a freer career to their sacred ministry.'[2]

The Church of Ireland was almost unanimously hostile to disestablishment. The Primate, Marcus Beresford (Archbishop of Armagh 1862–85), expressed the general view when he said, 'We are not here to amend Mr Gladstone's Bill or to throw out any suggestions respecting it. We condemn it utterly from first to last in principle. We look upon

1. *A History of the Church of Ireland* (1933), Vol. III, p. 326.
2. In 1861, the Church of Ireland had a membership of 693,357, 11.9 per cent of the total population.

it as a confiscation.' Yet, when the blow fell, the Church met
it with dignity, courage, and composure; and perhaps there
would be few today who would not agree that disestablishment was really the greatest blessing that has ever befallen
the Irish Church. The Church was not badly treated by
the government of the day. The rights of all bishops and
priests were safeguarded for their lifetime, and a sum of
£500,000 was paid over to the Commissioners of Church
Temporalities in compensation for the loss of the ancient
endowments.

The greatest evil which the State connexion had done to
the Church was that it had deprived it of all power of self-
government. The Irish Convocations had not met since 1711.
Every attempt of the archbishops to convene synods had
been frustrated by the stone-wall opposition of the government in Dublin. Everything had to be constructed from the
foundation up. On 14 September 1869 the Provincial
Synods of the Provinces of Armagh and Dublin met together
in Dublin, in what was in fact a National Assembly of the
Church. Its first resolution was 'That this Synod deems it its
duty to place on record a declaration that it is now called
upon not to originate a Constitution for a new Communion,
but to repair a sudden breach in one of the most ancient
Churches in Christendom'; its second, that 'the cooperation of the faithful laity has become more than ever desirable'. Nothing is more characteristic of the Church of
Ireland than the part which the lay people play in all its
affairs. One of the first decisions of the lay conference
that met in Dublin on 12 October 1869 was to the effect
that 'it is expedient that the number of lay representatives in the General Synod should be to the clerical in the
proportion of two to one'. The principle was accepted, incorporated into the Constitution, and has remained in force
ever since.

The legislative body of the Church is the General Synod,
of two Houses, which regularly sit together, avoiding the
cumbrousness of the English Convocations and the American General Convention, but can vote by Houses, when this
is held to be desirable. The Representative Church Body is

the legal body which holds the property of the Church. Each diocese has its own synod with full lay representation. Laymen take their share in the election of bishops and in the appointments to the parishes. There is no official assembly in which bishops or presbyters sit without the presence of laymen.

One of the first tasks of the General Synod was the drawing up of Constitutions and Canons Ecclesiastical. This was the period at which the ritual controversy in England was at its height; a strong party in Ireland was deeply suspicious of any innovation in doctrine or ritual, and succeeded in imposing on the Canons governing worship an extremely rigid and conservative character. At the Holy Communion the celebrant must stand at the north side of the Holy Table; and a decision of the General Synod of 1928 made it clear that 'north side' means 'north end' and nothing else. Such gaudy raiment as coloured stoles are not permitted in church; and the placing of a cross on or behind the Holy Table was forbidden until 1964. Many of the clergy would be glad to see a relaxation of some of these mid-Victorian rules; but, in that strongly Roman Catholic land, they had become as the ark of the covenant, and attempts on the part of the bishops and clergy to change them have always to reckon with the possibility of massive resistance on the part of the laity.

In 1879 the Irish Church put forth its revision of the Prayer Book, characteristically conservative and characteristically intelligent. Further revision has taken place at intervals since that date. The book has been enriched by some additional offices, such as that for the institution of a minister, and by a number of well-chosen special prayers. The Athanasian Creed is printed in the book as a witness to the faith, but is not used in public worship. The form of Absolution in the Visitation of the Sick has been replaced by the Absolution from the Order of Holy Communion. Lessons from the Apocryphal Books are not read in church. But the great glory of the Irish Book is its Psalter; Coverdale is still the basis, but the translation has been carefully revised where Coverdale makes nonsense; no longer in

Ireland does 'indignation vex him as a thing that is raw' (Psalm 58), no longer do the 'beasts of the people . . . humbly bring pieces of silver' (Psalm 68). The gain is immense. The Church of Ireland has one lectionary, one pointing of the Psalms, and one hymnal. They may not be the best that can be imagined; but in every church in the land the worshipper can count on regularity and continuity; he will meet the service that he knows.

A new crisis faced the Church of Ireland with the formation of the Irish Free State in 1922, and later with the development of the Free State into the Irish Republic outside the British Commonwealth. There was a momentary danger that the Church might take up a line of opposition as the Church of the continuing English loyalty. Whatever the opinions of individuals, the Church has steadily refused to take up any such position. Under the leadership of such men as T. A. Harvey (Bishop of Cashel and Emly, Waterford, and Lismore, 1935–58), the Church maintained that it was, as it has always been, the historic Church of Ireland, and that political changes could not affect the mission or the vocation of the Church. The Government of the Republic has been admirably and scrupulously neutral in religious affairs. The Church of Ireland is one of the few bodies that exist in both parts of the tragically divided country, and can work for appeasement and mutual understanding. The primatial See of Armagh itself has jurisdiction both north and south of the border.

In worship the Church of Ireland is monochrome. Yet the rigidity of the Canons governing worship has not hindered the development of movements, tendencies, and even tensions within the Church.

But what strikes an English visitor to Ireland, in his first contact with the Church, is the intensity of the loyalty of both clergy and lay people to their Church. Every priest of the Church knows that he will have to work under a clearly defined Constitution and Canons, and that Canons are there to be obeyed. The lay people over most of the country know that they form a minority Church, perpetually threatened by the preponderance of another organization of the

Christian faith. No other Province of the Church surpasses the Irish Provinces in vigour, devotion, and loyalty to the faith as it has been received from the fathers.

Canada

Canada is an immense country, and its ecclesiastical history has been in large measure conditioned by distance, difficulties of communication, and the problems of a sparse and widely scattered population.

Quebec was captured from the French in 1759, and Canada was ceded to Britain in 1763. That was in a sense the beginning of Canadian Church history; and serves as a reminder that one of the problems of the Church is the presence of a large French-speaking minority, almost all the members of which are devoted and fervent members of the Roman Catholic Church.

But we must look back rather further for the true beginnings of the Canadian Church. In 1699 the 'Planters of St John's Harbour' in Newfoundland petitioned the Bishop of London for 'a sober Clergyman, whose first task will be the rebuilding of the Church which was here previously, but which was destroyed by the French'; and we have noted that one of the first acts of the S.P.G. was to extend help to the priest who was living a solitary and arduous life on the bleak shores of the island.[1] The great change came when after the American Revolution Nova Scotia was flooded with loyalist refugees from the seceding Colonies. As early as 1758 the Church of England had been established in Nova Scotia; thus the Anglican immigrants and the clergy who accompanied them found themselves at least ecclesiastically at home in a strange land. It was to care over this vast region that the first overseas bishop of the Church of England, Charles Inglis, was consecrated in 1787. The Crown gave him all such authority and power as were

1. Newfoundland became part of the Dominion of Canada only on 1 April 1949. At first it was ecclesiastically a part of the diocese of Nova Scotia, but as an independent diocese from 1838 to 1949 it was not part of the Church of Canada. It acceded to the Canadian Church in 1947, and became a part of it in 1949.

enjoyed by bishops in England. It was not found possible to make him *ex officio* a member of the Legislative Council of the State; but Bishop Inglis and his successors were for some years personally summoned to sit as members of the Council.

In 1793 the needs of Canada properly so called were recognized by the consecration of Mountain Jacob (1749-1825) to be Bishop of Quebec. At that time there were in the whole wide area no more than nine priests, six in Upper Canada and three in Lower Canada. In the mainly French-speaking area of Lower Canada the Anglican Church has had a hard time of it, and has remained small. In Upper Canada immigration from the British Isles was rapid, and everywhere, especially perhaps in Toronto, one of the greatest cities and one of the greatest church-going cities in the British Commonwealth, the Church is strong. The original Province of Canada was formed in 1862. The greater strength of Upper Canada was recognized when the separate province of Ontario was formed in 1912. The Province of Canada includes Quebec, the Maritime dioceses of the east, and since 1949, Newfoundland.

While all this was going on in the east, Canada west of the Great Lakes was a wide empty land, in which small groups of Indians wandered at their will. As early as 1822 the Church Missionary Society had sent a missionary to the Indians. In 1849 the westernmost bishopric of Rupert's Land was formed. At that date no one could possibly have foreseen the vast changes that half a century would bring. In 1867 the Dominion of Canada was constituted. It was discovered that the great prairies of Canada's Middle West offered the finest wheat-growing lands in the world. Immigrants, mostly of British speech, began to pour in; the future prosperity and greatness of Canada were assured.

The hour was matched with the man. In 1865 Robert Machray (1831-1904), a Scot born in Aberdeen and a Fellow of Sidney Sussex College, Cambridge, became Bishop of Rupert's Land at the age of thirty-four.[1] Machray

---

1. It is interesting to note that his first episcopal act was to ordain to the priesthood in London the Rev. W. C. Bompas, the 'Apostle of the North', who became Bishop of Athabasca in 1874.

was one of the greatest churchmen in the history of the
Anglican Communion, and his fame, outside Canada, is far
less than it deserves to be. His first task was to organize the
scattered missions as parishes, and to press upon his people
the duty of self-support. Foreseeing the great developments
that lay ahead – 'In the Providence of God I have been
present at the birth of a new people' were his words in 1888 –
he pressed steadily for the formation of new dioceses. In
1875 the Province of Rupert's Land was formed with four
dioceses; by the time of Machray's death the number had
increased to nine. For many years the Archbishop combined
with his other duties the office of Chancellor of the University of Manitoba.

Machray was an Evangelical, though the most tolerant
of men. Eastern Canada had been liberally helped by the
S.P.G. The Church in the Middle West was built up largely
through the efforts of the Evangelical Societies, the Church
Missionary Society, and the Colonial and Continental
Church Society. Its ecclesiastical colour is still prevailingly,
though not exclusively, Evangelical.

There is yet a fourth Canada, that of the Pacific Coast.
The first Bishop of Columbia was appointed in 1859; but
the beginnings of the Church were slow and difficult.[1]
Everything was changed when in 1885 the first train of the
Canadian Pacific Railway ran from the Atlantic to the
Pacific. The western colonies were brought out of their
isolation. They have developed as the most English part of
Canada. In 1914 the Province of British Columbia was
founded, and thus a new unity was created among the four
dioceses of the Far West.

From 1873 the whole of mainland Canada had been
united politically in one Dominion of Canada.[2] The question of a corresponding unification of the Church could not
but be raised. The answer was given in 1893, when the
General Synod of the Church of Canada was formed. The

1. As so often the Church Missionary Society was the first in the field
and had sent a missionary to the Indians in that area in 1856.

2. British Columbia joined the Federation in 1871, Prince Edward
Island in 1873.

provinces were left in existence, but all the existing provinces and dioceses were brought together in the unity of the 'Church of England in Canada'. A solemn declaration was made that the Church of England in Canada desired to continue as an integral part of the Anglican Communion, adhering to and upholding all the distinctive tenets and features of the Church of England. The office of 'Primate of All Canada' was created, but the tenure of the primacy was not tied to any particular see. Robert Machray was naturally chosen to be the first holder of the office. It was decided that in future the metropolitans of the various provinces should bear the title Archbishop, which up till that time had been unknown in Canada.[1]

A few points of special interest in the Canadian scheme may here be mentioned.

Usually the Primate of All Canada has been one of the four metropolitans, but this has not been invariably so. Derwyn Owen as Bishop of Toronto, and G. F. Kingston as Bishop of Nova Scotia were, paradoxically, inferior as bishops to the metropolitans of the Provinces in which they served, superior to them as Primates of all Canada.[2] A proposal was for a time under consideration for the creation of a small primatial see, perhaps in the vicinity of Ottawa the federal capital, but it now seems unlikely that this proposal will meet with acceptance.

When the General Synod was formed in 1893, though communications had greatly improved, a journey from one side of Canada to the other was still a long and costly business. Air travel has changed all that; in a few hours a delegate from Vancouver can be in Toronto. Inevitably the present tendency towards centralization is affecting the Church in Canada. As the General Synod increases, it

1. On the asperity with which this decision was received in England, see some amusing correspondence in R. Machray: *The Life of Robert Machray* (1909), pp. 391 ff. The Church of England has always found it extremely difficult to realize that the Anglican Churches overseas have really grown up.

2. Dr Barfoot, was (when elected) Bishop of Edmonton, but later became Archbishop of Rupert's Land (1953). He resigned because of ill-health in 1959.

seems that the Provincial Synods are likely to decrease; how far in this situation the Province can remain continue to be an effective ecclesiastical entity must be regarded as at least in a measure uncertain.

In 1933 the decision was taken to gather into one the far-flung missions among the Indians and the Eskimos in the far north of Canada, and the diocese of the Arctic came into being. This must be, in land surface, the largest diocese in the world – with almost the smallest population. The cathedral at Aklavik lies north of the Arctic Circle. The first bishop, A. L. Fleming, 'Archibald the Arctic', belonged to the heroic generation of missionaries who had endured the extremes of privation in order to bring the Gospel to the remotest of the inhabitants of the earth.

The Church in Canada has its own Prayer Book. This, in its present form, was accepted by the General Synod in August 1961, and came into use in the first Sunday in Advent of that year.

Canada is in some ways an uneasy country. It maintains an intensely strong loyalty to the British Crown, and is proud of its membership in the British Commonwealth. Yet it cannot but be profoundly influenced by its great neighbour to the south. It has before it the prospect of great wealth. Yet in many ways it is an undeveloped country, and over large areas the cultural level is low. The Anglican Church has to some extent failed to rise to the height of its opportunities; it is numerically smaller than both the Roman Catholic Church and the United Church of Canada. In a Dominion so proud of its independence, it could not be regarded as suitable that this independent Church should continue to be known as 'The Church of England in Canada'. In 1955 it adopted the new name of 'The Anglican Church of Canada' – the first of all the Churches officially to adopt the term 'Anglican' in its title.[1]

1. The history of the words 'Anglican', 'Anglicanism', etc., still remains to be written. The first citation for 'Anglicanism' in the *Oxford English Dictionary* is from Charles Kingsley in 1846! I do not know when the term 'Anglican Communion' was first used.

The Church of the Province of South Africa[1]

Some Anglicans were to be found in South Africa from an early date. Bishops of Calcutta were asked, during their long voyage to India, to stop at Cape Town to hold confirmations in that land without a bishop. But the first bishop for South Africa, Robert Gray (1809–72), was not consecrated till 1847. On his first long journey he discovered the extent of the neglect from which the Colony and the Anglicans in it had suffered: 'Since I left Cape Town I have met with *one* English Church, but I travelled nine hundred miles before I came to it. . . . The real wants of this Colony are far greater than I imagined in England; the Church population, too, is far larger, and yet they have been, and still are, in various places aiding the erection of dissenting chapels, for want of any effort upon the part of the Church. . . . We have suffered great spiritual destitution from the long neglect shown by the Mother Church. . . . People do not seem to be aware that up to this time the Church can scarcely be said to have had a footing in South Africa.'[2] It was largely through the devoted and tireless vigour of Robert Gray himself that this state of things was brought to an end.

In 1853 the Bishop was successful in securing the division of his vast diocese, and bishops were appointed for Grahamstown and Natal. To these St Helena was added in 1859. In 1853 the See of Cape Town was raised to be the Metropolitical See of South Africa, and fresh Letters Patent were issued to Gray in the capacity of Metropolitan of the Province.[3] Like Selwyn in New Zealand, Gray was convinced that government of the Church by its own synods and councils must be introduced. He held the first Synod of the Diocese of Cape Town in 1857, and the first Provincial Synod of the bishops of the Province in 1861. This meeting

1. The full official title is 'The Church of the Province of South Africa (otherwise known as the Church of England or the English Church or Church of the Anglican Communion in these parts)'.
2. C. Gray: *Life of Robert Gray*, Vol. 1 (1876), pp. 197 ff.
3. On the title page of his biography Gray is called 'Metropolitan of Africa'; but I have not found any authority for this form of his title.

gave the opportunity for the consecration of C. F. Mackenzie to be bishop in charge of the new mission on the Zambezi. This was the first occasion on which a bishop was consecrated to be a pioneer – to a region where no Church existed, and where therefore a diocese in the sense in which that term was understood in those days could not be formed.[1] Gray acted without a mandate from the Queen. The word 'charge' was used instead of 'diocese'. But 'Charles Frederick Mackenzie, chosen Bishop of the Mission to the tribes dwelling in the neighbourhood of the Lake Nyassa and the River Shire', did take the oath of canonical obedience to the Metropolitan Bishop and Metropolitical Church of Cape Town. This consecration constituted a precedent of immense importance, and greatly helped to free the whole missionary work of the Church of England from some of the trammels imposed upon it by the State connexion.

The exact legal status of an Anglican Church overseas was the problem underlying the most famous ecclesiastical trial in the Anglican world of the nineteenth century. The first Bishop of Natal, J. W. Colenso, had been a distinguished mathematician, and became one of the greatest of Anglican missionaries. He was one of the first to realize, and to maintain against all comers, the essential equality of the African with those of other races, and his potential though still undeveloped capacity. But Colenso was what would now be called a modernist. Like many mathematicians, he had a rather hard unimaginative mind; and this, when applied to the problems of the books of Moses and of the Epistle to the Romans, resulted in his putting forward views which were regarded by almost all Anglican bishops of the time as dangerously heretical. Bishop Gray decided to proceed against Bishop Colenso on a charge of heresy. The trial was held between 17 November and 14 December 1863. Colenso

1. As a specimen of the *legal* thinking of that period, we may cite a letter of Bishop Gray: 'The Foreign Office is at present in a flutter about the possibility of issuing a licence without defining the limits of the Central African Diocese. They have referred the question to the Law Officers!! Shall I name the Mountains of the Moon?' *Life*, Vol. 1, p. 468, footnote 1.

did not attend in person. The trial ended with a declaration by the Metropolitan that the articles of accusation had been proved, and with formal action, depriving Dr Colenso of the bishopric of Natal, and inhibiting him from exercising any divine office within the Ecclesiastical Province of Cape Town.

Colenso appealed to the Privy Council, which gave judgement on 20 March 1865. Without pronouncing on the spiritual aspects of the case, the Court declared that the judgement of the Metropolitan was null and void so far as the civil position and status of the Bishop of Natal was concerned, and that Dr Colenso was still therefore legally Bishop of Natal. Bishop Gray then proceeded to excommunicate Bishop Colenso. In 1868 a rival bishop was chosen and consecrated under the style of Bishop of Maritzburg. The Colonial Bishoprics Fund endeavoured to stop payment of Colenso's stipend; but this too was secured to him by a judgement of the English law courts. It may well be thought that, at this juncture, Colenso should quietly have withdrawn; but, like Gray, he felt that he was standing for a principle – in his case the principle of freedom of thought and of resistance to ecclesiastical tyranny. He held on to his bishopric of Natal until his death in 1883. Part of his diocese sided with him till the end, maintaining its independent position as the 'Church of England in South Africa'.[1]

The affair of Colenso caused immense scandal throughout the Anglican world, and its consequences will occupy us in a later chapter. The situation was evidently intolerable, and it was essential that the legal position should be cleared up. The good that emerged from these painful legal proceedings was the clear establishment of the fact that a Church in a self-governing Colony cannot be in any sense legally a part of the Church of England. On 24 June 1863 Lord Kingsdown delivered judgement in another South African case in the following terms: 'The Church in England, in places

1. Some churches in South Africa still maintain their position, in independence of the Province, as 'The Church of England in South Africa'. But they are not recognized by the Church of England as having any Anglican status.

where there is no church established by law [i.e. in Colonies with legislative bodies of their own, such as the Cape Colony had become in 1852], is in the same situation with any other religious body, in no better, but in no worse position, and the members may adopt . . . rules for enforcing discipline within their body, which will be binding on those who expressly or by implication have assented to them . . . in such cases the tribunals so constituted are not in any sense courts; they derive no authority from the Crown, they have no power of their own to enforce their sentences.' The loss of coercive jurisdiction based on Letters Patent issued by the Crown was in fact the charter of emancipation of the overseas Churches. The Church could now be regarded as being, from the human standpoint, a voluntary society, dependent on a common faith and a common loyalty; clergymen in all the independent Anglican provinces are bound only by engagements into which they have themselves, personally and with full knowledge, entered.[1]

The later years of Bishop Gray's troubled episcopate were much concerned with the building up of a Constitution for his Province, in which the duties of all members of the Church should be clearly set forth, and the rights of all should be carefully safeguarded. The Constitution was provisionally accepted in 1870; it was finally adopted in 1876, by which time Gray had died (1872), and had been succeeded as Bishop and Metropolitan by William West Jones (1838–1908). It provides for full synodical government, by a synod in every diocese, by a provincial synod, which meets regularly once in five years, and an episcopal synod which meets every year. In 1945 the Province stepped into line with the more progressive provinces by admitting women to membership of the provincial synod.

1. One bright spot in this melancholy story is that, on hearing of the death of Robert Gray, Colenso said in a sermon preached on 22 September 1872: 'in labours for this end, most unselfish and unwearied, in season and out of season, with energy which beat down all obstructions, with courage which faced all opposition, with faith which laid firmly hold of the Unseen Hand, he spent and was spent, body and soul, in His service. . . . We all "know that there is a prince, and a great man fallen this day in Israel".'

Inevitably the Church has to some extent been from the beginning the Church of the English-speaking element in South Africa. The Boers were there before the British; they had their own Churches, to which they have from the beginning remained faithful. But gradually and increasingly the South African Church has taken up work among the coloured people (those of mixed African and European blood) and among the African peoples of South Africa. Almost every diocese has its missionary work, and its staff of African priests. No African has yet become a diocesan bishop. A great step forward was taken when in 1962 Canon Alpheus Zulu became Suffragan Bishop of the diocese of St John's (Bishop of Zululand, 1966–75).

The burning question in the South Africa of the twentieth century is that of the relationship between the races of which its mixed population is made up. In general it can be said that the Church of the Province has steadfastly held to the policy of maintaining the equality of all men in the sight of God. Owing to differences of language, Europeans and Africans for the most part worship apart; but delegates of all races meet as equals in the synods and councils of the Church. All the official pronouncements of the Church are in opposition to any government policy that would fix on the African the stigma of a permanent inferiority. It may be that not all the lay people of the Church would uncompromisingly take up this attitude; not all the leaders are in exact agreement as to the policy that should be followed; as to the general and official attitude of the Church there can be no doubt.

South Africa would not wish to be regarded as a 'one colour' province; but it is just the fact that its churchmanship inclines in one direction rather than in another. Bishop Gray felt himself more in sympathy with the S.P.G. than with the C.M.S.; the S.P.G. has a long record of noble service to the spiritual needs of all classes and races of men in South Africa. Such religious communities as the Society of St John the Evangelist (Cowley) and the Community of the Resurrection (Mirfield) have worked in the Province, and have left their impress on its churchmanship. It is likely that

the 'high churchman' will feel more at home in its climate
than the 'Evangelical'.

The Province has followed other Provinces in providing
itself with a Prayer Book of its own. The 'South African
liturgy' was produced as long ago as 1924. This is a form of
the Communion Service which follows in the main the
order of the Scottish liturgy of 1637, and is in some respects
the most satisfactory of Anglican revisions along that line.
In putting forth its own liturgy, the Provincial Synod
affirmed its continued loyalty to the Order in the English
Prayer Book, 'as a sufficient and completely Catholic rite,
endeared to multitudes of churchmen by the most sacred
associations'; and in fact the English Order is still used in
many churches, though it is probable that the South African
liturgy is gaining ground. Since 1924 other forms of service
have been issued; and the Church in South Africa now has
a complete Prayer Book, drawn up by its own authorities,
and adapted to the conditions under which it works.

The Church of England in Australia and Tasmania

The beginnings of Church life in Australia were not exactly
auspicious. In May 1787 the first shipload of convicts was
sent off to Botany Bay. It is typical of the spiritual degen-
eracy of England in the eighteenth century that no provision
whatever was made for the religious needs of the convicts
either on the voyage or in their new home. At the last
moment, through the intervention of William Wilberforce,
one priest, the Reverend Richard Johnson, was allowed to
accompany the fleet. For six years after his arrival Mr
Johnson had to hold services out of doors, wherever he could
find a shady spot. When he succeeded in building a church
out of his own resources, it was promptly burned down by
the convicts.

In 1794 Johnson was joined by Samuel Marsden, whom
we have already encountered as the apostle of New Zealand.
In 1814 Australia was included in the rather extensive territories allotted to the newly-appointed Bishop of Calcutta.

In 1817 the Government appointed five chaplains to minister to a population of 17,000, of whom 7,000 were convicts. Gradually the appalling conditions prevailing in the colony were brought to the notice of the authorities at home;[1] as a first step towards alleviating the miseries of the convicts, it was decided to appoint an archdeacon. In 1829 W. G. Broughton arrived in Australia in that capacity. 'Imagine your own archdeacon,' he wrote, 'having one church at St Albans, another in Denmark, another at Constantinople, while the bishop should be at Calcutta – hardly more distant from England than from many parts of the Archdeaconry of Australia.' It was not much, but it was a beginning.

In 1836 Broughton returned to Sydney as the first Bishop of Australia. The Church was the established Church of the Colony, the salaries of the clergy being a charge on the public revenue, and Broughton himself had a seat on the Legislative Council. In 1842 a bishop was appointed for Tasmania. In 1847 Australia was divided between Sydney, Newcastle, Adelaide, and Melbourne. It was laid down in the Letters Patent issued by the Crown that the new bishops 'shall be subject and subordinate to the Bishop of Sydney as Metropolitan in the same manner as any Bishop of any See within the Province of Canterbury in our Kingdom of England is under the Metropolitical See of Canterbury and the Archbishop thereof'. In 1850 Broughton held his first gathering of bishops, at which the Bishop of New Zealand was also present. One of the acts of this gathering was to determine that the Australian Church as such should organize itself to undertake missionary work among the aborigines of Australia, and among the peoples of the islands adjacent to its shores.[2]

In Canada, the Provinces came first, and a single Church

1. There is an extraordinarily vivid account of them in *The Life and Times of Bishop Ullathorne* by Dom Cuthbert Butler (1926).

2. The last of the Tasmanians died in 1867. The Australian aborigines have experienced immense difficulty in adapting themselves to the new conditions of life introduced by the white man; but remedial measures seem to have stayed the decline in their numbers, and there are flourishing churches among them.

for the whole of Canada followed. In Australia it was the other way about. The General Synod of the Church was formed in 1872. Division of dioceses went forward rapidly, but the Archbishop of Sydney continued to be Metropolitan and Primate of the whole of Australia. At the beginning of the twentieth century, on the suggestion of the Lambeth Conference, separate ecclesiastical Provinces were formed for the civil Provinces of New South Wales, Victoria, and Queensland, to which later Western Australia was added.[1] Each Province had its own archbishop; it became necessary to provide for a Primate of all Australia. It would have seemed natural that Sydney should continue to be, as it had always been, the primatial see. But, if the bishop elected to the See of Sydney was to become also the primate of the whole of Australia, it seemed necessary that the other Provinces and dioceses should be in some way associated with his election. The churchmen of Sydney refused to accept any limitation on their freedom to choose their own archbishop, and therefore Sydney could not be the primatial see for the whole of Australia. It was settled (and here once again the difference from Canada is to be noted) that, on a vacancy, the Australian bishops should elect one of the four metropolitans to be Primate of Australia and Tasmania.[2]

In most respects, synodical government in the Australian Provinces and dioceses follows much the same pattern as that which we have noted in other parts of the world. Laymen have a full share in the government of the Church. Provincial Synods meet every three years, the General Synod once every five years. But there is a strange and paradoxical lack of authority in these higher bodies. 'Considerable weight naturally attaches to the deliberations of these bodies; but they have no real legislative powers. . . . Their acts are binding only on such dioceses as accept them. The

---

1. Four dioceses are ordinarily required for the formation of a Province. But in 1973 South Australia with only three dioceses was recognized as a Province. Tasmania still remains as an extra-provincial diocese.

2. From 1935 to 1947 Dr Le Fanu, Archbishop of Perth in Western Australia, was Primate. The Primacy then returned to Sydney.

Church does not govern the dioceses, but the dioceses the Church. This is clearly an inversion of the rightful order.'[1]

There is another paradoxical feature in the life of the Australian Church. It was slow in becoming an autonomous Church. Gradually self-government was granted to each of the civil Provinces, and later to the Commonwealth of Australia. In each Province the Church was disestablished, and ceased to have any connexion with the State. What then was the nature of the legal nexus, if any, between the various dioceses in Australia and Tasmania and the Church of England in England? This is a most complicated historical and legal question, and tied up with it is the question of legal ownership of a good deal of property which has been given for the work of 'the Church of England in Australia and Tasmania'. In 1905 the question was submitted to counsel; and the answer came back, that 'the Anglican Churches in Australia and Tasmania are all organized upon the basis that they are not merely Churches "in communion with" or "in connexion with" the Church of England, but are actual parts of that Church'. If this is legally so, then the Australian Church has no power, for example, to alter the Prayer Book and other formularies which it has inherited from the Church of England.

An immense amount of effort has been expended in the attempt to win for the Australian Church its spiritual freedom. A draft Constitution was put forth in 1926, since when it has been constantly under revision and discussion. Prior to 1948 it had twice been accepted by the General Synod. But no further action could be taken until the Constitution had been accepted also by at least eighteen out of the twenty-five dioceses. It was not until 1962 that the necessary number of acceptances was forthcoming. Dioceses which held an extreme form of churchmanship in one direction or the other (notably Sydney on the Evangelical side) feared that, under a Constitution giving such spiritual freedom to Australia as is enjoyed by all the other Anglican Provinces, their particular views might be endangered; and so the necessary consent·

1. J. L. C. Wand (ed.) *The Anglican Communion* (1948), p. 119. But see *The Church of England Year Book 1976*, p. 207.

was long withheld. But at last the great change took place. The Church was able to hold its first fully independent General Synod on 8 May 1962.

Three further points in the life of the Australian Church require special mention.

Australia, like Canada, is a land of immense distances, and the problem of ministration to the widely scattered communities has always been perplexing. A great step forward was taken when, under the inspiration of Bishop Westcott of Durham, the Bush Brotherhoods came into being. A Bush Brotherhood has a central Community House, at which the priests who are members of the Brotherhood spend part of their time. For the rest they are out on the road, endlessly travelling from place to place, and seeking out the members of the Church in localities where no established ministry is yet possible. It is a hard, exacting life; of the enormous value of the services of the Brotherhoods to the Church there can be no question.[1]

The Australian Churches have maintained since 1850 a large and expanding missionary work. Australia became politically responsible for Papua, for the larger part of New Guinea, and for the adjacent archipelagos; naturally the Church in these areas claims a considerable measure of support from the Australian Churches. But this is not the whole story. Australian missionaries are serving with distinction in India; and the diocese of Central Tanganyika is in the main staffed and supported from Australia and Tasmania.

The Australian Church is becoming increasingly Australian. There, as in all other parts of the Anglican Communion, shortage of man-power in the ministry is the greatest of all problems. Training for the ministry in Australia is all the time being strengthened, and efforts are being made to raise the standard. More of the bishops are being chosen from the ranks of the Australian clergy. One of the Metropolitans is English; but for the first time three out of the four are Australian-born. The Anglican Church in Australia is now

---

1. The non-Australian reader will obtain a vivid idea of the conditions involved from Nevil Shute's *A Town Like Alice*. Australia has pioneered in bringing the aeroplane into the service of the Church.

the largest and strongest of all the Christian denominations in the continent (about forty-four per cent of the population); and on it rests a heavy responsibility for the spiritual future of these great new lands.

The Church of the Province of the West Indies

One of the first Anglicans to entertain the strange idea that the Negro slaves in the West Indies might have souls was Bishop Joseph Butler. In the Annual Sermon preached on behalf of the S.P.G. in 1739 he spoke as follows: 'Despicable as they may appear in our eyes, they are the creatures of God, and of the race of mankind for whom Christ died, and it is inexcusable to keep them in ignorance of the end for which they were made, and the means whereby they may become partakers of the general redemption. On the contrary, if the necessity of the case requires that they be treated with the very utmost rigour that humanity will at all permit, as they certainly are, and, for our advantage, made as miserable as they well can be in this present world, this surely heightens our obligation to put them into as advantageous a situation as we are able with regard to another.' No record has survived of the effect of the Bishop's gentle irony on his immediate hearers; it is certain that very little was done in the eighteenth century by the Church of England for the welfare of the unhappy slaves.

In this field as in others the Church of England was slow to act, and appeared only when others had already made their presence strongly felt. From the middle of the seventeenth century several of the islands had had established Churches (though always with a tragic paucity of ministers), and provision for them out of public funds; but Claudius Buchanan, writing in 1813, affirmed that Anglican missionaries, viz. those of the S.P.G., were only six in number, whereas those of other Churches – the Moravians, the Methodists, and so forth – amounted to ninety-two.

The Church of the West Indies has to minister to a strangely assorted population. There are those of European

origin. There is a tiny remnant of the original Carib population that was almost exterminated by successive conquerors.
There is the very large Negro population, the descendants of
the freed slaves, and also the population of mixed African
and European origin. There are aboriginal Indians in
Guiana and British Honduras; East Indians and Chinese in
Trinidad and elsewhere. The greatest event in all the history
of the West Indies was the freeing of the slaves. But freedom
in itself solves no problems. Generations of oppression have
left their mark. For centuries regular marriage was impossible for the slaves; it is not surprising that disregard of
Christian rules of marriage is widespread in the islands.
Poverty has been and still is terrible, and is accentuated by
unusually rapid increases of population, and the overcrowding of almost all the islands. The West Indian Province is in
many ways, through no fault of its own, the Cinderella, the
problem child, of the Anglican Communion.

The first bishoprics to be created were those of Jamaica
and Barbados (1824). British Guiana and Antigua were
added in 1842, and four other dioceses in later years. The
Churches were all established Churches, and such generous
provision was made by the Governments that men of eminence were ready to accept what has rather unkindly been
called 'the congenial gloom of a colonial bishopric'.[1] For a
time the C.M.S. maintained work in the islands. The S.P.G.
raised very large sums for the Negro Education Fund.
Everywhere there were signs of rapid progress and development.

A heavy blow befell the Church in 1868, when the
Government of Mr Gladstone in London imposed upon the
islands the policy of the disestablishment and disendowment
of the Church. This was carried out everywhere (with a
limited exception in Barbados); and, though here as everywhere else, the separation of the Church from the State has
been to its lasting good, the loss of the civil emoluments was
a severe shock to Churches the natural development of which

1. In 1824 Bishop Coleridge invited John Keble, then aged thirty-two,
to accompany him as Archdeacon of Barbados. If he had accepted,
would there have been an Oxford Movement?

has ever since been hindered by their poverty. But this was not allowed to stand in the way of canonical and spiritual development. On 25 August 1880 the Archbishop of Canterbury signed the formal Instrument authorizing the formation of a West Indian Province. Thereupon the bishops, with the consent of the dioceses, declared the Churches in the islands to be constituted into the Province of the West Indies 'as a branch of the Church of England in full union and communion with the said Church'. In 1883 the first set of Canons for the Province was drawn up and promulgated.

The West Indian Province resembles others in the general lines of its canonical and synodical administration. But, as in each Province, there are some features which are worthy of special note. The Archbishop is chosen by the bishops from among themselves, and the custom has grown up, though there is no rule, of electing the bishop who is senior by consecration. A bishop consecrated to serve in any diocese in the West Indies makes a declaration that he 'will pay due honour to the Archbishop of Canterbury and will respect and maintain the spiritual rights and privileges of the Church of England and of all Churches in communion with her'. One diocese has obtained permission to use the Order of Holy Communion of 1549, which has thus been resurrected after centuries of neglect.

Nothing has more hindered the progress of the West Indian Church than the poverty of communications between the islands. This has been so grave that, when the bishops of the Province have desired to meet, it has often been easier for them to meet in London than at any point in the West Indies. At last there is hope of improvement, and of the ending of the long neglect under which this 'imperial slum' has suffered. The attempt to create one great independent dominion of the West Indies was unhappily frustrated by local jealousies and ambitions. Jamaica and Trinidad have become separately independent; and plans are on foot for the federation of some of the smaller islands. It is to be hoped that some day the plan for a wider federation may be revived. In the meantime, the fuller development of self-governing institutions will present the Church with new opportunities and responsibilities.

It is sad to have to record that, at the time of writing, the Church is ill-equipped to meet even the opportunities that are already before it. The shortage of clergy is very grave indeed. Codrington College, Barbados, yet once again re-organized, is sending out a steady stream of men into the ministry, but not yet enough to fill the vacancies. The shortage of priests in England inevitably affects recruitment for the overseas Churches. There is also a feeling that the episcopate is still too much a foreign episcopate, and that full opportunity is not yet being given to the West Indian clergy to exercise to the full their gifts and powers.[1]

In this area we encounter a recurring Anglican problem – the astonishing failure of Anglican forces drawn from dif-ferent parts of the world to work together and to coordinate their efforts. All that has so far been written has relevance only to those Caribbean countries which are within the British Commonwealth. But there are also missionary dis-tricts of the American Church in Haiti, in Cuba, in Puerto Rico, in the Dominican Republic, and in the Panama Canal Zone. Work in these French- and Spanish-speaking areas, where the population is mainly Roman Catholic, is very different in nature from that in the islands which have centuries of a British tradition. Yet it is paradoxical that Anglican dioceses in the same region of the world should go on almost unaware of one another's existence, and without any regular and official instrument (short of the Lambeth Conference) for mutual counsel and enrichment.

The Church of the Province of Wales

The appetite for disestablishment grows with success. Once the Church in Ireland had been disestablished, it was certain that what had been thought good for Ireland would be deemed good also for Wales.

1. The situation in 1976 is very different from that which obtained in 1956; almost all the bishops are of West Indian origin.

The ancient British Church in Wales retained its independence until the twelfth century; but then, with the English conquest of Wales, it was merged in the Province of Canterbury. English treatment of Wales was generally scurvy. Many bishops appointed to Welsh sees were non-resident. A Welsh bishopric was often regarded as no more than a stepping-stone on the way to a better bishopric in England. Translations were unreasonably frequent. Even among the best and most devoted bishops, many could not speak Welsh, and were therefore cut off from intimate ministrations to their flocks; for in Wales the Welsh language has remained a living instrument, as is well known to anyone who has ever heard Welsh people sing in Welsh. In consequence of all this the Church lost ground, and the non-Anglican bodies, notably the Calvinistic Methodists, gained where the Church had lost. The Church had become no more than a privileged minority.

The agitation for disestablishment became strong about 1885. The Act of Parliament separating the four Welsh dioceses from the State was passed in 1914, and came into operation in 1920. The Welsh Church was treated less generously than the Irish. It retained all the ancient cathedrals and churches; but the ancient endowments were transferred to secular bodies. The poverty of the Welsh Church since disestablishment has been one of its main difficulties.

Since disestablishment two new dioceses have been created, and the Province now consists of six dioceses. This Province has in certain matters made arrangements different from those in force in other Provinces. A bishop is chosen not by the diocese over which he will preside, but by the Electoral College of the Province; in this all the bishops are members, and meet with six clerical and six lay representatives from the diocese for which the bishop is to be elected, and three clerical and three lay representatives from each of the other dioceses. For the election of an archbishop, who must be one of the Welsh diocesan bishops, a two-thirds majority is required in a College consisting of the bishops, and three clerical and three lay representatives from each diocese. If after three days no candidate receives the necessary majority, the appointment lapses to the Archbishop of Canterbury. The Province has a Supreme Court, consisting of the Archbishops of Canterbury, York, Armagh, and Dublin, and the Primus of the Episcopal Church in Scotland; its only function is to hear charges against the Archbishop, and it has not yet met. The Welsh bishops, though no longer bishops of the Church of England, are invited to attend meetings of the English bishops. The lay folk are at every level associated with the government and administration of the Church. In Wales, layman does not mean lay *man*; women are eligible for election to Church bodies on exactly the same terms as men.

The chief problem of the Welsh Church is the pressure exercised upon it by its much larger neighbour. Every year the Province of Wales loses to England a number of clergymen, attracted by wider spheres of service, better salaries, or what is believed to be a greater freedom of opinion and expression. This is serious, and there is no easy remedy for this bleeding wound. But, in Wales as elsewhere, the Church has gained more by disestablishment than it has lost. Most of the bishops of the dioceses where Welsh is widely spoken are themselves Welsh-speaking. Services in Welsh are provided wherever they are needed. The Church is the Church of the country, and no longer an alien importation ruled from far away. But much lost ground has to be regained before the Church in Wales can claim to play the same part in the life of Wales as is played in the life of England by the Church of England.

The rest of the world

The Church has always regarded it as its duty to follow its members, wherever they may go, and to provide for them the ministrations of the Church in their own language. One of the most important spheres has naturally been Europe; in every capital and in every major port there are considerable British colonies. And the rich Englishman pursued so successfully what Bishop Gore accurately described as his habit of wandering over the Alps in search of bacon and eggs for breakfast and eleven o'clock Mattins that there are almost a hundred Anglican churches and chapels in Switzerland.[1]

These chaplaincies have been organized as the diocese of Gibraltar (1842) for the Mediterranean area, and the Jurisdiction of Northern and Central Europe, under the Bishop of Fulham, who is a suffragan of the Bishop of London. But here again we meet with striking examples of Anglican lack of coordination and cooperation. At certain points the Americans have not unnaturally entered into the field, but in total independence of the Church of England, and at least in a number of cases without any attempt at joint planning or pooling of resources. Thus it comes about that in Geneva there is an English Anglican Church and an American Anglican Church a few hundred yards apart, each officially unaware of the existence of the other, and belonging to wholly separate jurisdictions.[2]

The Church of England has always taken the line that in nominally Christian countries it is not a proselytizing Church – it is there to care for its own children only. It is for this reason that services have been held usually in English, and in no other language; and many of the chaplains have been almost aggressively uninterested in the existence of other forms of Christianity in the area in which they ministered. But it is becoming increasingly doubtful whether this rigid attitude can be maintained. In every European country there are Christians in considerable numbers who cannot find a home either in the Church of Rome or in the local varieties of Protestantism, but are intensely interested in the Anglican combination of freedom with liturgical order. It is said that eighty per cent of the Anglicans in Holland are Dutch by origin and speech. Circumstances may be too strong for the Anglican chaplaincies, and may impose a change of policy and practice.

1. Only seven or eight regular chaplaincies, and a variable number opened in the holiday seasons.

2. The Lambeth Conference of 1878 discussed this problem. See *The Six Lambeth Conferences 1867–1920*, pp. 92–3.

One possible solution of the problem has been found in
Spain and Portugal. During the nineteenth century the
Anglican chaplains advised Roman Catholics who felt that
they could no longer remain in the Roman Church not to
become Anglicans but to form their own national Churches.
Thus the Reformed Episcopal Church of Spain and the
Lusitanian Church in Portugal came into being. Both these
Churches are extremely small, and the Spanish Church
suffers under all the disabilities with which Protestants have
to wrestle in Spain. But each has its own indigenous and
vigorous life; each has its own Prayer Book, based on the
Book of Common Prayer, but with most interesting en-
richments from the old Mozarabic sources.[1] The Church of
Ireland has taken these small and isolated Churches under
its wing. On 23 September 1894 three Irish bishops consecrated one of the Spanish pastors, Sr Cabrera, a former
Roman Catholic priest, as bishop of the Spanish Church.
The second bishop was consecrated in 1956 by one Irish
and two American bishops. The consecration of a bishop for
the Portuguese Church followed in 1958.

One other region may be cited as an illustration of the
astonishing range of Anglican adventures and responsibilities
in the world.

In 1844 a pious and eccentric retired naval officer,
Captain Allen Gardiner, founded a mission on the bleak and
cheerless shores of Patagonia. On this third visit, in 1850, he
and all his companions starved to death. But in spite of this
tragic beginning the work was developed, among this very
simple and primitive people, to the point at which Charles
Darwin wrote of it: 'I certainly should have predicted
that not all the missionaries in the world could have done
what has been done.'[2] In other parts of South America
also, the South American Missionary Society has worked

1. The ancient Latin Prayer Books of the Iberian Peninsula. A brief
account of these books is to be found in *Liturgy and Worship* (1932), pp.
816–17. See also Resolutions 51–2 of the Lambeth Conference of 1958,
and the recommendations of the Committee of the Conference on Church
Unity and the Church Universal (*Lambeth Conference, 1958*, pp. 2. 56–7).

2. Quoted in F. C. Macdonald: *Bishop Stirling of the Falklands* (1929),
p. 70.

among the aboriginal Indian inhabitants with considerable success.

True to its tradition of non-interference with Christians of other allegiances, the Church of England has not attempted to convert nominal Roman Catholics, but has limited itself to the care of the Indians, and to chaplaincies for the large groups of English people resident in all the great cities of the continent. A number of these chaplaincies were instituted by the Foreign Office, and maintained out of public funds. The smallest of all British colonies, the Falkland Islands, gave its name to the bishopric which was created in 1869 for the oversight of these two classes of work. But here a familiar difficulty reproduced itself. When the heroic missionary Bishop Stirling started to go round his enormous diocese, he found that some of the chaplains and churchwardens considered that their connexion with the Foreign Office was quite sufficient, and that the arrival of a bishop was an intrusion rather than a blessing; episcopal supervision is by no means always welcome to those who have accustomed themselves to doing without it.[1]

Such difficulties have long since been overcome. But another was lying in wait. The United States have always felt a special interest in and responsibility for Latin America; and the Protestant Episcopal Church has adhered less strictly than the Church of England to the rule of not proselytizing in nominally Christian countries. That Church has built up a considerable work in Brazil, partly among the Japanese immigrants, partly among Brazilians of other national origins. The Brazilian Church has now (1965) been recognized as an independent national Church within the Anglican Communion. Here again was the problem of concurrent and overlapping jurisdictions, of dioceses which are all Anglican but entirely independent of one another.

1. In 1871, the Churchwardens at one port wrote to the Bishop 'saying that they would not trouble him to remain'! One chaplain, as late as 1880, wrote: 'My letter of appointment from the Foreign Office expressly states that as Consular Chaplain I do not require a licence from any Bishop. As the omission to obtain the Bishop's Licence was quite unintentional, I hope his Lordship and you will overlook the irregularity, and kindly accept the will for the deed.'

Through discussions between the Archbishop of Canterbury and the Presiding Bishop of the Protestant Episcopal Church, progress has been made towards straightening out these anomalies.[1] The Lambeth 1958 Committee on Progress in the Anglican Communion reported that 'South America offers a challenge and an opportunity to the Anglican Communion as a great field for evangelistic work. There is no reason why it should not extend and strengthen its work in this continent. There is every reason why it should assume larger responsibilities there.'[2]

1. Jurisdiction in Brazil was ceded to the American Church in June 1955, with the exception of the British Chaplaincies. A similar adjustment of jurisdictions in Central America was brought into effect in June 1957.

2. *Lambeth Conference, 1958*, p. 2.72. Following upon this resolution, a notable conference of all the Anglican forces interested in Latin America was held, its recommendations were followed up by the formation of a new diocese for Chile and Peru, by an invitation to the Australian C.M.S. to take up work in Peru, and by the formation of a new American Missionary District for the Republics of Colombia and Ecuador.

CHAPTER 12

A Missionary Church

UNTIL the end of the eighteenth century the missionary efforts of the Anglican Churches had been few, weak, and intermittent, and mainly carried on through the agency of those who were not themselves Anglicans. When a small group of Evangelical clergymen met in 1799 to found the Church Missionary Society, they gave a new dimension in the Anglican world to the words 'I believe one holy, catholic and apostolic Church'. At that time the S.P.G. had limited itself almost exclusively to the work among English-speaking peoples, though progressively throughout the last century and a half it has played an increasing part in Anglican missionary work among the non-Christian peoples and in the formation of new Anglican dioceses and Provinces. The work of the S.P.C.K. in India was languishing. This really was a new beginning.

Like most new things in the life of the Church, missionary enterprise was not looked on with favour by those who were at ease in Zion. English candidates did not come forward, and almost all the missionaries of the C.M.S. in early days were Germans. Bishops would not ordain men for missionary work. The missionaries made almost every conceivable mistake. And yet the work went forward, until the Anglican Churches had to be reckoned with as one of the greatest missionary forces in the world. But it is well that those who today take this for granted should not forget the great effort on the part of a minority, and the great afflictions, through which this happy state of affairs has been attained.

India, Pakistan, Burma, and Ceylon

As we have seen, the chaplains in India in early days had taken little interest in the evangelization of the people among whom they lived. A change came, at the end of the eighteenth century, with the 'pious chaplains', the disciples of Charles Simeon, notable among them David Brown, Henry Martyn, Thomas Thomason, and Daniel Corrie. These men were debarred by their appointment as government chaplains from engaging in missionary work. But Martyn succeeded in translating the New Testament into Hindustani, Arabic, and Persian, a unique feat of Bible translation, especially when it is remembered that the translator died at the age of thirty-one; and in winning for Christ one convert, Abdul Masih, the first Indian ever to receive ordination at the hands of an English bishop (1825).

Gradually the English Church spread itself all over India and the adjacent countries. It has engaged in no less than eight different kinds of work.

The first three may be grouped together. Chaplains were provided to care for British residents in India, for British troops stationed in India, and for the Anglo-Indians, the people of mixed origin, many of whom are Anglicans. These chaplains constituted the Ecclesiastical Establishment; they were paid from public funds, and for obscure reasons were under the Defence Department of the Government of India.

On the missionary side, the Church inherited from the Lutherans the fruits of a remarkable movement in Tinnevelly, very near the tip of India, which just at the end of the eighteenth century brought more than five thousand people into the Church in a few months. In 1820 the C.M.S. sent a remarkable German, C. T. E. Rhenius, to Palamcottah. In 1829 the S.P.G. took over the old work of the S.P.C.K. The movement continued to develop under wise and patient guidance. A number of the faithful village catechists were ordained to the ministry. By the middle of the century Tinnevelly could boast an Indian Church of fifty thousand members, largely ministered to by men of Indian race, at that time by far the most remarkable 'younger Church' in the non-Roman Christian world.

The fifth enterprise was a mission of help sent by the

C.M.S. in 1816 to the very ancient 'Syrian' Church, which has existed in Travancore at least since the fourth century A.D.[1] The aim of the missionaries was not to make converts but to help this ancient Church to put its affairs in order, and to repair the harm that had come about through long centuries of isolation from the rest of the Christian world. Much good work was done, and the Bible was for the first time translated into Malayalam. But unhappily misunderstandings arose; a schism developed, and some 'Syrian' leaders left the Church of their fathers in order to become Anglicans – their descendants became the leaders in what was later the diocese of Travancore and Cochin. A later controversy brought into being the reformed 'Mar Thoma' section of the Syrian Church, which we shall meet in another connexion.

A new field opened up with the growth of immense movements towards Christianity among the 'depressed classes', the under-privileged outcastes, who make up one-sixth of the population of India. The movement was strongest in the Telugu country north of Madras, where between 1890 and 1940 perhaps a million people were brought into the Churches, the Anglican share being about one-fifth of that number. Similar movements took place also in the Punjab and in the United Provinces (Uttar Pradesh), but on a smaller scale. The entry into the Church of such large numbers of people – in their pre-Christian state dreadfully poor, illiterate, and demoralized – brought with it many problems with which the Churches are still wrestling.

Seventhly, the Church moved out into the field of the aboriginal, pre-Hindu peoples of India. A not very attractive process of 'sheep-stealing' from the Lutherans of the Gossner mission brought the Anglicans into the uplands of Bihar, and their work among the Mundas, the Oraons and the Hos laid the foundation for the present diocese of Chota Nagpur. The C.M.S. worked among the Gonds, the Bhils, and the Santals, with varying degrees of success. The Todas,

1. Every member of that Church believes with absolute confidence that it was founded by the Apostle Thomas himself. See Bishop L. W. Brown: *The Indian Christians of St Thomas* (1956).

surely the smallest people in the world (not more than 800 all told), have yielded about 250 members to the Church, and have their own church, All Saints, Toda Colony. In Burma the Karens had strangely been prepared for the Gospel by an ancient tradition that one day the white man would come and bring back to them the Word of God, which they had once possessed, but had lost in the course of their wanderings. The largest mission in Burma is that of the American Baptists; but the majority of the Christians of the Anglican diocese of Rangoon are Karens.

Finally, the Church has persisted in the endlessly difficult task of witnessing to the high-caste Hindu and the Muslim. Here one of the greatest instruments has been the Christian College. Robert Noble, following the example of the Presbyterian Alexander Duff in Calcutta, founded in Masulipatam (1843) the college which still bears his name, and was privileged to win a small group of outstanding converts. The blind Anglo-Indian Cruikshank at the C.M.S. Institution in Tinnevelly established the names of some of the best-known Christian families in South India. Colleges at Trichinopoly, Calcutta, Agra, Gorakhpur, and in other centres, carried on with varying degrees of success the dual task of educating the future leaders of the Indian Church and making the Gospel available to hundreds of non-Christian students of the higher castes.

But from the beginning the Indian Church has been in the main a village Church. If today it stands high among Indian communities in its educational level, this is due to the immense labour poured into the village schools by the missions during the nineteenth century.

India had to wait even longer than America for its episcopate. At last, in face of strenuous opposition from the East India Company and from conservative forces at home, Wilberforce and his friends managed to push the necessary measure through Parliament, and the first Bishop of Calcutta was appointed in 1813. Bishop Thomas Fanshaw Middleton (1769-1822) found himself bishop of the whole of India and of Australia, with a measure of jurisdiction over all the countries in between.

The first bishop's greatest contribution was the founding of Bishop's College, Calcutta, which, after many vicissitudes and several reconstructions, is today the strongest Anglican centre for theological training in India. His most perplexing problem arose from the two-fold nature of the ministry that he found working under him. With the chaplains he could deal easily enough; their ministry was only a variation of the parochial ministry in England. But the work of the missionaries corresponded to nothing in the parochial organization of the Church in a Christian country. 'I must either license them or silence them,' said the bishop in his perplexity. His successor, the much loved and deeply lamented Reginald Heber (1783–1826), had no such doubts; without hesitation he licensed the missionaries, and thus brought them within the ecclesiastical system in India. But the sharp distinction between chaplains and missionaries persisted until the end of British rule in 1947.

The dioceses of Bombay and Madras were founded by Act of Parliament in 1833, and in that year the Bishop of Calcutta was given the title of Metropolitan.[1] Thereafter the State connexion caused endless difficulties in the extension of the episcopate. It was almost impossible to get any ecclesiastical acts through Parliament. New bishoprics in India were brought into being by what may almost be termed subterfuges. For Travancore and Cochin (1879) which were 'native States', recourse was had to the 'Jerusalem Act' of 1841, which provided for the setting up of dioceses outside the British dominions. Tinnevelly (1896) and Dornakal (1912), though functioning in every way as dioceses, were legally parts of the diocese of Madras. The Church in India was still simply a part of the Church of England, with no rights of self-government, and bound to every letter of the English Prayer Book.[2] The Church

1. India has never adopted the title 'Archbishop'; the Bishop of Calcutta is the only Metropolitan in the Anglican Communion who habitually used, till 1970, the title 'Metropolitan' and no other.

2. So much so that the Tamil Prayer Book contained unaltered the requirement that the bishop is to satisfy himself that the candidate for ordination is 'learned in the Latin tongue'. We did teach our students some Greek; we did not teach them any Latin.

made greater progress than might have been expected, but the situation was in every way intolerable.

A great step forward was taken with the consecration, on 29 December 1912, of the first Indian bishop. Vedanayakam Samuel Azariah (1874–1945) was the son of one of the early village pastors in Tinnevelly. He had found his special field in the great mass movement area in the Telugu-speaking country. In his thirty years' episcopate he proved himself a great teacher, evangelist, and organizer, and became the beloved and trusted friend and counsellor of all the English bishops in India. He was perhaps the greatest leader as yet produced by any of the third-world Churches.

As early as 1877 the bishops in India agreed that the time had come 'for taking steps to provide a system of synodal action, both Diocesan and Provincial'. But it was only in 1930 that the Church in India, Burma, and Ceylon obtained its freedom. In that year the Church became an independent Province, and was set free to realize its own expressed intention of developing a form of Christianity that would be Asian and not European. In 1883 the bishops had declared that 'we do not aim at imposing upon an Indian Church anything which is distinctively English or even European'. Declaration 21 of the Constitution of the Church includes the words: 'As the Church of England, receiving Catholic Christianity from the undivided Church, has given a characteristically English interpretation of it, so the Church of India, Burma, and Ceylon aspires to give a characteristically national interpretation of that same common faith and life.' A national interpretation of a universal religion involves an element of paradox, even of danger; but, if the nation really is one of the natural orders appointed by God, we may think that, provided that the 'national' is kept strictly subordinated to the 'catholic', such a national interpretation may be permissible and even desirable.

The *Constitution* of the Church of India, Burma, and Ceylon (to which, in 1947, Pakistan had to be added) was perhaps the best of all Anglican constitutions. It was in the main the work of three of the ablest Anglican minds of this century,[1] and provides for almost every contingency that can occur in the life of a Church. Among other things, it offers a method for the election of bishops which avoids almost all the difficulties experienced in other Provinces.[2] The Constitution is democratic throughout. Laymen take part at every level, and women are not excluded from participation.

A peculiar feature of the Indian Constitution is that it recognizes an episcopal synod which is not simply the House of Bishops of the General Council; but, when the bishops meet in synod, it is laid down that they must have with them as assessors an equal number of presbyters and four laymen[3] – an admirable safeguard against any episcopal tendency to talk nonsense.

The most important decision taken so far by the Indian Church was that which, in 1947, made possible the formation of the Church of South India, through the union of the four South Indian dioceses of Madras, Dornakal, Tinnevelly, and Travancore, with the Methodist, Congregational, and Presbyterian bodies in South India. In that area, nearly half a million Anglicans lost their Anglican existence in order to find a new life in a larger united body.

The greatest problem for the Church has been the tragic division between India and Pakistan.[4] As in similarly and tragically divided Ireland, the Church is one of the few bodies which have preserved their unity in spite of the political divisions. The difficulties have been many; the

1. Edwin James Palmer, Bishop of Bombay (1908–28); Edward Harry Mansfield Waller, Bishop of Tinnevelly (1914–23) and Madras (1923–41); and Frederick James Western, Bishop of Tinnevelly (1928–38).

2. The diocesan Council, with clerical and lay membership, may elect a panel of candidates for the bishopric, and at the same time elect two bishops to sit with the Metropolitan and to make the final choice from among the candidates put forward by the diocese.

3. Miss G. I. Mather, who typed these pages, informs me from personal experience that women also are eligible to serve as assessors.

4. The separate independence of Burma and Ceylon seems to have caused no difficulty, since not accompanied by such feelings of bitterness as divide India and Pakistan; though the formation of the Church of North India in 1970 made advisable the creation of a separate Province for Burma.

division ran right through the heart of at least one diocese. It is the special vocation of the Church in that area to show that there is a unity in Christ which stands and remains in spite of all the strains and tensions of separate political existence.

China – The Chung Hua Sheng Kung Hui

Three times the Christian forces had entered China, and three times they had been thrown out. The Nestorians came in through Central Asia in the eighth century – and little more than a century later their work died away in persecution. In the thirteenth century came the Franciscans – but their work died out through lack of support from Europe. The Jesuits followed in the sixteenth century, and for a time won great and wonderful successes – but once again persecution and martyrdom reduced the Church to a hidden remnant. China remained a closed land, until at last in 1844 certain ports were opened to the foreigner. Then for the fourth time the foreign Christians came in. Just over a century later for the fourth time the foreign missionaries were withdrawn or expelled – but this time they have left behind them a great and glorious Chinese Church. Will it be able to survive in isolation and under the constant pressure of the Communists? To that question history alone will give the answer.

Life in China has never been very safe or easy for Christians of any race. From the Chinese point of view the Gospel was brought in at the point of the bayonet; and some missionaries, particularly the French, engaged in political activities beyond the limits of what was wise. The Boxer troubles at the end of the nineteenth century, in which hundreds of missionaries and thousands of Chinese Christians lost their lives, seemed a decisive check to Christian progress; yet this was in effect the beginning of the greatest advance, since in the period of reconstruction that followed the troubles countless young Chinese came to the conclusion that the Gospel alone had the moral power that was needed for the regeneration of their country. The First World War was followed by another time of anti-foreign feeling. Then came the first period of Communist aggression, followed by the disastrous days of Japanese invasion and occupation; now, finally, have followed the victory of the Communists, the liquidation of all the foreign missions, and the almost total isolation of Chinese Christians from their friends in other lands. Few Churches in the world have lived through so exciting and dangerous a history.

Almost as soon as China was opened in 1844, an American mission arrived at Shanghai, with Bishop Boone at its head. An English diocese was founded in 1849 with its headquarters in Hong Kong, a second at Ningpo in 1872, and a third for North China in 1880. Here, as elsewhere, the British and the Americans worked without any common plans or consultation, and without clear delimitation of dioceses. Further complications were added when the Anglican Church of Canada entered the field. Gradually these difficulties were overcome, and a regular diocesan system established. But, although Shanghai was the centre of a diocese supported from America, the large cathedral in the middle of the foreign area of the city was an English cathedral.

The Chinese Church was the fruit of the work of no less than twelve missionary societies. Of the fourteen dioceses that existed in 1948, five had connexions with the C.M.S., two with the S.P.G., three with the American Church, one with the Church in Canada, one with the China Inland Mission, and two were the local missionary enterprise of the Chinese Church itself. China is an immense country, with, until very recently, primitive communications. It has the advantage of one common classical language, but the local dialects are mutually unintelligible. The difficulties, geographical, psychological, and ecclesiastical, of producing unity out of such a situation can well be imagined. Symptomatic of the difficulty was the fact that almost every diocese had its own Prayer Book, with bewildering affiliations to several of the existing Anglican Prayer Books and some features that were peculiar to the Chinese books.

The Holy Catholic Church of China (*Chung Hua Sheng Kung Hui*) came into being in 1912, as a result of a conference attended by delegates from all the dioceses. In 1930 it was recognized by the Lambeth Conference as an independent Province within the Anglican Communion.[1] The Province has no archbishop, but the Chairman of the House of Bishops is recognized as the head of the Church, and is accepted by other Provinces as a metropolitan.

One meritorious feature in the Chinese constitution is the position accorded to assistant bishops. In many Provinces, e.g. India, the assistant bishop is no more than an episcopal curate, wholly dependent on the diocesan bishop who has appointed him, and with a commission which expires on the death or resignation of that bishop. Not so in China. The appointment of an assistant bishop requires the consent of the whole Church, and he has a seat and a vote in the House of Bishops. The first Chinese assistant bishop was appointed as recently as 1918.[2] But, once the first step was taken, the number was rapidly increased. Assistant bishops were of two types – the elderly Chinese pastor with little knowledge of the outside world, and the promising young man educated in Europe or America. The system worked so well that it was in many cases found possible to replace the foreign diocesan bishops by those who had been their assistants. The first Chinese Chairman of the House of Bishops, Bishop Lindel Tsen of Honan, was present in that capacity at the Lambeth Conference of 1948; his former diocesan, the Canadian Bishop W. C. White, was present at the same Conference in the capacity of his assistant bishop.

The Chinese Church holds that responsibility for missionary work rests upon every single member of the Church.

1. On the help given in this development by Archbishop Davidson of Canterbury, see G. K. A. Bell: *Randall Davidson*, (1935), Vol. 11, pp. 1227–9.

2. There was at this time no Province in China. Archbishop Davidson's judgement was very important: 'I am ready to agree that ... a Bishop who is a native of China shall on consecration make his profession of canonical obedience to the Chung Hua Sheng Kung Hui, and the laws, canons, doctrine and discipline thereof, rather than to any other ecclesiastical authority.' Letter of 26 February 1918, op. cit., p. 1228.

One of its first corporate acts was to organize its own mission field in the province of Shensi. A central theological college in Shanghai, and a national office of the Church, also in Shanghai, serve to bring the widely scattered dioceses into closer fellowship with one another.

China had long been threatened by the Communists; but the final collapse of the Kuomintang and the passing of power to the Communists came very suddenly. There can be no doubt that to the majority of Chinese Christian leaders the event brought a sense of liberation and relief. They were weary of the corruptions of the régime of Chiang Kai Shek (whose Government also contained a number of eminent Christians) and its failure to introduce the most urgently needed reforms. Some were, no doubt, sorry to see their missionary friends depart; yet almost all seem to have felt that to be free from dependence on foreign money and foreign Churches gave the Chinese Church for the first time liberty to be itself, to be truly a Chinese Church.

For some time friends of that Church were anxious as to whether it would survive at all. It has lost almost all its institutions, and can carry on no educational, medical, or social work. It has been thrown entirely on its own resources. It lives under a regime which makes no secret of its hostility to all religion. And yet it has managed to carry on. The churches are open, and within the church building a measure of freedom is permitted. The theological colleges are open, and training for the ministry is going forward. One Chinese bishop has been allowed to visit the West. Until 1959 delegations of Christians from the West were allowed to visit China. It is clear that, though the Chinese Church will stand to the last for its own absolute independence of any foreign control, it desires to remain in fellowship with all other Churches throughout the Christian world.

There seems to have been no movement in China in recent years for the amalgamation of existing denominations, as might perhaps have been expected. Chinese Anglicans seem to feel that they have a special witness to bear, and that for the time being this witness can be better borne as a separate Church than through absorption into organic union with the Protestant denominations that so greatly outnumber them.

Japan – the Nippon Sei Ko Kwai

The history of Christianity in modern Japan is even shorter than that of the Church in China. Persecution had succeeded in almost completely obliterating the earlier mission of the Jesuits. The modern period began with the arrival in Japanese waters of the ships of the American Commodore Perry in 1853. The missionaries were not far behind. The first Roman Catholics arrived in 1858, the American Episcopalians in the following year. English missionaries of the C.M.S. entered Japan in 1869, those of the S.P.G. and of the Canadian Church a little later.

The difficulties under which the missionaries worked were considerable. Government edicts against 'the evil sect of the Christians' were still in force. Until 1889 any Japanese who became a Christian was technically liable to the punishment of death, though after 1873 the edicts were in fact generally disregarded. There has never been anything like a mass movement in Japan; conversions have been individual, and numbers have always been small. In 1882 it was reported that all the converts of all Anglican societies together numbered only 761. But this small Church has produced a remarkably high proportion of leaders; and from the beginning the national pride and independence of the Japanese have made it difficult for them to accept as much direction and control from the foreigner as have been customary in other fields. Unlike the Church of India, the Church in Japan has always been a city Church; the evangelization of rural Japan has as yet hardly even begun.

The ordination of the first Japanese clergyman did not take place till 1877, the first ordination in the English mission not till 1884. After that, progress, though still slow, was more rapid. But much confusion was caused here, as in China, by the presence of a number of bishops owing allegiance to separate Provinces of the Anglican Communion, with no clearly defined jurisdiction, and with no common policy. Japanese Christianity owes a great debt to the English bishop Edward Bickersteth (1850–97; South Tokyo 1886–97), son of the Evangelical Bishop of Exeter, who first saw that the only remedy for this state of affairs was the formation of an independent Episcopal Church of Japan, in full communion with the rest of the Anglican Communion, but with perfect liberty to manage its own affairs. At a conference of all the Anglican missions held at Osaka in February 1887 the decision was taken to form such a Church, with its own synods and local councils, and on the principle of full self-government. Agreement was reached on the name the 'Nippon Sei Ko Kwai', the Holy Catholic Church of Japan, which the Church still bears. The conference, looking even further ahead, 'wishes to place on record the desire for the establishment in Japan of a Christian Church which, by imposing no non-essential conditions of communion, shall include as many as possible of the Christians of this country'. The hope that the Sei Ko Kwai might prove the rallying-point for such a larger union has not yet been fulfilled.

Even this conference did not find it possible to deal with the problem of overlapping and concurrent jurisdictions. This was at last straightened out in 1894, and six regular dioceses (later increased to ten) were formed. The question of a Japanese episcopate was repeatedly shelved. It was not till 1923 that, through the consecration of Dr Motoda and Mr Naide to the dioceses respectively of Tokyo and Osaka, a Japanese episcopate began to come into being.

The later history of the Japanese Church is closely linked with the political fortunes and policies of the country. In 1940 the Japanese bishops suddenly announced that all the foreign bishops would resign and be replaced by Japanese. There was no personal ill-will in this decision. It was simply felt that, in the situation of the country then, the Church would be in a stronger position if it cut itself off from everything that suggested dependence on the West either for financial support or for leadership.

In the same year, the Government, under the Religious

Bodies Law, made a strenuous attempt to bring all the non-
Roman Christian Churches together into a single Church.
The Nippon Kirisuto Kyodan, the Church of Christ in
Japan, came into being. Almost all the non-Roman bodies
joined it. The Anglicans pressed constantly for separate
registration; but this was always refused, and in 1942 the
Sei Ko Kwai was dissolved, thus ceasing to have any legal
existence as a corporate body. About one-third of the
Church, with three bishops, at this time decided to enter the
Kyodan, some because they were genuinely in favour of a
wider union on such a basis as the Kyodan offered, others
because this seemed to them to be the only way in which
any kind of corporate existence for the Sei Ko Kwai could
be maintained. In order to maintain a regular episcopal
succession, the three bishops who had joined the Kyodan
raised seven priests to the episcopate.

As soon as the war ended, the Kyodan began to break up.[1]
It was clear that the divided elements of the Sei Ko Kwai
would wish to recover their lost unity. But this was not at all
easy. Those who had not entered the Kyodan, some of whom
had suffered for their faith, persisted in regarding and treat-
ing those who had entered it as mere schismatics. Party
feelings were to some extent aroused, as in the main the
Evangelicals had joined the Kyodan, and the Anglo-
Catholics had remained aloof from it. Above all there was
the problem of the bishops consecrated during the separation. Some held that these consecrations, having lacked a
number of essential elements, were entirely invalid. Those
who had taken part in the consecrations maintained that
they had acted in a time of crisis, and that the consecrations,
though in some respects irregular, must necessarily be held
to be valid. It was wisely decided to remit this burning
question to the judgement of the Lambeth Conference of
1948.

That Conference found that those who had carried out,
and those who had received, consecration, had acted in good

1. It has, however, succeeded in maintaining its existence, with
elements drawn from the Methodist, Presbyterian, and Congregationalist
traditions, and is the second largest Christian body in Japan.

faith. Fully regular election was impracticable, but as far as possible all due forms had been observed. The crisis had come to an end much more quickly than anyone had expected; but this could not be foreseen in 1941, and the episcopal succession would have been gravely endangered if the crisis had been of long duration. In the circumstances, the consecrations, though irregular, were to be reckoned valid, and the bishops, though without jurisdiction, should be regarded as eligible to perform any episcopal function for which the Nippon Sei Ko Kwai might later authorize them. This wise decision was readily accepted by the Church in Japan.

After the war, the Japanese Church was in so enfeebled a condition that the bishops resolved to go back on the decision to maintain complete financial independence of the West: 'While we have no thought of going back upon the declaration of autonomy already made, we shall hold it in abeyance for the time being, in view of the abnormal inflation and for the sake of rehabilitating the Church, and hereby appeal to the sister Churches of England, America and Canada to renew the friendly relationships so unhappily interrupted during the war and to help us both spiritually and materially.' In point of fact, since the war the Japanese Church has received considerable financial help, and a number of missionaries from various Anglican societies are at work. All the diocesan bishops are Japanese, and all the leadership is rightly in Japanese hands; but the Church has given clear expression to its complete freedom from anti-foreign prejudices by inviting one American and one Canadian priest to accept appointment as assistant bishops under the Church.

The Japanese Church has one standard Prayer Book, which in some respects represents a conflation of the English and American books. A process of liturgical experiment has been inaugurated since the war. The first draft of a revised eucharistic liturgy suggested a considerable departure from the Anglican tradition, and by Anglican standards a rather grave impoverishment of eucharistic worship. It is right that every independent Province should have full freedom to make its own liturgical experiments. But the unity of the Anglican Communion is so largely a unity of liturgical experience that any wide departure from its hallowed traditions must be viewed with some anxiety by the other Churches in the fellowship.

The Japanese Church is extremely small. With between 30,000 and 40,000 Christians, its ten dioceses together have within them a far smaller Christian population than many a single diocese in India and Africa. Yet strength is not to be reckoned in terms of numbers alone, and the Sei Ko Kwai exercises a greater influence in Japan than its numbers would suggest. The Kyodan now has a confession of faith of its own; but since its inauguration in 1940 it has suffered from the indefiniteness which often adheres to a Church that is an amalgamation of several distinct traditions. The Sei Ko Kwai has a firm hold on the historic faith of the Church. It is deeply attached to episcopacy. It has a clear understanding of the significance of liturgical worship in the life of the Church. It may well be that it will have a great part to play in the midst of denominations which lack these elements in the Christian tradition, but are coming increasingly to appreciate their value.

Above all else, the Sei Ko Kwai is called to a vigorous policy of evangelization. In Japan far less than one per cent of the population is Christian. The hope of a great Christian movement after the war has not been fulfilled, but there is a readiness to hear the Gospel. The number of baptisms recorded each year has about doubled, as compared with the years before the war. Here is the responsibility and the opportunity of the Holy Catholic Church of Christ in Japan.

Africa South of the Sahara

For centuries Europeans and Christians have been familiar with the coasts of Africa; for centuries Africa's heart has resisted European penetration. Trackless deserts, tropical diseases, the tsetse fly – these things combined to bar the way into Africa to all but the hardiest travellers. In 1857, when

Livingstone reported on his famous journey across Africa from coast to coast, the word 'unexplored' could still be written across the greater part of the surface of the continent. Even fifty years ago there were wide regions that remained unknown. It is only the coming of the aeroplane that has made it possible to treat Africa as a unity; but today it is conscious of its unity, and pressing forward relentlessly to its place in the sun.

In 1864 there were only two wholly independent countries in Africa north of what is now the Union of South Africa; by 1976 every country in tropical Africa had attained complete independence. A century ago Christians could be counted in a handful of thousands; in 1964 there are more Christians in tropical Africa than there were in the Roman Empire when Constantine made Christianity the religion of the State. If the Churches do not lose heart, it may well be that in fifty years' time tropical Africa will be mainly a Christian country.[1]

(i) WEST AFRICA

West Africa is one of the oldest mission fields of the Anglican Communion.

We have already noted the courageous enterprise of Thomas Thompson on the Gold Coast in the eighteenth century. But this was not followed up; and, when at last the Anglicans returned, they found that territory strongly occupied by the Methodists and the Basel Mission. The dioceses in Ghana have a vigorous life of their own; but Anglicans still form a minority Church in the area.

The C.M.S. arrived in Sierra Leone in 1804, and carried on a wonderful work among the Creoles, the freed slaves of the Colony. The cost was terrible; in twenty years fifty-three members of the missionary parties perished. But the work went on, and with such success that in 1852 Sierra Leone became a diocese, and that in 1860 the Sierra Leone

1. In some areas such as Northern Nigeria, Islam is already firmly established; in others it is spreading rapidly; but, if the Churches are alert, the fields are open for a wonderful harvest in what is still pagan Africa.

Native Pastorate Church was founded – the first self-governing Anglican 'younger Church' in the world. For reasons
that will become clear in another context this was a prema-
ture step, and hindered rather than helped the development of the Church. The Creoles of the Colony had little
sympathy with the inland tribes of the Protectorate, and
have never reached out towards them in vigorous evangelistic work.

The diocese of the Gambia owes its origin to the missionary enterprise of the West Indies. In 1851 the Reverend
R. Rawle, at that time principal of Codrington College
(later Bishop of Trinidad, 1872–89) wrote: 'We want to
leaven the West Indian dioceses with missionary feeling. . . .
A great reaction is to be stirred up, opposite in direction as
in character to the traffic by which these colonies were
peopled, sending back to Africa, as missionaries, the descendants of those who were brought over here as slaves.' It was a
noble idea; but, though it has never entirely been lost sight
of, the extreme poverty of the West Indies has hindered the
development of the work on any large scale, and the Gambia
remains the Cinderella of the West African dioceses.

Liberia is the only African mission of the Protestant
Episcopal Church in America. This was founded in 1835,
and in 1851 placed under the supervision of a bishop. Here,
as in Sierra Leone, the divergence between the English-
speaking descendants of the slaves and the inland tribes has
been a perpetual difficulty; the work has progressed only
slowly and with many setbacks.

Nigeria is the greatest of West African countries, and it
has become one of the greatest of Anglican mission fields.
In 1846 Henry Townsend entered Abeokuta, and the Gospel
began to take root among the vigorous intelligent Yoruba
people. In 1857 the freed slave Samuel Adjai Crowther,
whose name stands first on the roll of the great Fourah Bay
College in Sierra Leone, was established as a missionary at
Onitsha on the banks of the lordly Niger. In 1864 Crowther
was called back to England to be consecrated in Canterbury
Cathedral as the first Anglican bishop of non-European
race.

It was a great venture. But it was ill-conceived and ill-conducted, and it ended almost in disaster. Half a century later, the great Bishop Azariah in India had as helpers three exceptionally able European archdeacons and a very strong missionary staff. Azariah knew (though many leaders have still failed to realize) that the handing over of new power and responsibility to the indigenous Church should be accompanied by a strengthening rather than by a diminution of the missionary personnel. Crowther was crippled from the start by Henry Venn's false theory that the formation of the 'native Church' is the moment for the disappearance of the 'mission'. Crowther's colleagues were English-speaking Sierra Leoneans, who regarded themselves much more in Crowther's phrase as 'black Englishmen' than as Africans, and lacked both natural sympathy for the Nigerians and the experience necessary for the successful building up of a pioneer mission. When the old bishop died in 1891, everything was in a state of terrible confusion. The great labours of three African long-timers, Bishops Herbert Tugwell (1894–1919), F. Melville-Jones (1919–40), and Bertram Lasbrey (1922–45), were needed to put things in order again. Though a number of Africans have rendered distinguished service as assistant bishops, the experiment of an African diocesan bishop was not to be tried again till two generations had passed (1952).

The question of a West African Province was taken up at a meeting of bishops in Lagos as long ago as 1906, but nothing came of it. It was raised again in 1944 with better success; and in 1951 the West African province was inaugurated in Freetown by the Archbishop of Canterbury, who personally paid a visit to West Africa for the purpose. Five dioceses of British or West Indian origin were brought together in the Province. It is typical of Anglican lack of logical coherence that the Missionary District of Liberia, which is dependent on the Church in America, does not form part of the Province though it has a common frontier with Sierra Leone.

Until the year 1963, the Province was under a

temporary Constitution, which granted to the Archbishop of Canterbury certain continuing rights of supervision. The work of constitution-making went steadily forward. When it had been completed, the Province was able to assert its complete independence of Canterbury, and its full equality with other Anglican Provinces.

One of the immediate results of provincial autonomy has been the creation of new dioceses. Where there were two dioceses in Nigeria, there are now eleven, and all of these are presided over by African diocesan bishops. Even earlier, most of the dioceses had gone a considerable way in the direction of self-government, and were familiar with diocesan constitutions. To create a Province on paper is not difficult. To fill it with reality is harder, when the dioceses lie strung out over nearly two thousand miles of African coast-line, and are separated from one another by considerable stretches of French territory. The unity of the Province is spiritual; but that unity will find in the aeroplane a singularly useful ally.

(ii) CENTRAL AFRICA

The work of the Universities' Mission to Central Africa, that most Anglo-Catholic of societies, had as its patron saint the Scottish Congregationalist David Livingstone. There is a unity in the faith that transcends the tragic differences between denominations. It was the challenge of Livingstone, on his return from his first great journey, that led to the formation of the U.M.C.A., which after a century of separate existence, decided in 1964 on amalgamation with the S.P.G.

Most Anglican missions have been begun by priests or laymen. The bishop has followed comfortably at a much later stage, when the work has already been well established. The Universities' Mission rightly decided to reverse this order, and to make the bishop the pioneer. We have already read of the consecration of C. F. Mackenzie in Cape Town in 1861, to be the head of the missionary party. Such a method at once solves one of the recurrent and perplexing problems of missionary work. As Bishop Steere wrote in 1881: 'Our missionaries are not dictated to by any home committee.

The Church has been a missionary body from its foundation, and its Episcopate are by the very nature of their
office the chiefs of its missions.'

In the early days disaster followed disaster. Within a year
Mackenzie had died, and other members of the mission did
not long survive him. To Livingstone's intense indignation,
Bishop Tozer decided to withdraw from the mainland and
to settle on the island of Zanzibar. At the moment this may
have seemed pusillanimous; but no one who has stood in
the great cathedral, built on the very site of the old slave-
market of Zanzibar, is likely to doubt the wisdom of this
temporary withdrawal, to be followed by renewed advance
inland from a base that could be securely held.

Within a few years the mission was pressing forward into
what is now Tanzania. Another party was established
on the shores of Lake Nyasa and on Likoma Island. A third
had moved south to found what later became the diocese of
Masasi. The number of converts has never been overwhelming. The mission has believed in steady development,
careful preparation of candidates for baptism, very careful
training of priests, and a steady attempt to transform the old
African ways after the pattern of the Gospel, retaining and
sanctifying all in those old ways that is susceptible of baptism into a new and Christian life. This aim was clearly set
forth by Bishop Hine in 1899; 'What this Mission has always
proposed to aim at is the building up of a native Church . . .
native in the true sense of the word; the Church of the
people of the land, irrespective of European influence, and
adapting itself to the special circumstances of the race and
country in which it exists.' The greatest specialist in putting
this into practice was perhaps W. V. Lucas, the first Bishop
of Masasi. It is interesting, however, to note that, in contrast to the C.M.S., the Universities' Mission to Central
Africa did not put forward its first African candidates for
the episcopate until 1963.

*The Church should not be dependent on the civil power;
yet all Church history shows that what happens in the world
of government has its influence also on the life of the Church.
One of the great events in recent African history was the

formation in 1953 of the Federation of Southern Rhodesia, Northern Rhodesia, and Nyasaland, a noble experiment which has ended in tragic failure. South Africa has its doctrine of *apartheid*, the separation of the white and the black races; West Africa is so exclusively an African country that the question of partnership between the races does not come up in any burning fashion. But the Rhodesias are a white man's country and a black man's country – a land of wonderful opportunity and peril. If partnership could really be made to work here, a precedent would have been set for all lands in which men of different races have to learn to live together; failure here would make the outlook for the future, here and elsewhere, dark indeed.

*It was natural that the Church should think it well to adapt its organization to that of the new country which was coming into being. The three dioceses of the Rhodesias, together with Nyasaland (shorn of its Tanganyika territories, which then became the diocese of South-West Tanganyika) joined on 8 May 1955 to form the new Province of Central Africa. This is in many ways a unique Province. In the first place, its numbers, both of clergy and of communicants, are very small. And, more even than the Church of India in the days of British rule, it is called to be a Church which cares faithfully for the white man, and at the same time is a truly missionary Church among the Africans. It has drawn up a Constitution, mainly on the pattern already familiar to us in so many other areas, but naturally with a specially close affiliation to the Constitution of the Province of South Africa. But, just because the Province is so new and small, rather more rights of oversight and guardianship are left to the Archbishop of Canterbury than is the case in the more fully independent Provinces.

(iii) EAST AFRICA

The first missionary of the C.M.S. in East Africa, the German Dr J. L. Krapf, reached Mombasa in 1844. Almost his first task was to bury his wife and child; but he wrote to the Society: 'Tell our friends that there is now on the East

African coast a lonely missionary grave. This is a sign that you have commenced the struggle with this part of the world; and as the victories of the Church are gained by stepping over the graves of her members, you may be the more convinced that the hour is at hand when you are summoned to the conversion of Africa from its Eastern shore.' In the following year, Krapf was joined by J. Rebmann. So little was then known of Africa that, when the two discovered Kilimanjaro, much of the learned world refused to believe in the possibility of snow on the Equator.[1] These two pioneers hung on for a generation under immense difficulties; there was no indication then of the mighty harvest that was to follow.

This harvest was to be reaped six hundred miles from the sea in a then entirely unknown land. On 14 April 1875 H. M. Stanley wrote a famous letter to the *Daily Telegraph* in which he challenged the Churches to send Christian teachers to Uganda. The ever-adventurous C.M.S. took up the challenge; a mission was dispatched, and held on in face of the deaths and difficulties attendant on all African missions in those days. In 1878 the famous Scottish engineer Alexander Mackay arrived, and the Church began to take root. Then, in 1879, the real disaster happened; two French Roman Catholic missionaries of the White Fathers arrived in Uganda. Nowhere in the world have the rivalries of Christians been more shamefully and more harmfully displayed than in Uganda. It was not a missionary historian who wrote: 'Dr Cust of the C.M.S. travelled to Algiers in the vain attempt to beg the Cardinal [Lavigerie] not to send these missionaries, with such vast stretches of pagan Africa still untouched, to work in the very place where the C.M.S. had begun. . . . There is no need to emphasize the depth of bitterness aroused in the hearts of those already there by the arrival of the new missionaries who made claims to higher authority and truth, nor the bewilderment and disunity created among the Baganda by the tragic divisions of Christendom. . . . The evils of Christian disunity were thus seen

1. Rebmann on this quietly remarked that, having spent some years of his youth in Switzerland, he knew snow when he saw it.

at their maximum, as both tended to compete for the support of the King and the ruling class.'[1]

The accession of the weak and vicious Mwanga to the throne of Buganda in 1884 was followed by a bitter persecution, in which at least thirty young converts, Anglican and Roman Catholic, were burned alive. The Bishop, James Hannington, approaching from the east, was caught and speared to death with many of his company. Far more harmful was the civil war (1892), waged under the rival banners of Anglicans and Roman Catholics, but really between the pro-French and the pro-British parties which was brought to an end only with the proclamation, in 1894, of the British protectorate.

Yet nothing seemed able to stay the progress of the Gospel among the vigorous and intelligent Baganda, and later among the neighbouring tribes.[2] It is probable that not less than half the population of this country of seven million inhabitants is by now nominally Christian, and that the twenty Anglican dioceses together have to care for much more than a million Christians.

This is not all. Fifty years ago J. J. Willis, later Bishop of Uganda (1912–34), sailed down the lake and started a flourishing work in the Kavirondo country of Kenya. The C.M.S. pressed up from the coast, and developed missions round Nairobi and among the Kikuyu. The Mau Mau crisis has shown both how superficial much of the work has been – inevitably superficial as missionary work must be when the missions are kept so dreadfully short of missionaries – and how splendidly good and solid some of it has been. In Central Tanganyika, the area taken over by the Australian C.M.S., progress has been less rapid, and the land is less thickly populated. But there, too, a growing African Church has been coming into being among peoples of varying race and language.[3] The consecration of four African assistant

1. Margery Perham: *Lugard: The Years of Adventure 1858–1898* (1956), pp. 214–15. The whole of this early period in Uganda is brilliantly and impartially lighted up by this experienced writer.

2. It must be remembered that the Kingdom of Buganda, ruled over by the Christian Kabaka, formed only one part of the Uganda protectorate.

3. But almost all, fortunately, able to use the *lingua franca*, Swahili.

bishops in Uganda Cathedral by the Archbishop of Canterbury on 15 May 1955 drew world-wide attention to the progress that had been made. And the great Revival movement, which, starting in the Ruanda area of the Belgian Congo, has brought new life to so many African Christians and Churches, has found echoes far beyond the shores of the African continent.

*With nearly a million Anglican Christians in East Africa, it might seem that the time should long ago have come for the formation of a Province. The idea was, in fact, discussed as early as 1927; but, though the East African bishops met regularly (and unofficially) for consultation, progress towards a Province was slow in coming. For this there are several reasons. One is ecclesiastical. The very strongly Evangelical dioceses that stem from the work of the C.M.S. found it very hard to picture themselves associated in one organization with the very strongly Anglo-Catholic dioceses of the U.M.C.A. As African Christians travel more widely and see more of the world, this is a difficulty that comes to be less acutely felt. Far more serious is the political difficulty. Suspicion of Kenya, and the policy loudly proclaimed by some of the settlers of keeping Kenya as a white man's country, aroused such intense suspicion in other parts of Africa, that any suggestion of provincial union which would involve closer fellowship with the Churches of Kenya met with fierce opposition in the Church of Uganda. The effort to unite the three territories, Uganda, Kenya, and Tanganyika, had to be abandoned as impracticable.

*Since the single province could not be formed, the decision was reached to have one province of East Africa for Kenya and Tanganyika, and a second for Uganda. The former was inaugurated on 3 August 1960, the latter on 16 April 1961. Seventeen dioceses have been constituted (9 and 8); of these, at the time of writing (1964) seven are under the care of African bishops; but there is no doubt that this proportion will rapidly increase. Political changes give cause for some anxiety; but there is good hope that the Church will be able to maintain itself, and to play a leading part, here as elsewhere, in the quest for Christian unity.

*The Church in Africa has suffered under its own gigantic successes. Those who love it most are best aware of its weaknesses and failings – the persistence of old pagan customs, disregard of the Christian standard of marriage, belief in witchcraft, worldliness, and ambition. All these things are true. The great need is not for better organization, but for the deepening of spiritual life. In this the lead must be taken by the sons of Africa themselves – they alone can perfectly understand their own people – but there is still a tremendous task to be undertaken by the African and the European together.

Other Lands

*It is impossible to write in detail of all the areas in which the Anglican form of the Christian faith has taken root. A mere catalogue of dioceses would be dreary and uninspiring. The Anglican Communion makes no claim to world-wide jurisdiction or to universal mission. Yet there are few large areas of the world in which it has not been led to pitch its tent; and a few specimens may be added to complete the impression of its world-wide expansion and far-flung responsibility.

Korea is a land which must not be confused with either China or Japan. It has undergone deep influences from both these larger countries. But the Koreans have their own racial type, their own language, and their own culture, No people in the world rejoiced more greatly over the recovery of their independence than the Koreans at the end of the Second World War.

Anglicans entered Korea late in time, when American Methodists and Presbyterians, and the Roman Catholics, were already entrenched. The first bishop was consecrated in 1889. The mission took clear shape under Bishop Trollope (1911–30). Trollope gave the young Korean Church three great gifts – the most beautiful Anglican cathedral in the East; a deep and abiding interest in Korean culture; and the sense that it should serve as a link between, a rallying-point for, the various Christian forces that were striving for the soul of Korea. He was one of the first to discern the perils of nationalism, and to draw a clear distinction between a 'native' and a 'national' Church. 'It is quite possible to build up a "native" Church, racy of the soil, without attempting to organize a "national" Church, with all its undesirable political implications; the surest way of stifling and perverting the healthy and legitimate development of a "native" Church lies in clamping it to the wheels of a "national" chariot.'[1]

The work of the diocese was disrupted by the Second World War. Hardly had it begun to reorganize itself, when it was again disrupted by the disaster of the Communist invasion. The heroic Bishop A. C. Cooper suffered long imprisonment. Several members of the small staff of the mission died on the long march into captivity. The effects of these losses have not yet been made up. But the small Korean Church stands as a witness to the reality of the Catholic tradition in a country where in the future its value is likely to be increasingly appreciated.

Madagascar[2] lies near to Africa, but it is not part of it. Its peoples seem to be in the main of Polynesian stock, and their common language, Malagasy, bears no relationship to the languages of the African continent.

The Malagasy Church has had a varied and exciting history. The missionaries of the London Missionary Society who arrived in 1818 found a ready hearing, translated the Scriptures, and laid the foundations of a Church. Then followed the period of the fearful persecution under Queen Ranavalona, which lasted for a whole generation (1828–61). Nothing could exceed the heroism of the Malagasy Christians, many of whom died as martyrs. When the persecution at last ended, it was found that the Christians were more numerous than when it had begun.

1. Quoted in J. McL. Campbell: *Christian History in the Making* (1946), p.177.
2. On this, see a useful article by T. B. Hudson in *The East and the West Review*, April 1957, pp. 54–60.

When the island opened up again, both the C.M.S. and the S.P.G. entered the field. With the consecration of the first bishop in 1874,[1] the C.M.S. withdrew, and since then the field has been held by the S.P.G. At that period British influence was strong; but with the coming of the French everything was changed. In 1865 France had agreed with Britain that both countries should respect the independence of the island; but in 1895 the French stormed the capital, Tananarive, Malagasy royalty was brought to an end; and, worse than this, the French Government proved for a time to be violently anti-Christian and anti-clerical, causing the closing of hundreds of Christian schools and of a certain number of churches. The end of the Second World War was marked by a terrible insurrection, in which plans had been made for the extinction of every foreigner in the country.

In all this tormented history the S.P.G. held quietly on its way, maintaining friendly relations with the Protestant forces in the island, but finding relations with the French Jesuits extremely difficult. The Anglican community in the island has always been very much a minority Church, including not more than five per cent of the non-Roman Christians of Madagascar. The great missions are those of the Lutherans in the south of the island, of the Congregationalists and the Friends in the centre and north. Movements for Church union in Madagascar have been very strong in recent years; but from these the Anglican diocese has not unnaturally held aloof – its characteristic witness could hardly be rendered in close fellowship with bodies the extreme Protestantism of which has sometimes rendered co-operation difficult. A great step forward in the life of the Church was the consecration, in 1956, of the first Malagasy assistant bishop Jean Marcel, who has since become diocesan bishop.

1. It is interesting to note that Bishop Kestell-Cornish was consecrated, at the request of the Archbishop of Canterbury, by that ever-serviceable body the bishops of the Scottish Episcopal Church. At the last moment, the royal mandate for the consecration of a bishop for Madagascar had been withheld by the government of which Mr Gladstone was the head.

Mauritius, with the Seychelles, is a small and remote colony; but nowhere in the world is the task of the Church more complex and difficult. Mauritius was taken over from the French in 1814; French it remains to this day in language and to some extent in religion. But the population is an astonishing mixture of French and British colonists, of Africans, Malagasys, Hindus, Malays, and Chinese. No Anglican bishop had ever visited it until Bishop Chapman of Colombo appeared in 1850, confirmed a number of candidates, and consecrated three churches. The diocese was constituted in 1854; and a bishop with about twenty priests attempts to meet the needs of a flock of many races and tongues in the other islands as well as in Mauritius, and to carry on missionary work among the non-Christians who still form the majority of the population. (The Seychelles became a separate diocese in 1972.)

Rather more detailed consideration must be given to two large areas to which religious and political factors give a certain homogeneity of their own.

The Muslim World stretches from Albania to Indonesia, and from the heart of Africa to Western China. Its intellecttual centre is in the ancient Christian lands of Egypt, Palestine, Syria, and Iran, with its spiritual centre in Arabia. Nothing in the history of the world since the First World War is more significant than the resurgence of independent Arab States, fired by a fanatical renaissance of Islam, and lying across many of the greatest trade routes the world. Christian missions among Muslims have always been difficult – the deep hatreds generated by the Crusades have not yet died away. But from the time of Ramon Lull onwards, some of the best and noblest spirits in the Christian world have felt led to devote themselves to the presentation of Christ to the Muslim.

The Church of England officially entered this field with the strange plan, launched in 1841, for a joint Anglo-German bishopric in *Jerusalem.* Many motives were involved. The King of Prussia seems to have had in mind a German protectorate over Protestants in the Levant parallel to that exercised by the French on behalf of Roman Catholics, and by the Russians on behalf of the Orthodox. He seems also to have had plans for the introduction of episcopacy into the Old Prussian Union, which had been formed in 1817 from the Lutheran and Calvinistic Churches in his dominions. It was agreed that the bishop should be alternately nominated by the British and by the German authorities, but that he should always be in Anglican episcopal orders. Never in Church history has there been a more unpropitious attempt than this to build with untempered mortar. The scheme was welcomed and supported by many leading Anglican churchmen; the difficulties do not appear to have been foreseen.

The first bishop, Michael Solomon Alexander, was a converted Rabbi of considerable distinction; but he survived only four years. Then in 1846 the German choice fell on Samuel Gobat, a Swiss from the Jura, a brilliant linguist, who had served the C.M.S. as a missionary in Abyssinia. Until his death in 1879 Gobat was the centre of stormy controversy. One of the tasks of an Anglican bishop in the Middle East must be to maintain good relations with the ancient Christian Churches of that region. Gobat was accused of making proselytes from among members of these Churches; and, though most of the charges were baseless, the accusations continued to pour in.[1]

When the third bishop, Barclay, died in 1881, and the Germans decided to withdraw from the alliance, many people felt that the bishopric had better cease to exist; but in 1887 Archbishop Benson decided to reconstitute it. There were a number of specific tasks for the bishop – the spiritual care of English people resident in the Levant, including Cyprus; oversight of the missions of Church Missions to Jews; oversight of C.M.S. work among the Muslims; and the development of friendly relations with the ancient Churches of the East. Under a succession of wise bishops, all these aims have been satisfactorily fulfilled .The forma-

1. A thorough and objective study of this tangled history is a great need. Almost everything that has been written on it is marred by passion and prejudice on one side or another.

tion of the State of Israel has immensely complicated the bishop's task; but the Anglican missions have been a centre of fellowship and goodwill for Christians of many races and Churches, and St George's Cathedral, Jerusalem a centre of study and research in the concerns of the three great religions that have their meeting-point in Jerusalem.

In *Egypt*, where a separate Anglican diocese was formed in 1920, some of the Protestant (especially Presbyterian) missions had followed the policy of detaching Christians from the ancient Coptic Church, believing that Church to be far too corrupt ever to be reformed. The Church Missionary Society was in Egypt for the evangelization of the Muslims, and only in exceptional cases has it received Copts into the Anglican fellowship. The results have been a widespread movement of reform within the Coptic Church, especially among the educated laity; the conversion of a small number of Muslims; and the creation of an atmosphere of cordiality and mutual trust among Christians hard to parallel in that region of ancient and envenomed hatreds. Cairo could at one time be regarded as being the most ecumenical city in the world; even the Roman Catholics held membership in the committee on religious cooperation.

In *Iran*, the Anglicans were late in entering. Henry Martyn had passed through Persia, as it was then called, on his last and fatal journey homewards in 1811-12, and had argued with the Moulvies. A generation later the German C. G. Pfander, the great controversialist against the Muslims in India, had worked in the country and laid the foundation of his marvellous knowledge of Islam. But it was not until 1869 that an Anglican missionary was stationed in the country. Robert Bruce, an Irishman, who had been a C.M.S. missionary in India for ten years, obtained permission to spend a year in Persia for study of Islam. In 1871, just as he was preparing to leave for India, nine Muslims in Isfahan asked for baptism. This was an event without precedent in the Muslim world, and it seemed clearly right that Bruce should remain permanently in the country where his work had been so remarkably blessed. In 1881 he returned to England, bringing with him a revision of Henry Martyn's

Persian New Testament. Later he completed the whole Bible in Persian. What he wrote of his work is characteristic of mission work in all Muslim lands: 'I am not reaping the harvest; I scarcely claim to be sowing the seed: I am hardly ploughing the soil; *but I am gathering out the stones.* That, too, is missionary work.'

Episcopal supervision was provided for Persia in an unusual way. Thomas Valpy French, the most distinguished missionary who has ever served the C.M.S., after resigning the bishopric of Lahore which he had held between 1877 and 1888, decided to launch out on a one-man assault on Islam in Arabia, its very heart; after three months at Muscat, he died on 14 May 1891. Inspired by this example, Edward Craig Stuart, who had been a colleague of French in India, and Bishop of Waiapu in New Zealand from 1877 till 1894, decided to resign his bishopric and to offer himself for missionary work in Iran. The Anglican diocese, however, was not formed on a regular basis till 1912. Muslims in Persia seem less inaccessible to the Gospel than those of some other areas, and conversions have been more numerous than in, for instance, Egypt or Arabia. The Church, though tiny, is a living part of the great Anglican fellowship.

The happily temporary expulsion of the Irish bishop in Iran at the time of the controversy over Iranian oil, and the apparently permanent exclusion of the bishop in Egypt as a result of the Suez Canal episode, may serve as a reminder, if one were needed, of the vital necessity that the Church in every region should become genuinely indigenous, with bishops who cannot be expelled on the ground that they are foreigners.[1] With the consecration in 1961 of the first Persian bishop of the Anglican Communion, the Reverend Hassan Dehqani-Tafti, himself a convert from Islam, the Church in Iran came of age, and many beyond its boundaries joined in its rejoicings.

South-East Asia is becoming ever more conscious of itself as

1. Just after these pages were written, the formation of a new Anglican Archbishopric of Jerusalem, was announced. The first bishop for Jordan, Lebanon, and Syria, an Arab, was consecrated on 6 January 1958.

a separate and distinct region of the world. The area is a
meeting-point of races and religions. Thailand is one of the
great centres of conservative Buddhism. Indonesia and
Malaya are mainly Muslim. The Philippines are the one
country in the East which is mainly Christian, though many
would feel that the kind of Christianity professed by the
seventy per cent of the population which is Roman Catholic
is little more than skin-deep. It is a disturbed part of the
world. Vietnam became the scene of violent international
conflict. Malaya endured the trials of a nine-year emergency. The situation in Indonesia is highly unstable. If this
region as a whole were to go communist, the balance of
forces in the world would be radically changed.

The great cities of the region, which rank among the
great centres of activity of the world, are Singapore,
Djakarta, Bangkok, Manila, and Hong Kong. In Djakarta
the Anglican Communion hardly exists, since Indonesia has
been left mainly to the great missionary work of the Dutch
Churches, with their unusual success among Muslims.
There was a promising S.P.G. mission in Bangkok, but it has
not been followed up. In the other three centres the Anglican Churches are present in strength.

The work in *Borneo* goes back a long way. In 1841 that
astonishing man Sir James Brooke became Rajah of the
fantastic Kingdom of Sarawak. He wished for a clergyman,
and his need was met by the sending of the Rev. F. T.
McDougall, a devoted man who combined the offices of
doctor and parson. In 1855 Brooke thought that it would be
nice to have a bishop of his own; the authorities obliged, and
McDougall was consecrated in Calcutta as Bishop of Labuan
– the first Anglican consecration to take place in Asia.[1] In
1869 the Straits Settlements were added to McDougall's
jurisdiction. In 1909 the separate diocese of Singapore was
formed. In 1949 the old diocese took the name of Borneo,
and has since been divided into the two dioceses of Jesselton
and Kuching.

1. The third consecrator (with Calcutta and Madras), was not Bombay, but Bishop G. Smith of Victoria, Hong Kong, who happened to be
on leave in India at the time. This was the first *English* consecration to
take place outside England.

Progress in Borneo was for a long time slow. Work was begun among the head-hunting Dyaks in 1867, and among the Chinese a little later. But the extreme difficulties of communication were a great hindrance, and the climate took a heavy toll of the workers. Since the Second World War, there has been a quickening of the tempo in all directions. With a Church population that now numbers more than 15,000, the two dioceses give the impression of an expansive, almost explosive, vigour.

In *Malaysia*, the Churches have hardly begun the work of evangelizing the Malays, who are Muslims. The Christians are almost all Indian or Chinese immigrants. A vast new opportunity of work has been offered by the government's policy of grouping the inhabitants in new villages, as one means of bringing the communist emergency to an end. In many of these villages missionaries have been welcomed, and some of those expelled from China have found a second life opening out before them in Malaya. The Anglican body cooperates with Methodists and Presbyterians in a joint theological college for the preparation of the clergy. If there is any place in the world where there is a case for a united Church, if it can be attained without sacrifice of principle, it would seem to be Malaysia and Singapore.

In the *Philippines*, most of the Protestant missions have devoted themselves, with very considerable success, to detaching individuals from the Roman Catholic Church. Under the leadership of the great Bishop C. H. Brent, the American Episcopal Church set itself to break through into the world of the still pagan tribes in the mountains. Though small, as compared with some other bodies, the Church is influential; and by training in its seminary the candidates for the priesthood of the (Aglipayan) Independent Church of the Philippines, it is rendering an ecumenical service that may be of great importance for the future.

Hong Kong is the great listening-post for the vast world of Chinese Communism. Fearfully overcrowded with refugees of all kinds, it offers an almost unparalleled situation of opportunity and peril. Well led by bishops of long experience in Chinese affairs, who have known how to make themselves dearly loved by the Chinese, the Anglican Church in the Colony holds a strategic place amid the sometimes competing and sometimes cooperating forces that have flowed in from almost every part of the Christian world.

It might seem well that these South-East Asian dioceses, the special responsibilities of which mark them out so clearly from those of Australia to the south-east, of Japan to the north, and of India to the west, should be brought together in one Province. The bishops of that area have met for consultation, but have not gone further. The conferences of all the non-Roman Churches of the area held at Prapat in Indonesia in 1957 and at Kuala Lumpur in Malaya in 1958 are an indication that the movement is steadily in the direction of integration, closer understanding, and cooperation. It is desirable that, in such a situation, the Anglican forces should not lag behind the movement of the times.

*

In this chapter it has not been possible to depict more than a fragment of the missionary work of the Anglican Communion, and of the younger Churches that it has been its privilege to call into existence in many lands. Perhaps enough has been written to show something of the range of opportunity and peril with which the Churches are confronted, and of the flexibility of the responses that the Provinces are making. This chapter is a record of steady growth in independence, in self-government, and in the increase of the indigenous episcopate. The Anglican Communion is a living, growing entity, so rapidly changing and growing that, by the time this chapter is in print, some of the things written in it will already be out of date.

CHAPTER 13

A World-wide Communion

Two hundred years ago there was not a single bishop of what is now the Anglican Communion outside the British Isles. A hundred years ago, the Anglican Communion was beginning to take shape; the number of bishops was steadily increasing, and almost every year was marked by the formation of a new diocese. Parliament had conferred on a number of bishops the title of Metropolitan, though without any very clear idea of what a Metropolitan might be, or what he could do. But this growing Communion had no central organization of any kind whatever. Bishops sent out from England acknowledged the primacy of the Archbishop of Canterbury, though in many cases they were much too far away from him to consult him or to receive guidance in any but cases of the most extreme difficulty. But the bishops of Scottish and American Churches acknowledged no such primacy. In 1857 New Zealand had become a fully independent Province. The Archbishop of York in every generation has tended to stress his privileges as 'Primate of England',[1] and to deny to his brother of Canterbury any rights in the northern Province.

While all Anglicans are united in the great veneration in which they hold the successor of St Augustine, a considerable number of them are equally united in their determination to see to it that his authority is kept within strict limits. Any suggestion that a second Vatican is in process of development at Lambeth rouses most vigorous feelings; and even in 1964 the Anglican Communion has taken only the first steps towards such a central administration as is provided by the Lutheran World Federation for the Lutheran Churches.

The Anglican Communion had grown up without planning and without observation. One of the most

1. The Archbishop of Canterbury is 'Primate of all England'.

important steps in its development took place almost
fortuitously.

The case of Bishop Colenso had awakened strong feelings
and intense anxiety in every part of the Anglican world.
Apart from the doctrinal and personal issues, the whole question of the relation of the colonial churches to England had
been raised. A clash had been revealed between the Constitution of a Church and the law of the land. Bishop Gray of
Cape Town had declared that there could be no appeal
against the sentence of deposition and excommunication
which as Metropolitan he had pronounced against one of his
suffragans, the Bishop of Natal. Others felt that this claim
could not be allowed to pass unchallenged; some of Gray's
own clergy wrote in protest to the Bishop of London (A.C.
Tait): 'If the claim put forward by the Bishop of Capetown, to
have his decision as Metropolitan regarded as final be allowed,
if, in other words, as he affirms, there is no appeal to any
court on earth from a judgement which he may pronounce
as Metropolitan, it is evident that the Suffragan Bishops of
the Province are in a far worse position than the humblest
Priest in Pre-reformation times; he at least had an appeal to
the Roman pontiff, whilst they are subject without appeal to
the sentence, however arbitrary, of the Metropolitan.'[1]

That this anxiety was widely shared is evident from a
paper of questions circulated by Bishop Tait to all bishops,
deans, and archdeacons of the colonial Churches, which included the following: '*First* – the desirableness, or otherwise,
of all Bishops in British Colonies receiving their mission
from the See of Canterbury, and taking the oath of canonical obedience to the Archbishop. . . . *Third* – How far the
Royal Supremacy, as acknowledged by the United Church
of England and Ireland, can be maintained in our Colonial
Churches; *Fourth* – What seems the best guarantee for maintaining unity of doctrine and discipline between the different
scattered branches of our Church in the Colonies?'

It was at this juncture that the all-important proposal
came from an unexpected quarter. The Provincial Synod of

1. Quoted in *Archibald Campbell Tait, Archbishop of Canterbury*, Vol. I
(1891), pp. 373–4.

the United Church of England and Ireland in Canada, meeting at Montreal in September 1865, on the motion of the Bishop of Ontario (J. T. Lewis) addressed a letter to the Convocation of Canterbury in which they declared that 'we believe that [this connexion] would be most effectually preserved and perpetuated if means could be adopted by which the members of our Anglican Communion in all quarters of the world should have a share in the deliberations for her welfare, and be permitted to have a representation in one General Council of her members gathered from every land'. This was accompanied by a letter to the Archbishop of Canterbury (C. T. Longley), in which they suggested, 'to obviate, as far as may be, the suspicion whereby so many are scandalized, that the Church is a creation of Parliament, we humbly entreat your Grace, since the assembling of a General Council of the whole Catholic Church is at present impracticable, to convene a National Synod of the Bishops of the Anglican Church at home and abroad, who, attended by one or more of their presbyters or laymen, learned in ecclesiastical law, as their advisers, may meet together, and, under the guidance of the Holy Ghost, take such counsel and adopt such measures as may be best fitted to provide for the present distress in such Synod, presided over by your Grace.'

This was a sensational proposal, and its results have been memorable. But it is important to note that what followed was not in the least what the Canadian Church had asked for. The Canadian proposal was for a General Council or Synod of the Church. The Lower House of the Convocation of Canterbury, in its discussion on 14 February 1867, seems to have taken much the same view of the possibilities. But the Archbishop, foreseeing the opposition and the difficulties that were certain to arise, took a much more modest view of what it would be within his power to achieve: 'It should be distinctly understood that at this meeting no declaration of faith shall be made, and no decision come to which shall affect generally the interests of the Church, but that we shall meet together for brotherly counsel and encouragement. . . . I should refuse to convene any assembly which pretended to enact any Canons or affected to make any decisions binding on the Church.'[1]

In these words, Archbishop Longley defined what the Lambeth Conference was and is today. It is not a synod; it is not a General Council. It is a purely informal gathering of bishops, met on the invitation of the Archbishop of Canterbury in consultation with his Convocation and with other bishops but essentially of his own volition, and without any power to command or to constrain either attendance or submission to any decision whatsoever of the gathering. This was made perfectly plain in the letter sent out by Archbishop Longley, signed by him alone and by no one else, notifying the bishops of the Anglican Communion of the holding of the Conference: 'I request your presence at a meeting of the Bishops in visible communion with the United Church of England and Ireland,[2] purposed (God willing) to be holden at Lambeth, under my presidency, on the 24th of September next and the three following days. . . . Such a meeting would not be competent to make declarations or lay down definitions on points of doctrine. But united worship and common counsels would greatly tend to maintain practically the unity of the faith; whilst they would bind us straiter in bonds of peace and brotherly charity.'

This letter was sent out by Archbishop Longley to all the bishops of the Anglican Communion, who then numbered 144. Six English prelates, including the Archbishop of York, refused on principle to attend such a gathering, which seemed to them to lack all canonical precedent. Bishop Thirlwall hesitated long, believing that the tendency of the meeting would be 'to modify the Constitution and Government of the Church'; he and Tait withdrew their opposition, on receiving the Archbishop's assurances as to the nature of the Conference that was to be held; and in particular a private undertaking on the part of the Archbishop to Thirlwall that the stormy question of the position of Bishop

1. *Chronicle of Convocation*, 15 *February* 1867, p. 807.
2. At this period, no bishops outside the Anglican Churches were in visible communion with the 'United Church of England and Ireland', as many are today.

Colenso should not be discussed. In the end seventy-six pre-lates accepted the invitation, though some of those who had accepted were prevented by illness from being present at the Conference.

Four days is a brief period for a Conference that had so many things of importance to discuss. But it is clear from the records that its proceedings were on the whole businesslike and to the point. The first principle established, and one to which the Conference has happily adhered ever since, was that the Press should not be admitted to the discussions. The further principle was accepted that a great deal of the business must necessarily be done through committees. It was impossible entirely to exclude the affair of Colenso from the discussions, and at certain points tempers flared up. But most of those present would probably have been in agree-ment with the summing up of Bishop Tait: 'I feel that the tone of the Conference generally was very good. And as the Bishop of Argyll said, we must consider that the poor Bishop of Capetown has had a most difficult position – that it is very difficult to know what anyone would have done with such a Suffragan as Colenso. On the whole, let us be thankful for the kindly spirit of the Conference – for the essential amity.'[1]

The Conference had met, prayed, debated, and dispersed. Would it ever meet again? There was no special reason to think that it would. Yet, with surprising unanimity, the Anglican Provinces decided that it had been a good thing and should be repeated. As early as 1872 the Canadian bishops assembled in synod addressed to the Upper House of the Convocation of Canterbury the request that 'the Convocation of Canterbury will take such action as may seem most expedient to unite with us in requesting the Archbishop of Canterbury to summon a second meeting of the Conference'. The expression 'a second meeting of the Conference', rather than 'a second Conference', is specially noteworthy. In 1873, the West Indian bishops associated themselves 'in the request lately made to the Archbishop of Canterbury by the Bishops of the Canadian Province, that

1. *Archibald Campbell Tait, Archbishop of Canterbury*, Vol. 1 (1891), pp. 381-2.

he would summon another meeting of the Bishops of the Anglican Communion throughout the world at as early a date as may seem to his Grace practicable and expedient.'[1] In 1874 forty-three out of forty-six American bishops warmly agreed to the idea of another Conference, though some of them wisely 'wished that any action of the Bishops should be preceded by some expression from the clerical and lay deputies that would prevent any thought that the Bishops were acting for themselves alone, and not also for and with the clergy and laity'.[2] On 20 February 1875 the Convocation of York, which had held aloof from the first Conference, expressed 'the wish of this Synod that all necessary steps may be taken for the assembling of a second Conference at Lambeth'. Tait, who was now Archbishop of Canterbury, after consultation with the English bishops, sent out on 28 March 1876 a letter 'to all the Anglican bishops', asking for their advice.

At last, on 19 July 1877, an invitation was issued by Tait to 'a Conference of Bishops of the Anglican Communion' to be held at Lambeth beginning on 2 July 1878. This letter was not sent direct to the bishops of organized Provinces but was communicated to them through their metropolitans. Much more careful preparations had this time been made, and it was arranged that the Conference should last for not less than four weeks.

The term 'Bishops of the Anglican Communion' necessarily raised the question as to the identity of the persons to be included under this description. In 1878 the view seems to have been taken that all who had received episcopal consecration were eligible, and four venerable prelates who had resigned their sees were present at the Conference – McDougall (Labuan), Perry (Melbourne), Ryan (Mauritius), and Claughton (Colombo). At later dates the view seems to have prevailed that bishops attending the Conference

1. Note that the West Indian Bishops misquoted the letter of their Canadian brethren.

2. Readers of ecclesiastical documents will appreciate this sentence from the letter of the Presiding Bishop, dated 3 Nov. 1874: 'As some of the signatures may not be readily legible, I enclose a printed list of the names of the signers.'

should have some kind of jurisdiction as diocesans, as co-adjutors, as suffragans, or as assistants; but this was relaxed in 1920, when four bishops without jurisdiction of any kind were full members of the Conference,[1] and also in 1958.

One hundred and eight bishops accepted the invitation of the Archbishop of Canterbury, but, some having been prevented from attending, the number that gathered was exactly one hundred – 35 English, 9 Irish, 7 Scottish, 30 Colonial and Missionary, and 19 from the United States of America. The subjects discussed followed closely the lines of those that had occupied the Conference of 1867; but there was more ample time for discussion, and the arrangements had been far more fully thought out in advance. There was a great deal more public interest in the Conference than in 1867; but the rule of privacy and secrecy was observed.

There can be no doubt that this was Archbishop Tait's Conference. When it was over, the Bishop of Iowa wrote of him: 'First and foremost in rank, as he was unquestionably in his presence and "many-sidedness" of character, was the Primate of all England. . . . Fair and equitable in his address and rulings, and at the same time astute in feeling the temper of his auditors and brethren . . . the Archbishop's presidency was above praise. While avoiding all appearance of dictation, his presence and position were always felt; and the harmony and unanimity of the Conference were largely due to his uniform affability and good temper, and his masterly leadership.'

From this time onwards it came to be taken almost for granted that a Conference would be held every ten years or so. In 1886 Archbishop Benson, in a letter which he sent through metropolitans to all members of the Anglican Episcopate 'exercising superintendence over Dioceses or lawfully commissioned to exercise Episcopal functions therein', remarked that 'there appears to be a general desire that a Conference of the Bishops of the Anglican Communion should again be held at Lambeth within the

1. Browne (Bristol), Copleston (Calcutta), Montgomery (Tasmania), King (Madagascar).

next few years'. The third Conference was accordingly held, under his presidency, in 1888; the fourth, under Archbishop Frederick Temple, in 1897; the fifth in 1908, the sixth in 1920, both under Archbishop Davidson; the seventh in 1930 under Archbishop Lang; the eighth in 1948 under Archbishop Fisher. At every Conference there was an increase in the number of prelates taking part.[1]

It would be tedious to write in detail of each several Conference. In general the plan has been similar. The assembled bishops have carefully studied together one main theological problem. Much time has been spent on the purely domestic concerns of the Anglican Communion. All the problems presented by the rivals and enemies of the Christian faith – infidelity in its various forms, secularism, the decline of moral standards, and so forth – have come under consideration. The unity of the Church has been a recurring interest. The pronouncements of the Fathers have only rarely reached the level of classic inspiration; there has been a tendency to verbiage, to pious locutions, to superficial handling of grave themes. The Conference, like most ecclesiastical assemblies, has not always been able to resist the temptation to pass resolutions on every conceivable subject. Yet it is doubtful whether any similar assembly in modern times has put on record so much sense and so little nonsense, or has shown such sensitiveness to the movement of the times through which it has lived. Until Pope John XXIII announced his intention of convening a Council of the Roman Catholic Church the Lambeth Conference could be regarded as fulfilling more nearly than any other assembly in modern times the function of a General Council of the Churches.

*But the Conference of 1948 was the last of the series. In the preparations for the Conference of 1958 a new and revolutionary decision was taken. The principle that an invitation should be sent at least to all Anglican bishops holding any kind of jurisdiction as diocesans, coadjutors,

1. The numbers attending were as follows:

1867:	76	1888: 145	1908: 242	1930: 308	1958: 310
1878:	100	1897: 194	1920: 252	1948: 326	1968: 459

suffragans, or assistants was abandoned. There can be no question as to the canonical and constitutional propriety of this decision. The Lambeth Conference has no canonical existence at all; it is neither synod nor council. It has no constitution, and as we have seen, there has been a measure of fluctuation in its membership. It may well be that the smaller body will be more manageable and more efficient than the larger. But the Conference has lost its character of approximating to a Church Council; it has become no more than a selection of Anglican bishops, meeting at the behest of the Archbishop of Canterbury for consultation on themes of common interest. It would seem that a change in title might be desirable to mark this revolutionary departure from the traditions established since 1867.[1]

A continuing interest in the visible unity of Christ's holy Catholic Church throughout the world – this has been from the beginning one of the marks of the Anglican Communion. In 1920 the Lambeth Conference was inspired to give memorable and classic expression to this interest and this longing.

In almost every chapter the Anglican interest in union has come before us – from the ecumenical ideals of Cranmer, through the endless peregrinations of John Dury, the wide correspondence of Compton and Wake, to the fantastic dream of the Jerusalem bishopric. But the years after the First World War introduced a new period of more dramatic possiblities and of intenser strains.

The trouble started with the Kikuyu controversy. The Anglican missionaries in East Africa had met with others, mainly Methodist and Presbyterian, to consider whether a united Church of East Africa could be brought into being. They had not gone beyond preliminary discussions. The only unusual thing that happened was that, at the end of the session of 1913, the Bishops of Mombassa (Peel) and Uganda (Willis) held a Communion Service, at which those who were not Anglicans were invited to be present and to receive the Holy Communion. At once the whole Anglican world was in an uproar. Party warfare raged higher than at any

1. There appear to be at present 347 dioceses and missionary districts in the Anglican Communion. See Appendix II.

time since the 1860s. The air was thick with pamphlets on one side and another. A doughty Goliath marched into the field in the person of Frank Weston, Bishop of Zanzibar (1871–1924), who denounced his brother bishops, accused them of heresy, and threatened dire things if absolute assurance could not be given that the irregularity would never be repeated. Weston was a great man – scholar, saint, apostle; but the gentle side of his nature was likely to become evident to opponents rather later than the other and more aggressive side. The matter was handled by Archbishop Davidson with characteristic wisdom and patience; his judgement, issued at Easter 1915, was so balanced, so cautious, as to some extent to justify the caustic summary that in his opinion the events at Kikuyu 'were highly pleasing to Almighty God, but not in any circumstances to be repeated'. The more urgent preoccupations of the First World War had already descended upon the nations and for the moment the problems of Kikuyu were forgotten.[1]

But they were very much in the air at the Lambeth Conference of 1920. Frank Weston was there in untamed splendour. There were possibilities of head-on collisions between bishops of differing points of view. The Conference might have found itself unable to say anything at all. It might have lost itself in the pious platitudes to which such assemblies are singularly prone. Instead it was inspired to issue an 'Appeal to all Christian People', a noble and prophetic utterance on Christian unity, in the phrasing of which the skilful hand of Cosmo Gordon Lang, at that time Archbishop of York, is to be traced.

No citations can do justice to this classic statement of the Anglican position on unity; it must be read in full.[2] It opens by acknowledging the membership in the Church of all those who believe in our Lord Jesus Christ and have been baptized into the name of the Holy Trinity. It admits

1. The best account is that by J. J. Willis (Bishop of Uganda at the time of the crisis) in *Towards a United Church, 1913–1947* (1947), pp. 15–51. On the part played by Archbishop Davidson, see G. K. A. Bell: *Randall Davidson* (ed. of 1938), pp. 690–708.

2. In the Lambeth Conference (1920) Report, pp. 26–29. Or in G. K. A. Bell: *Documents on Christian Unity* (First Series, 1924), pp. 1–14.

sorrowfully that all Christians have shared in the causes that have led to our unhappy divisions. It sets forth the ideal of a Church, 'genuinely catholic, loyal to all Truth, and gathering into its fellowship all "who profess and call themselves Christians", within whose visible unity all the treasures of faith and order, bequeathed as a heritage by the past to the present, shall be possessed in common, and made serviceable to the whole Body of Christ'.

The Appeal then reaffirms, in a simplified form, what has come to be known as the 'Lambeth Quadrilateral' – the belief that the essentials for unity are the acceptance of the Holy Scriptures as containing all things necessary to salvation, the Nicene Creed as the sufficient statement of the faith, the Sacraments of Baptism and the Holy Communion as instituted by Christ Himself, and 'the Historic Episcopate, locally adapted to the needs of various regions and peoples'. The 'Quadrilateral' had already had a history of fifty years. It was first put forward in 1870 by an American Episcopalian, William Reed Huntington, in his book *The Church Idea, an Essay Toward Unity*. Affirmed by the General Convention of the American Church at its meeting held in Chicago in 1886, it was adopted with some modifications by the Committee on Home Reunion of the Lambeth Conference of 1888; and was now paraphrased rather than textually cited in the Appeal of 1920.[1]

In attempting tactfully to commend the Episcopate, the Lambeth Fathers accorded a generous measure of recognition to non-episcopal ministries; 'It is not that we call in question for a moment the Spiritual reality of the ministry of those Communions which do not possess the Episcopate. On the contrary we thankfully acknowledge that these ministries have been manifestly blessed and owned by the Holy Spirit as an effective means of grace.' They expressed a willingness to accept some form of commission from other Churches in cases where union had been achieved: 'If the

---

1. The varying forms of the Quadrilateral are conveniently set forth in *A History of the Ecumenical Movement 1517–1948* (1954), pp. 250, 264 ff. The divergences are not unimportant, though a concern for the specialist rather than for the general reader.

authorities of other Communions should so desire, we are persuaded that, terms of union having been otherwise satisfactorily adjusted, Bishops and clergy of our Communion would willingly accept from these authorities a form of commission or recognition which would commend our ministry to their congregations, as having its place in the one family life.' The Appeal ends with the words: 'We do not ask that any one Communion should consent to be absorbed in another. We do ask that all should unite in a new and great endeavour to secure and to manifest to the world the unity of the Body of Christ for which He prayed.'

In some respects the response to the Appeal in other Communions was disappointing. It was, however, widely recognized as the beginning of a new period in Church relationships, and may fitly serve as the starting-point for our consideration of the Anglican Churches today, and their position in the quest for Christian Union. It is impossible to deal in detail with negotiations that have spread out into every corner of the Christian world. It will be convenient to treat in order the different types of Christian body with which the Anglican Churches have at one time or another attempted to enter into closer fellowship.

With the *Roman Catholic Church* the Anglican Churches have had no official relationship for centuries.[1] An attempt was made, after the First World War, to break through the deadlock, mainly through the untiring optimism and enthusiasm of the second Viscount Halifax (1839–1934), an earnest Anglo-Catholic and a devout believer in the possibilities of Catholic union. Between 1921 and 1926 a number of conversations were held at *Malines*, in the archiepiscopal palace of the greatly venerated Cardinal Mercier. Halifax was most anxious that these should be more than private conversations between friends. A very cautious measure of approbation was secured from the Vatican. Archbishop Davidson was also very cautious. While not wishing to

1. The letter addressed by Queen Victoria to Pope Pius IX, sympathizing with him on his withdrawal from Rome in 1848, and addressed 'Most Eminent Sir', was probably the first letter addressed by a British sovereign to a Pope since the excommunication of Queen Elizabeth in 1570.

discourage the venture, he was most anxious to ensure that Anglicans who took part in the conversations should not regard themselves as in any way delegates of the Church, and that 'conversations should not be misunderstood to mean "negotiations"'. The phrase is Davidson's own: 'Needless to say, there has been no attempt to initiate what may be called "negotiations" of any sort.'[1]

A certain naïveté attached to all the proceedings. It was certainly good that eminent Anglicans found it possible to meet with eminent Roman Catholics in an atmosphere of charity and goodwill. But the position of Rome with regard to unity is perfectly clear and uncompromising – absolute submission to Rome in every detail of doctrine and discipline is required; and short of that there can be no question of unity. The Malines Conferences were valuable in exploring and revealing the profound differences that still separate the Romans and Anglican Communions; they led to no other practical result. With the death of Cardinal Mercier on 21 January 1926 the main inspiration dropped out of the conversations. Ere long it was decided that no more such meetings would be held.[2]

With the great *Orthodox Churches of the East* Anglican relations have become increasingly friendly. Closer acquaintance has revealed to many Anglicans the immense treasures of the Orthodox traditions of liturgy and of spirituality; and the Orthodox have been learning that it will not do simply to dismiss the Anglican Churches as 'another Protestant sect'. The friendship is so close that many Anglicans believe that there is some kind of inter-communion between the

1. Bell: *Randall Davidson*, p. 1284. Dr. Bell himself goes much too far in speaking of 'this, the first semi-official conversation'. Op. cit., p. 1260.

2. The Report, *The Conversations at Malines 1921–1925*, was published on 19 January 1928. There is a most valuable account in Bell: *Randall Davidson*, pp. 1254–1302. But I cannot go with the optimism of Dr Bell in the concluding words of his chapter: 'There has been progress in understanding, in charity, in desire. So far as the longed-for *rapprochement* was concerned, the fundamental difficulties remain unsolved. But channels of thought and methods of study have been started, from which in later days some great gain may result.'

Anglican Churches and the Orthodox;[1] and it is important to make clear exactly what has happend.

In 1920 an encyclical letter, commending the cause of the unity of the Church, was addressed by the Oecumenical Patriarchate in Constantinople 'to all the Churches of Christ, wheresoever they be'. Two years later the Oecumenical Patriarch and the Holy Synod of Constantinople put forward for the consideration of the Orthodox Churches a memorable statement of its judgement on the validity of Anglican ordinations: 'As before the Orthodox Church, the Ordinations of the Anglican Episcopal Confession of bishops, priests, and deacons, possess the same validity as those of the Roman, Old Catholic, and Armenian Churches possess, in as much as all essentials are found in them which are held indispensable from the Orthodox point of view for the recognition of the "Charisma" of the priesthood derived from the Apostolic Succession.' By 1935 four other Churches, those of Alexandria, Jerusalem, Cyprus, and Rumania, had expressed their agreement with the opinion of Constantinople.

The year 1948 semed to mark a setback in relationships. In that year a meeting of heads of autocephalous Orthodox Churches was held in Moscow. This conference took the view, akin to the Roman view, that complete agreement in faith and doctrine must precede inter-communion; and the judgement of the Conference on Anglican orders was expressed as follows: 'If the Orthodox Church cannot agree to recognize the correctness of Anglican teaching on the sacraments generally, and on the Sacrament of Orders in particular, neither can she recognize the validity of Anglican ordinations that have actually taken place.' This is not the judgement of a synod. Because of political difficulties, it is impossible at present that a general synod of the Orthodox Churches should be held; and there is a tendency for the Greek-speaking and the Slavic Churches to move in divergent directions. American Orthodoxy will probably in the

---

1. An idea probably strengthened by the Orthodox practice of distributing blessed, but not consecrated, bread to non-communicants at the end of the Orthodox Liturgy.

future exercise an increasing influence. But the lack of inter-communion has not hindered friendly intercourse and dis-cussion. An Anglican delegation was in Moscow in 1956, and discussions of value took place.

The position of the *Lesser Eastern Churches* – the fragments of the great Eastern Churches that are to be found in Armenia, Egypt, Mesopotamia, and elsewhere – has been carefully considered by successive Lambeth Conferences. But, though it is held that these Churches, so long regarded as heretical by the West, are probably not in any real sense heretical today, friendship and mutual service have not led on to inter-communion, with one single interesting excep-tion. An extension of the ancient 'monophysite' Church of Syria was the 'Syrian' Church which has existed in South India since at latest the fourth century A.D. As a result of the working of an Anglican leaven in that ancient Church, intro-duced through the work of the C.M.S., a division took place in 1887, and those who had accepted reforming ideas organized themselves as the 'Mar Thoma Syrian Church of Malabar', at that time the only completely autonomous Church in India. In 1936-7 this Church and the Anglican Church in India agreed on terms of limited inter-communion – the only example in the world of inter-communion between an Eastern and a Western Church. In 1974 the Church of England established full Communion with the Mar Thoma Church; other Anglican Provinces are in process of enter-ing into relations of full, or limited, Communion with this Indian Church.[1]

In 1931, by the Agreement of Bonn, the Church of Eng-land entered into relationships of full inter-communion with the *Old Catholic Churches* of the continent of Europe. When, in 1870, the Roman Catholic Church accepted the dogma of the Infallibility of the Pope, those who could not accept this new and previously unknown article of the faith organized themselves as the Old Catholic Churches in Germany, Austria, Switzerland, and other countries, and joined hands with the 'Jansenist' Church in Holland, which came into

1. *Lambeth Conference, 1958*, p. 2.52.

existence in the seventeenth century in separation from Rome.[1] This movement was followed with great interest and sympathy in England. The Lambeth Conference of 1878 had remarked that 'the fact that a solemn protest is raised in so many Churches and Christian communities throughout the world against the usurpations of the See of Rome, and against the novel doctrines promulgated by its authority, is a subject for thankfulness to Almighty God'. Increasing contact and theological discussion led on gradually to the Agreement of Bonn. This is so short that it can be cited in full:

1. Each Communion recognizes the catholicity and independence of the other, and maintains its own.

2. Each Communion agrees to permit members of the other Communion to participate in the Sacraments.

3. Inter-communion does not require from either Communion the acceptance of all doctrinal opinion, sacramental devotion, or liturgical practice, characteristic of the other, but implies that each believes the other to hold all the essentials of the Christian faith.

What this means in practice is that any Anglican on the Continent who finds an Old Catholic Church can without hesitation present himself to receive Communion if he so desires. Almost all the Anglican Churches have officially accepted the Agreement of Bonn. In 1946 the American Church established relations of inter-communion on similar terms with the Polish National Catholic Church of America, a body which is itself in communion with the Old Catholic Churches of Europe.

This raises an interesting point in Anglican relations with the Roman Catholic Church. That Church recognizes the Old Catholic succession of episcopal consecrations as valid but irregular. Old Catholic and Polish National Catholic bishops have often taken part in Anglican consecrations. Considerably more than half the Anglican episcopate now has the Old Catholic as well as the Anglican succession, and before long this is likely to be true of the whole episcopate. If at any time the Roman Catholic Church wished to move nearer to the Anglican Churches, this might clear the way

1. See C. B. Moss: *The Old Catholic Movement* (1948).

to happier relations. No Anglican imagines that anything is added to his consecration or ordination by Old Catholic participation; but from the Roman Catholic point of view such orders might be held to have regained something of that regularity and validity which the Pope's Bull of 1896 denied to them.[1]

In the Philippines, the American Episcopal Church came to the rescue of the *Philippine Independent (Aglipayan) Church*, a body which had come into existence at the beginning of this century through the separation of about a million people from the Roman Catholic Church. This body had retained episcopacy, but without any regular succession. In 1948 three American bishops consecrated three bishops of the Aglipayan Church, who in their turn consecrated the others, so that this body now has a regular Anglican succession. This was a purely American venture, carried out without any consultation with any other Province; the success of the Americans in helping the Aglipayans has provided justification for this courageous adventure, and in 1963 the Church of England, followed by other Anglican Churches, entered into full Communion with the Aglipayan Church.

As we have seen, the English Reformation in its early days owed much to Luther. Anglican exchanges with the Lutheran Churches in Germany have been friendly, but have led to nothing decisive. A more promising field for Church relationships has been found in Scandinavia.

As early as 1878 the Lambeth Conference had drawn attention to the possibility of closer relationships with the *Scandinavian Churches, and especially the Church of Sweden.* The five Scandinavian Churches hold closely together, but each has its own special character, and there are considerable differences between them. The Church of Sweden has retained the historic episcopal succession. This was, in a sense, an accident. In the sixteenth century the usurper Gustavus

1. The latest Roman Catholic discussion, that by Francis Clark S.J. in *Anglican Orders and Defect of Intention* (1956), comes to the conclusion that Anglican orders are so hopelessly defective that Old Catholic participation cannot do anything to rectify the defect. This is, however, a personal and not an official Roman Catholic view.

Vasa felt that it would be wise to add such lustre as he could to his crown through coronation by bishops of the ancient line, and so retained the episcopal succession; whereas the profoundly Lutheran King of Denmark called in Luther's friend Bugenhagen to ordain his new superintendents, as though to make it quite clear that an absolute break with the past was intended.

In 1920 the leisurely Anglo-Swedish discussions reached a conclusion. The Anglicans had been much concerned to be sure that the Swedish episcopal succession was in fact unbroken, and that the Swedish Church attached a real value to its episcopacy. The Swedes were anxious to be sure that the Church of England was sound on the doctrine of justification by faith. The Lambeth Conference of 1920 felt able to recommend a measure of inter-communion between the two Churches – not complete inter-communion, since the Church of Sweden maintains very close fellowship with other Lutheran Churches which are not episcopal, and in 1920 the missions of the Church of Sweden both in South India and in South Africa were without a bishop. The first outward sign of the new fellowship was the participation by two English bishops (Henson of Durham and Woods of Peterborough) in the consecration of two Swedish bishops on 19 September 1920. This has been followed up by many similar acts of fellowship.

The Church of Finland owed a deep debt, in its origins, to English missionaries, and has retained a vivid sense of its own continuity with the past as the national Church of Finland. Like the Church of Sweden it had retained the historic succession of the episcopate. But in 1884 this was almost fortuitously broken. The three Finnish bishops all died in the same year. The attitude of the Russian Government, which then controlled Finland, was such that it was impossible to secure for the new bishops any kind of consecration in the regular succession. However, in 1918 Finland became an independent country; and the Finnish Church set itself to recover through Sweden the succession which it had lost. In 1935 the Convocations of Canterbury and York passed resolutions in favour of relations between the Church of England and the Church of Finland similar to those which had beeen recommended by the Lambeth Conference in the case of Sweden, though at that time several of the Finnish bishops did not stand in the historical succession. Relations between the Anglican and Finnish Churches have become increasingly intimate and cordial.

This is true also of the Churches of Norway, Denmark, and Iceland; though the problem of the historic episcopate has so far stood in the way of any mutual recognition or inter-communion.

In *England*, the Lambeth Appeal of 1920 was followed by discussions between the Church of England and the *Free Churches*, and these led in 1938 to the publication of an *Outline of a Reunion Scheme*, in which a comprehensive Church in England was envisaged. But this led to nothing; and, under the preoccupations of the Second World War, the question of Church union slept.

A new impulse was given in 1946 in a remarkable sermon preached in Cambridge, on 3 November of that year, by the then Archbishop of Canterbury, Dr Fisher. The condition of the Church of England as an established Church sets grave difficulties in the way of corporate union. Would it be possible for the Free Churches to take episcopacy into their own systems, and thus take one step forward towards a common ministry and common sacraments?

The response of the Free Churches to this suggestion was immediate and friendly. A joint body of Anglicans and Free Churchmen was appointed to discuss the problem.[1] In 1950 their report appeared under the title *Church Relations in England*. This report was cautious in the matter of making proposals; but did point out six conditions which would have to be fulfilled if advance were to become posssible. It further emphasized the fact that each Free Church, if it so desired, would have to make its own approach to the Church of England, and that in each case there would be different problems to solve.

1. It is to be noted that the Anglicans were simply individuals appointed by the Archbishop of Canterbury on his personal initiative; they were in no way official representatives of the Church of England.

*So far only the Methodists have taken action on this Report, and have appointed a committee to engage in talks with the Church of England. As was to be expected, these discussions have revealed rather sharp divisions of opinion between the 'high church' and the 'nonconformist' wings among the Methodists themselves. It is as yet uncertain whether these discussions will lead to any practical result. The friendliness accompanying them is a welcome sign of the change that has taken place during this century in ecclesiastical relations in England.[1]

In 1952 the English Convocations recommended the reopening of discussions with *the Church of Scotland,* to which the Episcopal Church in Scotland and the Presbyterian Church in England should also be party.[2] These have turned out to be, from the theological point of view, among the most important discussions on Church Union that have ever taken place. All too frequently such discussions are bedevilled by words and names and by the associations attaching to them. Advance is sought without a real exploration of the theological implications of action.[3] We owe it to the Scottish Presbyterians that really serious theological study became the order of the day. A careful attempt was made to get behind words and phrases and current practice with regard to the ministry, and to work out the fundamental biblical concepts of oversight, shepherding, and service in the Church of Christ. In the light of these fundamental ideas, specific proposals were put forward as to the way in which the Presbyterian Churches might accept the office of bishop-in-presbytery, and as to modifications that might be made in the Anglican system in the direction of

1. The very important proposals for union were published in 1963 and were twice voted on by the Churches concerned.

2. An earlier set of discussions, in which the Episcopal Church in Scotland and the Presbyterian Church in England had been represented only by observers, had resulted in a report – *Relations between the Church of England and the Church of Scotland, a Joint Report* (1951).

3. In America Episcopalians and Presbyterians had moved a considerable way towards closer fellowship, when in 1946 the negotiations were broken off rather abruptly from the Episcopal side, in circumstances which left Presbyterian feelings deeply wounded.

recognizing more fully the priestly character of the whole Church, in which the laymen shares as well as the ordained minister.[1]

As was to be expected, the effect of the Report on the Churches was somewhat explosive. The Churches in the British Isles have still all the strength and all the weakness of insularity. Few English Anglicans realize the extent to which other Provinces, without losing any of their Anglican character, have made their own much of the tradition of synodical government and lay participation, which is the special glory of Presbyterianism. On the other side, as Professor T. F. Torrance so truly remarked, 'The word "bishop" in Scotland is not a nice word'. Many Presbyterians have failed to notice that the Reformed Church in Hungary, without ceasing to be Presbyterian, has had bishops for centuries. It is greatly to be regretted that in 1958 the General Assembly of the Church of Scotland rejected the Report and the proposals which it contained. But both sides have refused to accept defeat, and the way is still open for further discussions of the possibility of union on rather different lines.

In *Australia*, a different approach has been made to the problem of overlapping ministries. In the early days in Papua, in order to avoid competition between missions, the Government laid down strictly the area within which each might work. These geographical divisions still persist; but with greater mobility Papuan Christians in large numbers have moved beyond the area of the Church to which they belong. In 1937 the National Missionary Conference of Australia set itself to consider the problems of ministry involved. Was it possible that a minister, without ceasing to be a minister of his own Communion, could receive a wider commission such as would render his ministry acceptable to members of Communions other than his own? In 1943 proposals were put forward for the unification of separate ministries in the person of individual

1. The important report, *Relations between Anglican and Presbyterian Churches* (1957), appeared while this book was being written.

ministers, by a wider commissioning through a 'Mutual Formula', in which each minister would receive a commission from those Communions to which he did not already belong.

The Lambeth Conference of 1948 did not welcome such schemes with enthusiasm, through anxiety lest such partial unification might obscure the need for full and organic union, which in the opinion of the Conference must be the goal of all plans for closer fellowship. The Australian situation has changed materially since 1948, in that Congregationalists, Presbyterians and Methodists decided to go ahead with plans for union without the Anglicans. Two notable reports on the faith of the Church (1960) and on the structure of the Church (1963) have been published, the latter including the intriguing suggestion that the Australian Church should acquire episcopacy by way of South India. If this is accepted by the Churches concerned, the way will certainly be opened for Anglican participation later on.

A movement rather similar to that of Australia took place from 1943 onwards in *Canada*, where the same conditions of vast spaces, imperfect communications, and a rapidly growing population have to be met. The original proposals were for an enlarged commission like that proposed in Australia. But since 1949 the tendency in both the United Church of Canada and the Anglican Church of Canada seems to be in favour of movement directly towards organic union, without waiting at the half-way house of unification of the ministries. Nothing decisive has yet been done; but it seems likely that it is in this direction that future developments in Canada are to be expected.

The formation of the *Church of South India* in 1947 represented something new in the whole field of Church union. What has been talked about in England and Canada has already happened; for the first time since the Reformation episcopal and non-episcopal Churches have become one, and a new type of Church has come into being, for which there is no earlier precedent.

The bodies concerned were the Church of India, Burma,

and Ceylon, the Methodist Church of South India, and the South India United Church, the fruit of an earlier union between Congregationalists and Presbyterians (1908), which also included one area of the Basel Mission (Reformed and Lutheran) of Switzerland and Germany. The initiative was given by the Manifesto of the Tranquebar Conference of May 1919. The new Church was inaugurated in 1947. Almost a whole generation of human life and effort had gone into the preparation of the plans. It was agreed from an early date that the resulting Church would not be in any sense an Anglican Church; it would be an independent regional Church, like the Church of Sweden or the Church of Finland, and its relationship with the Anglican Communion would have to be gradually worked out. It was agreed that the Church of South India would be episcopal, and that from the date of union all ordinations would be by bishops and presbyters together. But a novel feature of the scheme was the provision of a period of thirty years for 'growing together', during which many anomalies would be permitted, and ministers of all the uniting Churches, whatever the nature of their ordination, would have equal rights.

From the Anglican side the acceptance of the scheme was a most complicated business. The scheme had to be voted on by all the fifteen dioceses of the Church of India, Burma, and Ceylon. Unless two-thirds of them were favourable, nothing further could happen.[1] In all the dioceses which voted except one (Nagpur) there was a large favourable majority; the four dioceses specially concerned were almost unanimously in favour. Then the matter came to the General Council of the Church. Here a simple majority was required in each of the three houses of bishops, clergy, and laity, and an over-all majority of seventy-five per cent of the three houses together. The scheme had twice received cautious commendation from the Lambeth Conference (1920 and 1930). The Consultative Body of the Lambeth Conference had twice been asked for its opinion, and had responded on the whole favourably. The Anglican authorities in India had

1. Some dioceses could not vote, owing to war-time conditions, but the necessary two-thirds majority was obtained.

been in touch personally with the most distinguished Anglican theologians all over the world, and officially with the metropolitans of all the Churches with which the Church of India was in communion. Thus the union was in no sense an act of the four dioceses, Madras, Travancore and Cochin, Tinnevelly, and Dornakal, which entered the new Church; it was a most carefully guarded synodical action of an independent Province, constantly in touch with Anglican opinion throughout the world.

The South India Church was and is the subject of intense controversy. Some Anglicans regard its formation as a conspicuous manifestation of the guidance of the Holy Spirit, others as a betrayal of every Anglican principle.

The Lambeth Conference of 1948 could not reach a clear decision about it, and, departing from its normal practice, stated without reconciling two divergent opinions. Many bishops felt that, in the first year of a new Church's life, it was impossible to pronounce upon it, and that time must be allowed to show what the Church of South India really is. Lambeth left it to each Province to determine its own relation to the new Church.

Shortly after the Lambeth Conference the continuing Anglican Church in India decided (1950) to recognize the regularity and validity of ordinations and consecrations carried out in the Church of South India. The English Convocations in 1950 postponed their decision for five years. In 1955, being satisfied as to the credal orthodoxy of the new Church and as to its intention to remain in the historic traditions of the Christian faith, the Convocations accorded to the Church of South India the same measure of recognition as had already been accorded by the Indian Province. A strong commission of the American Church, after visiting India, reported in general favourably on the Church; its report was discussed and received with general approval at the General Convention of the American Church in 1958.

The Church of South India has gone on its own way, eagerly desirous of fellowship with all Churches in the world, but at the same time jealous of its own independence. It has increased the number of its Indian bishops. It has produced a liturgy for the Holy Communion which has been commended by a Roman Catholic authority as being nearer to the Catholic ideal of Eucharistic worship than the Prayer Book of the Church of England. It has developed its evangelistic enterprise and the unity of its witness to the vast non-Christian population around it.

South India is important not only for what it is, but for what has sprung from it. The Churches in *Ceylon* and in *North India and Pakistan* have come forward with complete schemes for Church Union.[1] In each case, the Baptists have taken part in the discussions; this has opened up new problems, and has called for new and adventurous solutions. In North India there was the problem of reconciling the episcopacy of the Church of India with the other episcopacy of the Methodist Church in Southern Asia. A common feature of these two schemes is that each tries to avoid the anomalies of the period of 'growing together', and to provide from the start a unified and generally acceptable ministry. A form of wider commissioning is provided, through which it is hoped that the ministries of all the uniting Churches may be brought together into one in the united Church. Anglicans in these areas have been alarmed by the hostility manifested in certain quarters to the Church of South India, and have wished to safeguard their own future. For a variety of reasons these plans have run into difficulties, and it is not yet certain that union will be achieved in either of these areas.

*In the summer of 1957, *Nigeria* came forward with a complete scheme for the unification of the Anglican, Methodist, and Presbyterian bodies in that great country. It was gratifying to the scribes who wrote the South India scheme to find that page after page of their handiwork had been taken over exactly or with the minimum of change by the scribes in four other countries. After 1958 the Nigerians, under pressure from the Lambeth Conference, accepted a plan for the unification of all ministries at the time of union, rather different from that proposed in North India and

1. The third edition of the Ceylon Scheme was published in 1955; the third edition of the North India Scheme in 1957.

Ceylon. It is intended that this plan should come to fruition in December 1965.

This very brief and inadequate sketch has shown the Anglican Churches reaching out literally in all directions in the Christian world in search of closer fellowship. In doing so, they find themselves confronted by what is perhaps the largest problem of the Christian world in the mid-twentieth century. The movement towards unity is strong and growing. But what is to be the pattern of unity that emerges from these tendencies and strivings?

Two patterns are already visible. The first is the strengthening of denominational loyalties, and of the more intimate organization of the Churches on a denominational basis. Each of the great denominations now has a worldwide fellowship. By far the best organized and most effective is the Lutheran World Federation, with its headquarters in Geneva. We have seen something of this tendency in the gradual development of the Anglican Communion over the last century. The other pattern is that of the new regional Church, in which all previous denominational patterns are broken in pieces. Such a Church already exists in South India. In the next ten years, we may see similar Churches emerge in Ceylon, in North India and Pakistan, in West Africa, in Canada, each with Anglican elements, yet none a constituent member of the Anglican Communion. Which way does the future lie? And what ought an Anglican to think of these conflicting tendencies? This is a problem to which we shall return in our concluding chapter.

The search for the visible unity of the Church is only part of that growing tendency towards Christian fellowship which is the special characteristic of Christianity in the twentieth century; and the last section of this chapter must deal with wider aspects of what has come to be called the ecumenical movement.

It is generally agreed that the modern ecumenical movement began with the first World Missionary Conference held in Edinburgh in 1910. This was true in the sense that the first permanent organ of ecumenical cooperation, the

Continuation Committee, which prepared the way for the formation of the International Missionary Council in 1921, was an outcome of that Conference. It is true also in a deeper sense. The ecumenism of the nineteenth century had been in the main evangelical in a broad sense of the term, and its characteristic organ was the undenominational society. A typical example at the beginning of the nineteenth century was the British and Foreign Bible Society; later in the century followed the Young Men's Christian Association. In such bodies, formed with a view to good works and based on a common loyalty to Christ, the problem of varying denominational loyalties was, if not absolutely suppressed, at least not acutely felt. A change began when Anglican Anglo-Catholics began to be interested in the Student Christian Movement; could they conscientiously take part in a movement which was dominantly evangelical, and in which many of the leaders were Nonconformists?

Without knowing the importance of what they were doing, the leaders among the students worked out a new formula of cooperation, in which denominational differences, so far from being overlooked, were frankly recognized, and every student was welcome on the understanding that he would bring into the fellowship the riches of his own denomination, to which he was expected loyally to belong. 'The Student Christian Movement is inter-denominational, in that while it unites persons of different religious denominations . . . it recognizes their allegiance to any of the various Christian bodies into which the Body of Christ is divided. It believes that loyalty to their own denomination is the first duty of Christian students and welcomes them into the fellowship of the Movement as those whose privilege it is to bring into it, as their contribution, all that they as members of their own religious body have discovered or will discover of Christian truth.'[1] This was at the time a new and revolutionary idea of ecumenism.

It was the idea that underlay also the Edinburgh Missionary Conference. Some of the Anglican societies had

1. T. Tatlow: *The Story of the Student Christian Movement of Great Britain and Ireland* (1933), pp. 405 f.

hesitated to take part in a conference which seemed to be organized entirely under Protestant auspices. A weighty protest against this view was entered by Charles Gore, then Bishop of Birmingham, and Edward Stuart Talbot, then Bishop of Southwark: 'We think it would be widely harmful, and a great loss to ourselves, if the Anglican Church were left outside it . . . we desire to express our very earnest hopes that the S.P.G. will be strongly and powerfully represented.' In the face of some opposition the S.P.G. agreed to send the full delegation to which it was entitled. Anglican opposition was stilled when Archbishop Davidson, after long hesitation, himself agreed to be present and to address the Conference.

This was the beginning of many things.

By agreement the questions of Faith and Order were excluded from the programme of the Edinburgh Conference. But it was there that Charles Henry Brent (1863–1929), the Canadian-born bishop of the American missionary district of the Philippines, saw the vision of an international Christian Conference, in which Christians would meet precisely to discuss those things about which they differed. 'Faith and Order' had been born. For seventeen years Faith and Order was almost exclusively an Anglican, and an American Anglican, venture. The inspirer of the movement, Bishop Brent, the Secretary, Robert H. Gardiner, and the Treasurer, George Zabriskie, were all American Episcopalians. The faith and the money that kept the movement going through long and discouraging periods came mainly from the same source. When at last the first World Conference on Faith and Order met at Lausanne in 1927, the Anglican Churches were well and truly represented. By the time of the second Conference, Edinburgh 1937, William Temple (1882–1944), then Archbishop of York, had revealed himself as the greatest ecumenical personality of this age.

In the meantime, the other wing of the ecumenical movement, Life and Work, was going ahead. Here the first inspiration was Scandinavian, and no other man played a role in it comparable to that of Nathan Söderblom, Archbishop of Uppsala. Yet, when the first great Conference met at

Stockholm in 1925, the first sermon was preached by Frank Theodore Woods, Bishop of Winchester (1874–1932); and Dr G. K. A. Bell, then Dean of Canterbury, later Bishop of Chichester was asked to edit the record of the proceedings of the Conference.

When, in 1938, the decision was taken that the two wings of the ecumenical movement should join together in a World Council of Churches, no doubt at all was felt as to the person who should preside over the Provisional Committee appointed to bring the World Council into being. William Temple, Archbishop of York, was endowed with a capacious intellect and a large heart, unforced geniality and wide sympathy, on occasion with noble and moving eloquence, and with an astonishing – sometimes too astonishing – capacity for finding the formula in which irreconcilable points of view would find their reconciliation. His sudden death in 1944 was a severe blow to the ecumenical movement, and a grievous loss to the whole Christian world.

At last in 1948 the World Council of Churches was formed, and held its first assembly at Amsterdam in August 1948. It fell to Temple's successor, Geoffrey Francis Fisher, to declare the World Council of Churches duly and formally constituted, and to lead the great assembly in prayer. The first Chairman of the Central and Executive Committees was Dr G. K. A. Bell, Bishop of Chichester. The secretary responsible for the third great Conference on Faith and Order held at Lund in 1952 was another Anglican, Canon Oliver Tomkins, now Bishop of Bristol. The second General Secretary of the World Council of Churches, whose nomination as successor to Dr W. A. Visser 't Hooft was announced in August 1964, is a Scottish Anglican, the Reverend Patrick Rodgers.

Every regularly constituted Anglican Province is a member Church of the World Council of Churches.[1] This membership is far more than formal; these Anglican Churches feel that the World Council, with its original

1. It is irritating that, under the World Council's definition of automony, such extra-provincial dioceses as Singapore and Sabah (formerly British North Borneo), cannot obtain membership.

basis, as 'a fellowship of Churches which confess Jesus Christ as God and Saviour', is one of the greatest instruments yet given by God to the world-wide Church for the realization of the ideals, dear to every Anglican heart, of unity combined with freedom, and of comprehension without disloyalty to the faith once given to the Church.

In this ecumenical setting, the Anglican Churches are sometimes seen to advantage as *par excellence* the ecumenical Churches. With their own diversity and variety of tradition, they reach out in all directions, can find themselves at home with all manner of Churches, and can to some extent serve as interpreters to one another of widely divided Churches both within and outside the ecumenical fellowship. On the other hand, the non-Anglican Churches are sometimes driven to distraction and infuriation by the uncertainties of Anglican action and the indefinable quality of Anglican thought. With the Orthodox we know exactly where we are. But some Anglicans will receive the Holy Communion at a Lutheran or Presbyterian service, and others will not. How can this be? Comprehensiveness has its drawbacks as well as its advantages. But such difficulties are inseparable from membership in a group of Churches which the Roman Catholic uncompromisingly dismisses as 'Protestant', but which, as seen through the uncompromising Protestant eye, look disturbingly 'Catholic'.

Epilogue: *Present Situation and Prospects*

THE Anglican Communion, being a living entity and not an ossified institution, never abides in one stay, and is in a condition of perpetual change. This means that any description of its situation at a given moment will be out of date before it appears in print; the most that can be hoped for is an analysis of trends and an indication of the direction in which developments seem to be moving.

The first and burning question is, naturally, whether the Anglican Communion in anything like its present form can survive at all. The beginning of its dismemberment seemed to come with the formation of the Church of South India in 1947. It seemed likely that this process might go rapidly forward. But in point of fact nothing happened for a number of years. Then at last in 1970 the plans for a United Church in North India, the working out of which had begun in 1929, finally reached successful completion, and this new Church came into existence. Unlike the Church of South India, this Church provided an outward ceremony for the unification of ministries; and, the lawyers in England having declared that this could be taken as equivalent to episcopal ordination, the Church of England felt it possible to give full recognition to the new Church, a privilege which it has withheld from the Church of South India, since there a number of ministers who have never received episcopal ordination continue in office. Other Anglican provinces, not being bound by the Act of Uniformity of 1662, have felt less hesitation about entering into full communion with the Church of South India also.

The plans had been made for a Church of North India. But political changes have made the term obsolete. The unity of India, which the British had built up and maintained for just a century, was shattered by the formation of Pakistan in 1947, and again by the secession from Pakistan of Bangladesh. It is necessary, therefore, to speak of four Churches in the Indian sub-continent, all fully independent, all in full communion with one another, all desiring to maintain friendly relations with all other parts of the Christian world.

The expectation that similar Churches would come into being in other parts of the world has not been fulfilled. Union between Anglicans, Presbyterians and Methodists in Nigeria seemed very near, in fact, the date for union in 1965 had actually been fixed.

At the last moment opposition developed among the Methodists and the union has been postponed *sine die*. In England the plan for a fusion of the Church of England with the Methodists, accepted by the Methodists, has twice been rejected by the Church of England. In Canada, after years of somewhat desultory negotiation, the Anglican Church in Canada has decided (1975) that it does not wish to enter into union with the United Church of Canada. In the United States the consultation on Christian Union, moribund though not defunct, has failed to produce any result in actual church union. The Committee on Christian union in East Africa, in which Anglicans have been concerned since the beginning, has made progress in producing a common catechism and a united liturgy, but actual church union seems further off than it did in 1913, when through the Kikuyu Conferences Kenya gave a lead to the entire Christian world. It had been hoped that in 1975 a United Church of Sri Lanka (Ceylon) would have come into being, but at the last moment legal difficulties led to a postponement which has lasted to the present time. Some observers hold that the interest in Christian unity is no less than it was, but that the focal point has changed. The time of local unions has come to an end. The epoch has arrived in which world denomination should speak to world denomination. The major Christian confessions are more conscious of themselves as worldwide churches than they ever were before. It is on this basis, we are told, that Christian union should be sought, and not on that of merely local enterprise. The Congregational and Presbyterian bodies have set an example to the world by fusing their central organizations, even though this does not necessarily involve union at the local level. The Roman Catholic Church is engaged in discussion with a number of non-Roman confessions precisely at this level of worldwide involvement. The Anglican delegation which meets with Roman Catholic colleagues is not a deputation from the Church of England, but an international body on which Anglican churches in various parts of the world are represented on terms of perfect equality. There can be no doubt that this marks a change of emphasis; it may prove no less fruitful than the promotion of local unions, which was a central concern in the period between 1910 and 1960.

So the Anglican Communion survives, only slightly diminished from what it was in 1947. But it has found it necessary extensively to reorganize itself in the period under review. We have alluded to the effect of political situations on the development of Churches in one geographical area. Similar political changes have had marked effects on the proliferation of Anglican Provinces in many parts of the world.

In the 1930s plans were afoot for the formation of a single Province in East Africa. British colonial policy tended towards closer unity between the three territories of Uganda, Kenya and Tanganyika (now Tanzania). Dislike in Uganda of the settler regime in Kenya intervened to frustrate the plans that were being made. But the real division came after the granting of independence to the three countries. Instead of moving closer to one another the three territories have moved apart; it seemed natural that this political reality should find its reflection in the formation of three separate ecclesiastical Provinces – Uganda, Kenya and Tanzania (36 dioceses).

Mauritius and Madagascar did not seem to belong anywhere. It seemed right and good that these four dioceses, now with the Seychelles as the fifth, mainly French-speaking, not African but with an increasing tendency to look towards Africa, should come together in one body under one head, who rejoices in the charming title of Archbishop of the Indian Ocean (1973).

With political independence in 1947, Burma, which had been unnaturally linked with the Indian Empire, recovered its freedom. It was only a matter of time for Church to follow State, and for the Church in Burma to separate itself from the Church of India, Burma and Pakistan (1970; Archbishop enthroned 1973). The number of Anglicans in Burma is very small, less than the population of many single dioceses elsewhere. For political reasons the Burmese Province lives very much isolated from the rest of the Anglican world. But these four dioceses have managed to maintain themselves, and to carry on effective evangelistic work, especially among the mountain peoples of the north and east.

Political factors can influence the life of the Church adversely as well as favourably. For a brief period Malaysia and Singapore found themselves together in a federation engineered, with the best of intentions, by the British. Good intentions were frustrated by the inveterate dislike of the Malays for the Chinese. The Federation broke up. Tension between the two parts of the Federation continues to embitter the situation, so much so that it is almost impossible for theological students from Malaysia to be trained in Singapore. It would seem obvious that the four dioceses of West Malaysia, Singapore, Sabah and Kuching, once the home of Rajah Brooke and of the doughty Bishop Macdougall, should come together in a Province. For the time being the matter can hardly be so much as discussed. These four dioceses are, however, included in the Council of the Churches of East Asia, in which the three Korean dioceses have also found a home.

Melanesia had from the beginning of its Christian history been associated with New Zealand, but progress towards political independence indicated plainly that continued dependence of the Church in Melanesia on New Zealand would be undesirable; the formation of an ecclesiastical Province was achieved in 1975 (to be followed after only four months by the death of the first Archbishop). The same process, with the like results, is likely to be followed in Papua–New Guinea, a territory which is now self-governing, though not yet fully independent of Australia. The South Pacific Anglican Council is already in existence, and is likely to undergo extensive changes in the near future.

The Churches in Brazil and in the Philippines, which owe their origin to the activity of the American Episcopal Church, have both been recognized as independent Churches within the Anglican fellowship. The Philippine Church has a special responsibility to maintain and develop good relationships with the Philippine Independent Church. In other parts of Latin America and the Caribbean, similar processes are at work, though accompanied with certain difficulties arising from the co-existence of dioceses owing their origin to the work of the American Church with those of British origin.

The special elements which constitute an Anglican Province are the right to elect its own bishops, to draw up its own constitution, to work out its own liturgy, and to determine its relationships with other Churches in its area. Each Province is jealous of its independence, and, while remaining intensely loyal to the Communion as a whole, makes plain its intention in many matters to go its own way.

Evidently Churches will be affected by their own past, by local traditions, and by the political situations in which they find themselves. One area in which this is clear is marriage customs and rules. Newly independent nations make their own laws as to what can be accepted as a legally valid marriage; a Church must take account of such laws, even when they may not be entirely satisfactory from the Christian point of view. Anglican canon law, as it relates to marriage, is derived from Roman law and from the traditions of the west in the Middle Ages; it is impossible to hold that this law will be of universal application. In India, for instance, western canon law sanctioned marriages between cousins which Indian custom regarded as incestuous, since those whom the west calls cousins are in certain cases called brothers and sisters in India. The same law forbade marriage between uncle and niece, which in certain castes is not merely permissible but obligatory. In some African societies all marriages between cousins are regarded as incestuous. Provinces will naturally assert their right to work out their own rules, while retaining certain prohibitions held to be required by the law of God and not simply imposed by the laws of men.

Of recent years there has been a spate of activity in the field of liturgical revision. Provinces have tended to go their own way with little mutual consultation, and there has been a tendency to disregard or to repudiate what for four centuries have been regarded as accepted Anglican standards. This must be regarded as serious. Anglican unity has to a large extent been liturgical unity. Up till the year 1928 a wandering Anglican could be certain that, wherever he went, he would find the same service, within somewhat narrow limits of variation in different areas. Even if he did not understand a word of the language, he would know what was going on and would be able to participate intelligently. From 1928 onwards variations have multiplied, so much so that an Anglican moving from Province to Province never knows by what service he is likely to be confronted, and will find himself bewildered by the variety of uses. It is not merely that language has changed and been brought up to date; the very structure of services has been modified and much that was familiar as the Anglican pattern has disappeared.

The independence of the Provinces and their ability to adapt themselves to the most varied conditions are laudable and must be maintained. But the fissiparous tendencies which have become apparent suggest that attention has to be paid to structures of unity, such as had not been felt to be necessary in the past, when unity existed and did not have to be constructed.

On the whole Anglicans have felt suspicious of organization, and have managed to get on remarkably well without it. Until 1897 the Communion possessed no body charged with responsibility for regular consultation between its different parts. In that year the Consultative Body of the Lambeth Conference was called into being. This was a somewhat tenuous body, the eighteen members of which were appointed by the Archbishop of Canterbury after consultation with the heads of the various Anglican Churches. It was not stated that regular meetings were to be held.

Moreover, the functions of the body were strictly and rigidly defined. The Conference of 1920 stated that the title of the body accurately indicated its functions: 'The Consultative Body is of the nature of a continuation committee of the Lambeth Conference, and neither possesses nor claims any executive or administrative powers.' This was reiterated in 1930; it was further stated that its tasks were 'to advise the Archbishop of Canterbury or any other bishop who requests its help, and to assist the Archbishop of Canterbury in the preparation of the business of the ensuing conference'. The Consultative Body was, as a matter of history, frequently consulted by the authorities of the Church of India, Burma and Ceylon, over questions arising out of the formation of the Church of South India, as a means of maintaining contact with the rest of the Anglican Communion; but this had to be done by correspondence only, since meetings of the Consultative Body were rare, and none at all could be held during the crucial period of the second world war.

Very gradually further steps towards a regular organization were taken. The Lambeth Conference of 1948 brought into being an Advisory Council on Missionary Strategy, but failed to give this purely advisory body either adequate staff or convincing terms of reference. A much more important step was taken in 1958, when for the first time a whole-time officer was set apart, with the clearly stated duty of serving as a liaison between the scattered parts of the Communion, and of maintaining and promoting that unity which found brief and visible expression in the ordinarily decennial meetings of the Lambeth Conference. The first to hold this office was Bishop Stephen Bayne, who resigned the bishopric of Olympia in the United States in order to be available for this central and international responsibility. Even so the duties of this Executive Officer were not clearly defined, and a cloud of decent obscurity hung over the question of his relations with the Archbishop of Canterbury and Lambeth Palace. Yet this was a real advance in the direction of both the recognition and the expression of Anglican unity.

Similar processes had been going forward in other parts of the Christian world. Ecumenical activity has been followed, paradoxically but inevitably, by an increase of denominational self-consciousness. When we meet Christians of other confessions on terms of intimacy and mutual respect, we are bound to ask ourselves what it means that we are members of this or that denomination, and whether that denomination has any right to continued separate existence. One of the notable phenomena of the past quarter of a century has been the formation or the strengthening of the world confessional bodies. One of the most recent and by far the most powerful of these bodies is the Lutheran World Federation, which was formed as lately as 1947, but with its large and highly qualified staff in Geneva, in some points rivals, in others almost overshadows, the World Council of Churches itself. This movement throws into relief both the Anglican tendency to be caught up in world currents, and also its distinctiveness from other worldwide bodies. When an invitation was issued to the central organizers of world confessional bodies to attend a meeting in Geneva for mutual consultation and advice, the then Executive Officer of the Anglican Communion refused to attend, on the ground that the Anglican Communion was not a world confessional organization but an international and worldwide Church. He was right. The Anglican central organization creates nothing new, it merely gives clearer expression to something that has always existed. Nevertheless, when both the Roman Catholic Church and the Orthodox Churches had agreed to be associated in a consultative capacity with such meetings, there seemed little reason for the Anglican Churches to hold back, and, still with some hesitation, they have agreed to be represented when these new forms of ecumenical organization hold their meetings.

The Lambeth Conference of 1968 carried the process of central organization a step further. It recommended to the Churches the formation of an Anglican Consultative Council, on which all the Provinces would be represented, by clergy and lay people and not only by bishops, which would remain in permanent existence, and would meet ordinarily once in two years. The first meeting was held at Limuru in Kenya in 1971, the second at Dublin in 1973, the third, postponed by a year because of the meeting of the fifth assembly of the World Council of Churches in 1975, in 1976 in Trinidad. The Lambeth Conference defined the spheres of responsibility of the Council as being 'communication, mission, inter-church relations, and special studies'.

There can be no doubt of the value of such a small body, able to meet at frequent intervals, and capable of dealing without long delay with such important issues as come up from time to time. One specially valuable feature of the Council is the specially weighted representation granted to the Provinces in the third world. Though the Churches in that part of the Christian world are not numerically very strong, their representatives feel themselves to be the spokesmen for vast masses of human beings who have not yet found entrance into the visible fellowship of the people of Christ. Dramatic expression was given to this new reality in 1974, when the whole world saw on television the pause in the proceedings of the enthronement of the Archbishop of Canterbury, when the new Archbishop knelt to receive a special blessing from the Most Reverend Festo Olang, Bishop of Nairobi and Archbishop of the Church of the Province of Kenya.

For a century the Lambeth Conference of Anglican bishops has been the one clear expression of a growing and worldwide unity. It has always been stressed that the Conference is not a Synod; it is technically no more than a large committee of bishops, convened on the authority of the Archbishop of Canterbury for the time being, having no existence apart from his invitation, and with no guarantee that each Lambeth Conference will not be the last. Yet, although the Lambeth Conference has no authority, it has acquired great influence. Its best pronouncements have spoken to the Christian world far beyond the limits of the Anglican Communion. In recent years, however, the assets seem to have diminished. The Conference of 1948 did notable work in the field of Christian union, and dealt temperately and sensibly with the new problems arising out of the formation of the Church of South India. The Conference of 1958 produced a first-class report on the family, and dealt perhaps more successfully than any other Christian body so far with the new problems of family limitation and population control. The Conference of 1968 was so unwieldy that it produced nothing except the practical advance, already referred to, of the creation of Consultative Council. Naturally, the question has been raised whether any further conferences of this nature should be held. They are immensely expensive. All such meetings tend to duplicate those of other Churches, and of the World Council of Churches. Do they continue to serve any useful purpose?

To this question a resolutely affirmative answer must be given. The total product of a Lambeth Conference cannot be distilled in reports and resolutions. To a lonely bishop on the frontiers of the Church it is an incomparable gain to spend several weeks in close proximity with bishops from all parts of the world and from every kind of diocese. In the midst of many differences of point of view something like a common mind can emerge. An awareness grows of the greatness and uniqueness of the Anglican tradition. It is felt by many that some extensive changes in the planning of such a conference are overdue. Attendance should be limited to one bishop from each diocese (not necessarily in every case the diocesan bishop). Far more careful preparation is needed. The conference should limit itself to Anglican concerns and not attempt to pronounce on every Christian problem, especially on those with which the World Council of Churches is specially concerned. Except for the opening and closing sessions, it should avoid the crowded and overheated atmosphere of London. If such changes can be made, the conference may have a long and useful future worthy of its distinguished past.[1]

Not the least interesting feature of change is the position of the Archbishop of Canterbury in the new world of Anglicanism. Clearly he has far less power than he had. Until independent Provinces came to be formed, all the outlying parts of the Anglican Communion, with the exception of those which were under American control, were legally part of the Church of England. This meant that the Archbishop of Canterbury was their Metropolitan. He had more say than anyone else in the appointment of their bishops, except in so far as these were controlled by the Crown. A steady stream of letters from every part of the world flowed in to his desk, and the answers he returned to the questions addressed to him had almost the force of law. Now all that has changed. The Archbishop of Canterbury has no jurisdiction outside his own Province. Chaplains to the British armed forces hold a licence from the Archbishop. When such chaplains arrived in India during the second world war, it was a considerable shock to them to learn that the Archbishop's writ did not run in India and that they could do nothing on Indian soil unless they had received a licence from the local Anglican bishop, whoever he might be.

But, if the Archbishop's authority is a good deal less than it was, it is possible that his influence is even greater. In past times an Archbishop hardly travelled outside England, except for the occasional continental holiday. Now that he has so much less work to do than his predecessors, it is possible for him to become something like a travelling apostle. The last two Archbishops of Canterbury have travelled extensively throughout the world, perhaps a little more than was good for their special work as responsible heads of the Church of England. Archbishop Ramsey's visit to the Pope attracted worldwide attention. It is taken almost for granted that the Archbishop will be present at an Assembly of the World Council of Churches. His personal gifts are likely to make it possible for him to contribute something of exceptional value. But apart from this, no less than the Pope, he is the visible

1. This does not, of course, exclude the desirability of other representative gatherings in the intervals between Lambeth Conferences.

symbol of something very great—of a fellowship which has spread to almost every part of the world, which has maintained itself, in the face of extensive and continued hostility, has shown itself flexible and adaptable in many environments, has called out the devotion of men and women of every race and clime. It is likely that this role of the Archbishop as a travelling messenger of unity may increase rather than decline in importance, as increasing independence lessens the official bonds, and therefore adds to visible symbols an importance greater than they had before.

Like other Churches, the Anglican fellowship is a bewildering amalgam of strength and weakness.

In England, its homeland, its influence appears steadily to be losing its hold on that majority of the population which is officially classed as adhering to it. No power and no movement has yet succeeded in staying the decline in the number of confirmations and of those who make their communion. Recent figures, given in the *Church of England Year Book* for 1976, suggest that not more than five per cent of those who are called Anglicans are regularly in church; the percentage of regular communicants is even smaller. Until 1960 statistics for the United States and Canada were much more encouraging; but at about that date the powers of materialism seemed to begin to take charge, and there too the Church has to work much harder just to maintain its ground and to retain its hold on those who traditionally have been faithful to it. The general feeling is that the Church no longer has any power in those places where decisions are made that affect the life of men and of nations.

All this is true. But it is very far from being the whole picture. No one has yet managed to develop an organon for the measurement of diffused Christianity; no one can doubt that it is an element of great potency in the life of the once Christian nations. Certain signs can be recognized. An astonishing number of people listen to Christian services either on radio or television. In a recent opinion probe, almost two-thirds of those who answered stated that they engaged in some kind of prayer every day. Apart from the convinced Marxists, secularists and humanists, who oppose every form of religion, perhaps eighty per cent of the population would be prepared to assert that on the whole they wish England to remain in some sense a Christian country. Eighty per cent of that eighty per cent, or roughly two-thirds of the population retain some kind of connexion, albeit often a rather shadowy one, with the Church of England. The cry for the disestablishment and disendowment of the Church is still heard, but usually without the venom which attended on such demands in the nineteenth century.

A number of new movements within the Church indicate that it is very far from dead.

The most sensational is probably the charismatic movement, which twenty years ago was hardly known within the sober Anglican tradition. In more recent times, crossing the Atlantic from the United States, it has attracted a considerable number of the younger clergy, and made itself felt as an effective movement for renewal. Once empty churches are full, and lay people have borne witness to an experience of Christ of a kind previously entirely unknown to them. In some places the movement has shown itself to be divisive; in others it has remained well within the framework of the Church, and has made itself available to those who desire it rather than forcing itself on those who do not.

Party labels and distinctions are less a feature of English church life than they were half a century ago. A variety of emphasis does persist, though now as then the majority of Anglicans firmly resist classification and refuse to identify themselves with any party cry. The most notable change in the balance of emphasis, not only in England, has been the increasing strength of those who call themselves conservative Evangelicals. The theological colleges associated with this part of the Church are full. With a simple, definite and demanding message, they have managed to make themselves extensively heard, especially in the so-called middle class. Representatives of this group have reached the episcopate; it is not clear that they have so far been effective in the great conurbations, among the trade unions, and in the so-called working class.

As always it is necessary to guard against the error of identifying Anglicanism with the Church of England. The scene may be gloomy in England. This is far from being true of the Anglican Communion as a whole. In some areas, especially in Africa, the Church is staggering under the weight of its own success. In one diocese in Tanzania, the Church claimed to have quadrupled its numbers in sixteen years; this claim was based not on guess-work but on very carefully maintained statistics. In one area of Kenya, progress is at the annual rate of ten per cent. This means that the Church will double its numbers in seven and a half years. The Anglican Church in Australia is entering into missionary work in areas, such as Indonesia, with which it had never previously been concerned. Recently the whole of Canada saw on television the ordination of four young Eskimos to the diaconate. Eskimo Christians are confident that the next Bishop of the Arctic will himself be an Eskimo.

Like any true picture of the Church in any period and in any area the contemporary picture of the Anglican Communion is made up both of lights and shadows. This is no occasion for pessimism. But certain deeper questions do arise. In this ecumenical age, in which so many barriers have been broken down and the Churches have in so many ways drawn nearer to one another, do the Anglican Churches still stand for anything specific, and do they regard themselves as making a contribution to the sum total of Christian experience, which is not being made in exactly the same way by any other Church?

In the past the Anglican, if challenged to state the nature of his Church's faith and the authority on which it is based, was at once ready with his reply. The sources of Anglican doctrine were:

The Scriptures of the Old and New Testaments, as containing all things necessary to salvation. (Article VI.)

The three ecumenical creeds (Apostles', Nicene, Athanasian).

The dogmatic decisions of the first four general councils of the Church.

The Book of Common Prayer together with the Ordinal, as setting forth not only the Anglican order of worship, but also the theology of which that worship is the expression.

The Thirty-nine Articles of Religion as the official summary of the Church's standpoint in relation to central issues of theology and to a number of controversial points of doctrine and practice.

It is doubtful whether any of these five points can be cited with equal confidence today.

At the time of the Reformation all the Churches of the Christian world, with some minor variations, held what today would be called a fundamentalist understanding of the nature of biblical inspiration. The words of the Bible are the very words of God himself. They may require interpretation but may not be questioned. Some Anglicans believe themselves still to hold this view; but in point of fact almost all have made some concessions to the spirit of criticism in recognizing, more fully than even the teachers of the nineteenth century, the human element in Scripture. The mere quotation of proof texts without elucidation is no longer accepted as a valid method of theological argument. Some would go much further. What is meant by the inspiration of Scripture, other than that these are ancient documents which over the centuries have spoken to the hearts and consciences of men?

It is recognized that creeds and dogmatic decisions were conditioned by the situation which existed at the time of their formulation, though, as dealing with questions which are of perennial concern to the Church, they have a certain timeless quality. But it is no longer self-evident that these are the only possible formulations or that they best express the truth of the Gospel as it is relevant to the changed conditions of the modern and industrialized world. The creeds seem to present the Church as a static existent rather than as a growing and developing reality, constantly changing in relation to the changing world by which it is surrounded.

A number of the Anglican Provinces no longer accept the Thirty-nine Articles as an accepted statement of the Church's doctrine, binding on the ministers of the Church. Such Provinces have their own confession of faith, though they may also print in their constitutions the Thirty-nine Articles, as an historic document of value and as a point of contact with other Anglican Churches elsewhere in the world. No one has yet produced a satisfactory revision of these ancient formularies; it is unlikely that any theologian would feel able to accept every syllable of a sixteenth-century document, much of which is wholly admirable, but parts of which are no longer relevant and other parts state truth in forms which are no longer acceptable to the modern mind.

What are the acceptable limits of variation in doctrine? This is a question which has agitated the Church from the beginning, and no finally satisfying answer has ever been reached—truth is something which can never be wholly captured in a formula. On the whole the Anglican Churches have avoided becoming involved in persecution of dissentients and in heresy trials. The affair of Bishop Colenso and the agitations resulting from the publication of *Essays and Reviews* in 1860 have perhaps served as a permanent warning. In both cases a number of the views which at the time seemed shocking have come to be generally accepted as in no way inconsistent with essential Christian truth. Willingness for a certain amount of what appears to be error to continue to exist in the Church is not necessarily a sign of indifference to truth; it may arise rather from an awareness that underneath the appearance of error what is really a new discovery of truth may be concealed, that truth shines by its own light, that in the history of the Church innumerable aberrations have proved unable to maintain themselves, and that far more harm is done to the life

of the Church by the appearance of persecution and the making
of martyrs than by enduring for a time in the confidence that
orthodoxy and heresy will in course of time sort themselves out
and that in the end truth will prevail.

Yet anxiety cannot altogether be stilled. When Bishop J. A. T.
Robinson published his now famous book, *Honest to God* (1963), the
theological best-seller of all time, it was received by many not
as the Bishop had intended, but as a clear indication of how little
it is necessary still to believe in order to call oneself a Christian.
The Cambridge volume called *Soundings* (1967), called forth a
satirical rejoinder by Professor E. L. Mascall called *Up and Down
in Adria*. More recent pronouncements by some eminent Anglicans
have given the impression that the Church of England no longer
knows what it believes, and does not stand for anything particular
at all.

Confusion in the field of theology is not helped by confusion in
the area of liturgy to which allusion has already been made. Some
have maintained that there are liturgical principles, existent in
their own right, in the light of which liturgical reformation can
be carried out, and that no change in theology is necessarily
involved in such purely liturgical changes. This is an illusion.
Every form of liturgical expression has theological implications.
Every change in liturgy involves also a change in theology,
though naturally some changes are much more serious than
others. The wide variety of liturgical forms already existing within
the Anglican Communion makes it impossible to refer to any of
these liturgies as providing the same kind of theological standard
as was provided for four centuries by the Book of Common Prayer.
It is not always clear what form of theology these variant liturgies
are intended to express. Pelagius was the first of English (Welsh)
heretics. Pelagianism is the form of heresy to which Englishmen
and their Anglican colleagues of other races have been most
inclined. It is possible that our Anglican revisers, in a perhaps
excessive reaction against what was held to be the excessive
Augustinianism of the older Anglican tradition, may have produced what are in fact characteristic Pelagian liturgies. While so
much as yet remains undecided, no more than a provisional
opinion can be expressed. Note must be taken of the anxiety
which is felt in many quarters as to the direction in which the
work of liturgical revision seems to be carrying the Anglican
world.

If all this is true, if so many elements of uncertainty disturb what used to be the perhaps over-complacent tranquillity of the Anglican world, can it be affirmed that the Anglican tradition still exists, and that it stands for anything in the general framework of the Christian Churches?

The answer may not be obvious. Yet others besides Anglicans may feel that this group of Churches does stand for something significant in the Christian world, and does represent a kind of unity which is not found elsewhere.

Anglicans throughout the world are bound together by a strong sense of mutual loyalty. This is not altogether easy to explain to those who are not themselves Anglicans. It may be felt to be rather like the unity of that other mysterious body the (British) Commonwealth of Nations. With the dissolution of the British Empire, a few peoples withdrew from the fellowship; but the great majority decided to remain, perhaps because of certain advantages, but in most cases because they simply liked being there. The formal links are few, other than the recognition of the Queen of England as the Head of the Commonwealth. But the fellowship is a reality. A member of a Commonwealth country who crosses the border between the United States and Canada, two countries closely linked by very much more than propinquity, at once finds himself in a world in which he feels himself very much more at home than in that which he has left. The Commonwealth does seem to stand for certain ideals of human dignity, of democratic freedom and of mutual tolerance which must be kept alive if the human race is to advance in the direction of a universal community of nations. In the same way, Anglicans in all parts of the world do find one another, do realize that they are involved in a community far wider than the Church of England as by law established. They do feel that this community enshrines certain values which must be maintained if the Church is ever to recover the unity in which all Christians profess to believe.

The Anglican Churches on the whole have shown themselves accessible to the acceptance of new truth. This willingness to change is at the time of writing being severely tested by the question of the ordination of women. The Anglican middle way is once again becoming visibly recognizable. The Orthodox Churches have declared that never in any circumstances can a woman be ordained to the priesthood. Roman Catholic theologians seem to take the view that, though there may be no theological objections to such ordination, the practical difficulties are so great that proposals for action cannot be seriously entertained. Many Protestant Churches have gone forward and ordained women, without much theological consideration, on general grounds of the equality between the sexes and without much attention to difficulties that might later have to be faced. The Anglican attitude has been typically cautious, marked not so much by a refusal even to consider the question, as by a desire for theological clarity and for a careful weighing in advance of all that would be involved in action. The Anglican Church in Canada has decided that ordination of women to the priesthood may begin on Advent Sunday 1976. The Episcopal Church in the United States is almost certain to follow suit in the course of the same year. No one Province is bound by the action of any other; but it seems likely that the practice may spread throughout the Anglican world. The Church of England may be the last to make up its mind. But, when the diocese of Hong Kong, now an isolated diocese since it is at present practically unable to live out its fellowship with the Holy Catholic Church of China to which it once belonged, moved into action with the ordination of two women as priests, it was made plain that by doing so it would not forfeit its relations of communion with the other dioceses of the Anglican Communion, though women so ordained would not automatically be recognized as capable of ministering in other Provinces which had not followed the Hong Kong example.

This may be regarded as typical of a spirit which can be condemned as mere compromise or expediency, but which Anglicans would prefer to identify as combining intense loyalty to the past, together with a recognition that the past is not to be taken as a strait-jacket, making impossible any change or development in what is always and necessarily an undisclosed future.

Anglicans feel that their fellowship does hold in certain respects a central position within the Christian world. The Second Vatican Council, in one of its documents, referred to the Anglican Communion as standing in a special relationship to the Roman Catholic Church. It was not specified in what this special relationship consisted, but it is easy to guess what was intended—the Anglican fellowship has always set itself in a unique way to combine what are generally referred to as the Catholic and the Protestant elements in the Christian tradition.

As a result of this, the Anglican Churches have been able to gather around them a group of Churches in what has come to be known as the Wider Episcopal Fellowship. Relations with the Old Catholic Churches and the Philippine Independent Church, to which reference has been made in an earlier chapter, have been strengthened and developed. A number of Provinces now have fellowship with the Mar Thoma Syrian Church of South India, the Reformed section of the ancient Church of the Thomas Christians of that area.[1] The tiny Episcopal Churches in Spain and Portugal have been drawn in, the latter having contributed a bishop for an Anglican diocese (Lebombo) in South-east Africa. Some hold the view that the Lambeth Conference should be replaced by a conference of this wider fellowship.

It is not only within the episcopal world that the Anglican Churches have taken a lead in the work of Christian union. They have reached out in all directions, and to Churches of many different types.

The editors of the *History of the Ecumenical Movement 1517–1948* were at times embarrassed to find that over considerable stretches of the history the record was almost exclusively that of Anglican efforts to promote Christian unity. These efforts have sometimes produced embarrassment in others, through the Anglican insistence on certain historic traditions unknown to and undesired by those who have never had experience of them and cannot understand why Anglicans lay such stress upon them. This has led at times, unfairly, to a doubt as to the sincerity of the Anglican desire for unity. It does make plain the fact that the working out of effective plans for a workable Christian unity is more difficult than is often supposed by those who have never been involved in such work.

Any confession which takes seriously the possibility of visible and organic Christian unity must reckon with the possibility of its own disappearance, or of such radical modifications in its life and structure as would make it hardly recognizable as that which it had been in the time of separation. Denominations may be necessary, but they should never be regarded as a permanent feature of the Christian landscape. They justify their existence on the grounds of the maintenance of certain aspects of Christian truth which would otherwise be imperilled. But, if Christian truth can be adequately safeguarded in a united Church, is there any valid argument in favour of continuing in separation? Must not the denomination be prepared to lose its individual existence by becoming merged in a larger whole?

1. The term Mar Thoma Church is sometimes used incorrectly of the whole body of 'Syrian' Christians in India; it should be used only of that group, now about 300,000 strong, which in the nineteenth century, under Anglican influence, carried out an extensive reform of its life and worship.

The Anglican Churches have been the first in the world to consider seriously the possibility of their own demise. This possibility was soberly and cautiously put forward by the bishops assembled in the Lambeth Conference of 1948:

> Here we desire to set before our people a view of what, if it be the will of God, may come to pass. As Anglicans we believe that God has entrusted to us in our Communion not only the Catholic faith, but a special service to render to the whole Church. Reunion of any part of our Communion with other denominations in its own area must make the resulting Church no longer simply Anglican, but something more comprehensive. There would be, in every country, where there now exist the Anglican Church and others separated from it, a united Church Catholic and Evangelical, but no longer in the limiting sense of the word Anglican. The Anglican Communion would be merged in a much larger Communion of National or Regional Churches, in full Communion with one another, united in all the terms of what is known as the Lambeth Quadrilateral.[1]

This is a splendid vision. It must not be misunderstood. Such a Church would not be Anglican *in the limiting sense of the word.* But it would still be Anglican, in the sense that such a united Church would accept and incorporate and express all that the Anglican Churches hold to be essential in the way both of Christian truth and Christian order. Short of this, the formation of a united Church would mean impoverishment and not enrichment for all concerned.

It is this consideration that leads to the caution expressed in the *Encyclical Letter* of Lambeth 1948:

> It is well to keep this vision before us; but we are still far from its attainment, and until this larger Communion begins to take firmer shape, it would only be a weakening of the present strength and service of the Anglican Communion if parts of it were severed from it prematurely. If we were slow to advance the larger cause, it would be a betrayal of what we believe to be our special calling. It would be equally a betrayal of our trust before God if the Anglican Communion were to allow itself to be dispersed before its particular work was done.[2]

Christian Churches, like Christian individuals, if they desire to follow their Master, must be prepared to die for His sake; but it

1. *Lambeth Conference, 1948*, p. 22.
2. *Lambeth Conference, 1948*, p. 23.

may be incumbent on them, as on their Master, at certain moments to say, 'My time is not yet come'. Churches, like nations, are precious things; and, though a Church should not strive 'officiously to keep alive' things that in the providence of God were better dead, it has no right gratuitously to sell its life away without any assurance that the sacrifice will have been worth while. Churches cannot enter into union with one another except by dying to their existence as separate Churches; they ought not to do so, unless they are assured that God himself is calling them to death with a view to a better resurrection.

It was in this hope and expectation that four Anglican dioceses in 1947 entered into the Church of South India. They willingly accepted a kind of death. They knew that their bishops would no longer be invited to the Lambeth Conference. They knew that, though still related to the Anglican Communion, they would no longer be a part of it. They believed that the sacrifice would be worth while. After twenty-nine years of existence in a united Church, almost every one of those who have lived throughout that period would agree that their hope and expectation have been fully justified. They have lost nothing that was essentially Anglican; they have gained by finding themselves heirs of other traditions that were not theirs in the days of separation. In the years between 1910 and 1960 fifty united Churches came into existence in various parts of the world. Hardly one of them was formed without having to pay the price of some new schism; the Church of South India lost the Churches in the area of Nandyal. But not one such united Church has been dissolved. Those involved seem almost all to have been agreed that such losses as have been experienced have been more than compensated by the gains that have accrued through union.

It is to be hoped that other Anglican Churches will not unduly or selfishly cling to their Anglican life in separation. But the years have shown that the way to union involves a long slow journey, which cannot be carried out in haste. It seems plain that, for a long time yet, the Anglican Communion will have to continue to be, since God still has work for it to do in separation, with a view to that blessed union of all Christian people, which all the Anglican Churches firmly believe to be the will of God for the Church which is the body of his Son.

Select Bibliography

General

THE literature on Anglicanism is immense and grows from year to year. There is as yet no satisfactory bibliography.

Good short bibliographies are to be found in the *Oxford Dictionary of the Christian Church* (2nd ed., 1974) under the headings 'Anglican Communion', 'Anglicanism', 'Church of England'.

The *Dictionary of English Church History*, edited by S. L. Ollard, G. Crosse and M. F. Bond (3rd ed., 1948) is out of date at a number of points, but contains information not easily accessible elsewhere.

Articles on most of the leading figures in the history are to be found in the *Dictionary of National Biography*. These are of uneven value. Some of the notices in the Supplementary Volumes are excellent, as are the short biographies in the more recent editions of the *Encyclopaedia Britannica*.

Only one full-scale history of the Church of England is in existence:

A History of the English Church (edd. W. R. W. Stephens and W. Hunt, 9 Vols., 1899–1910; some volumes several times reprinted). The volumes are not all of equal value, and naturally all are now at some points out of date.

We are now promised, by A. and C. Black,

An Ecclesiastical History of England (General Editor, J. C. Dickinson), to be completed in nine volumes. So far only three volumes have appeared:

M. Deanesly: *The Pre-Conquest Church (1961)*;

O. Chadwick: *The Victorian Church, Part I, 1829–1859* (1961);

Part II, 1860–1901 (1970).

Each of these has received warm commendation from the experts.

The Oxford History of England (ed., Sir G. N. Clark, 15 Vols.) is now complete. Each volume contains chapters on the Church. These, if read continuously, will provide the student with a clear picture of developments within the Church. The lay historians, having no axe to grind, have at some points done a better job than the ecclesiastics.

Of one-volume histories, the most practically useful is that by Bishop J. R. H. Moorman: *A History of the Church of England*

(2nd ed., 1967). H. O. Wakeman: *An Introduction to the History of the Church of England* (1896; 2nd ed., with additional chapter by S. L. Ollard, 1927), can still be consulted with profit.

C. H. E. Smyth: *Church and Parish* (1955), though strictly the history only of one parish, St. Margaret's, Westminster, admirably opens the door to an understanding of what the Church of England is really like.

On the Anglican Communion as a whole there is as yet no comprehensive work.

J. W. C. Wand (ed.), *The Anglican Communion* (1948) shows fairly well the situation as it was at the date of publication.

The successive issues of the *Church of England Year Book* (Ninety-second Edition, 1976) give detailed and reliable information not only about the Church of England but about all the Churches which together make up the Anglican Communion, and also about other Churches with which Anglican Churches have special relationships.

CHAPTER 1 : *Beginnings*

For the earlier period,

M. Deanesly: *The Pre-Conquest Church* (1961) gives all that is needed.

For the middle ages, two admirable books are still indispensable:

J. R. H. Moorman: *Church Life in England in the Thirteenth Century* (1948);

W. A. Pantin: *The English Church in the Fourteenth Century* (1955).

We still await a great book on the fifteenth century. In the meantime,

H. Maynard Smith: *Pre-Reformation England* (1938) will serve.

For one aspect of Church life, we now have a complete and magisterial survey:

M. D. Knowles, O.S.B., *The Monastic Order in England* (2nd ed., 1963);

The Religious Orders in England (3 Vols., 1948–59).

CHAPTERS 2–5 : *The Reformation Period*

We are fortunate in now having a really good book on the Reformation period in England:

A. G. Dickens: *The English Reformation* (1964).

Professor Dickens, although a layman, deals competently with the theological issues involved. There is no bibliography, but 'the notes are also designed to provide bibliographical material on many topics' (p. 341).

C. Beard: *The Reformation of the Sixteenth Century in Relation to Modern Thought and Knowledge* (1883, reprinted, 1927) is still indispensable.

The classic work of

R. W. Dixon: *The Reformation in England from the Abolition of the Roman Jurisdiction to 1570* (6 Vols., 1895–1902) contains a great deal of information not available elsewhere.

From the Roman Catholic side,

P. Hughes: *The Reformation in England* (3 Vols., 1950–54) may be consulted.

For *Henry VIII* we now have an honest and impartial biography by J. J. Scarisbrick (1968). For *Elizabeth I* the best life is still that by J. E. Neale (1954; and see also his other writings on the reign). For *Thomas Cranmer*, A. F. Pollard's *Life* (1904) is still unsurpassed. The more recent life, by J. G. Ridley (1962), though full is not wholly satisfactory; Mr. Ridley has done better work on his ancestor, *Nicholas Ridley* (1957).

For the reign of Elizabeth I, see:

R. G. Usher: *The Reconstruction of the English Church* (1910);
P. Collinson: *The Elizabethan Puritan Movement* (1967);
H. C. Porter: *Reformation and Reaction in Tudor Cambridge* (1958);
J. A. Froude's *History of England* (1856–70), for all its imperfections, is a work which still repays careful study.

CHAPTERS 6 AND 7: *The Seventeenth and Eighteenth Centuries*

For the seventeenth century no one book can be recommended. The later chapters of:

S. C. Carpenter: *The Church in England, 597–1688* (1954) are learned and vivid.

Dr. Carpenter followed this up with:

Eighteenth Century Church and People (1959);

P. E. More and F. L. Cross: *Anglicanism* (1935) is important, giving extracts from many writers, mainly of 'the seventeenth century.

Among recent biographies:

Antonia Fraser: *Cromwell, Our Chief of Men* (1973) stands out, and throws much light on the complexities of Church life in the period.

On the Restoration:

R. S. Bosher: *The Making of the Restoration Settlement* (1951) is illuminating.

From this point on the story is carried forward in

J. W. Legg: *English Church Life from the Restoration to the Tractarian Movement* (1914).

From a different angle:

G. F. Nuttall and O. Chadwick (edd.), *From Uniformity to Unity* (1962) is valuable.

One notable book brought the eighteenth century back into true perspective:

N. Sykes: *Church and State in England in the XVIIIth Century* (1934).

With this should be read the same writer's massive life of an eighteenth-century Archbishop:

William Wake, Archbishop of Canterbury, 1657–1737 (1957).

Among the innumerable works on the Wesleys, the best introduction is:

M. Schmidt: *John Wesley* (2 Vols., German, 1953, 1956; Engl. trans., 1962 ff).

See also a delightful study of eighteenth-century hymnody:

B. L. Manning: *The Hymns of Wesley and Watts* (1942).

Much work remains to be done on the Anglican Evangelicals. The best book up to date is:

L. E. Elliott-Binns: *The Early Evangelicals* (1953).

CHAPTERS 9 AND 10: *The Nineteenth Century*

S. C. Carpenter: *Church and People, 1789–1889* (1933) gives a good overall picture.

O. Chadwick: *The Victorian Church* (2 Vols., 1966, 1970) deals more than competently with more than the purely Anglican scene.

R. P. Flindall (ed.), *The Church of England, 1815–1945* (1972) is a useful collection of documents.

For theological thought in the period:

V. F. Storr: *The Development of English Theology in the Nineteenth Century, 1800–1860* (1913);

L. E. Elliott-Binns: *English Thought, 1860–1900: The Theological Aspect* (1956) give most of what the student will require.

For the Tractarian Movement:

R. W. Church: *The Oxford Movement, 1833–45* (1891)

is still indispensable, but needs to be supplemented by:

E. A. Knox: *The Tractarian Movement, 1833–45* (1933);

Y. Brilioth: *The Anglican Revival* (1925);

O. Chadwick (ed.): *The Mind of the Oxford Movement* (1960);

E. R. Fairweather (ed.): *The Oxford Movement* (1964).

Much of the history of the Evangelicals in this century remains unknown. The work of:

G. R. Balleine: *A History of the Evangelical Party in the Church of England* (1908)

is reliable but incomplete.

For the so-called Broad Church movement, no satisfactory book exists. The movement is best studied in the biographies of such men as F. D. Maurice, F. W. Robertson, C. Thirlwall, A. P. Stanley, B. Jowett.

Many more recent biographies are filling in gaps in our knowledge, e.g.:

The Life and Letters of Mandell Creighton, Bishop of London, by his wife (2 Vols., 1904), universally recognized as one of the best of British biographies.

W. G. Fallows, Bishop of Sheffield, *Mandell Creighton and the English Church* (1964), adds little to what we already knew.

Lord Furness: *Wilberforce* (1975). (A fuller life by J. C. Pollock is in preparation.)

M. Trevor: *Newman* (2 Vols. 1962).

G. Battiscombe: *John Keble* (1963).

D. Newsome: *The Parting of Friends* (1966; deals with the second generation of Wilberforces).

CHAPTERS 8, 11, 12 AND 13: *A Developing Communion*

The general background can be found in

S. C. Neill: *A History of Christian Missions* (1964);

R. Rouse and S. C. Neill: *A History of the Ecumenical Movement, 1517–1948* (1954).

Anglicans have played a leading part in movements for Christian unity through the centuries.

On the specifically Anglican side,

J. McL. Campbell: *Christian History in the Making* (1946), is useful.

Details must be sought in the histories of the various missionary societies, among which

E. Stock, *History of the Church Missionary Society* (3 Vols., 1901) is outstanding. Also in the histories of the various Churches of

which the Anglican Communion is made up. A good example
of this *genre* is
    J. T. Addison: *The Episcopal Church in the United States* (1951).

EPILOGUE: *Present Situation and Prospects*

Naturally, much of the history of the twentieth century remains
as yet unwritten.
    R. B. Lloyd: *The Church of England, 1900–65* (1966)
is a solid work of 623 pages, but is limited by the limitations of
the writer's sympathies.

Indispensable are the Reports of the Lambeth Conferences of
1910, 1920, 1930, 1948, 1958 and 1968.

Much of the history is recorded in a series of notable biographies:
    G. K. A. Bell: *Randall Davidson, Archbishop of Canterbury*
(2 Vols., 1935);
    F. A. Iremonger: *William Temple* (1948);
    J. G. Lockhart: *Cosmo Gordon Lang* (1949);
    C. H. E. Smyth: *Cyril Foster Garbett, Archbishop of York* (1959);
    R. C. D. Jasper: *Arthur Cayley Headlam, Bishop of Gloucester*
(1960);
    R. C. D. Jasper: *George Bell, Bishop of Chichester* (1967);
    A. Paton: *Apartheid and the Archbishop; The Life and Times of
Geoffrey Clayton* (1973);
    F. W. Dillistone: *Charles Raven: Naturalist, Historian, Theologian*
(1975).

A life of Geoffrey Francis Fisher, Archbishop of Canterbury,
is being prepared by Dr. E. F. Carpenter, Dean of Westminster.

Index